Neoplasms: Principles of Development and Diversity

Jules J. Berman

JONES AND BARTLETT PUBLISHERS
Sudbury, Massachusetts
BOSTON TORONTO LONDON SINGAPORE

World Headquarters
Jones and Bartlett Publishers
40 Tall Pine Drive
Sudbury, MA 01776
978-443-5000
info@jbpub.com
www.jbpub.com

Jones and Bartlett Publishers
Canada
6339 Ormindale Way
Mississauga, Ontario L5V 1J2
Canada

Jones and Bartlett Publishers
International
Barb House, Barb Mews
London W6 7PA
United Kingdom

Jones and Bartlett's books and products are available through most bookstores and online booksellers. To contact Jones and Bartlett Publishers directly, call 800-832-0034, fax 978-443-8000, or visit our website www.jbpub.com.

Substantial discounts on bulk quantities of Jones and Bartlett's publications are available to corporations, professional associations, and other qualified organizations. For details and specific discount information, contact the special sales department at Jones and Bartlett via the above contact information or send an email to specialsales@jbpub.com.

Production Credits
Chief Executive Officer: Clayton Jones
Chief Operating Officer: Don W. Jones, Jr.
President, Higher Education and Professional Publishing: Robert W. Holland, Jr.
V.P., Sales and Marketing: William J. Kane
V.P., Design and Production: Anne Spencer
V.P., Manufacturing and Inventory Control: Therese Connell
Publisher: Cathleen Sether
Acquisitions Editor: Shoshanna Goldberg
Managing Editor: Dean W. DeChambeau
Associate Editor: Molly Steinbach
Editorial Assistant: Caroline Perry
Senior Production Editor: Karen C. Ferreira
Senior Marketing Manager: Andrea DeFronzo
Composition: Northeast Compositors
Text Design: Anne Spencer
Cover Design: Kristin E. Ohlin
Associate Photo Researcher and Photographer: Christine McKeen
Cover Image: Cytologic preparation, lung. Adenocarcinoma cells, lower right. Ciliated bronchial epithelials, upper left. Courtesy of G. William Moore, MD, PhD, Pathology and Laboratory Medicine Service, Baltimore VA Medical Center, Baltimore, MD.
Printing and Binding: Malloy, Inc.
Cover Printing: Malloy, Inc.

Library of Congress Cataloging-in-Publication Data
Berman, Jules J.
Neoplasms : principles of development and diversity / Jules J. Berman.
p. cm.
ISBN 978-0-7637-5570-6 (pbk.)
1. Cancer. I. Title.
RC261.B622 2008
616.99'4—dc22

2008014872

6048
Printed in the United States of America
12 11 10 09 08 10 9 8 7 6 5 4 3 2 1

Brief Contents

Contents

Preface

The most important questions are the ones that children ask and, getting no answers, stop asking.

—Anonymous

Cancer occurs in all mammals, all types of fish (including sharks), insects, and plants. Basically, if you happen to be a multicellular organism, you can add cancer to your anxiety list. No other disease seems to play such a constitutive role in terrestrial life. Our inability to eradicate cancer has only served to strengthen our resolve to conquer this enemy. To this end, the field of cancer research has enlisted the help of cell biologists, chemists, physicists, engineers, statisticians, and virtually every subdiscipline of human and veterinary medicine.

Although cancer research is a multidisciplinary field, there is limited exchange of ideas across the disciplines. Molecular biologists ask a different set of questions than anatomic pathologists. Clinical trialists ask a different set of questions than cell biologists. Often, each discipline speaks in a jargon that is impenetrable to outsiders. Although biology has become data intensive and often requires advanced computational techniques to analyze huge quantities of experimental data, it is a rare biologist who can hold a productive conversation with a computer scientist. With very few exceptions, each discipline is unaware of the fundamental questions posed by the others.

Most cancer books either overgeneralize cancer, discussing the "cancer cell" as though one concept applies to all neoplasms, or they focus on one specific neoplasm and ignore the others. This book describes all of the different forms of neoplasia and develops a neoplasm classification that simplifies the way scientists from many different fields may think about complex cancer phenomenology. As we develop the cancer classification, a great number of questions are discussed. The premise of this book is that these questions, if answered intelligently, will hasten the eradication of cancer (List 1).

This book is written for students, researchers, and medical professionals who have a particular interest in the field of cancer research and oncology. Cancer research is multidisciplinary. Smatterings of anatomy, embryology, genetics, histology, biochemistry, computer science, informatics, and probability theory all appear in these pages. Mark

List 1. Some questions posed in this book.

What intrinsic cellular properties are common to all cancers?
Are there properties of neoplasms that are not found in normal cells?
How can we stop the progression of a normal cell into a cancer cell?
How can we kill cancer cells without harming normal cells?
Can we reverse the cancer process and transform cancer cells into normal cells?
How have studies on rare tumors led to new cures for common tumors?
What are the inherited tumors, and how do they differ from non-inherited tumors?
What is the relationship between inherited cancers and inherited malformations?
Why do chemical carcinogens need to be activated by cells in our bodies before they can cause cancer?
Are there some carcinogens that cause only benign tumors and others that cause only malignant tumors?
What carcinogens are responsible for the majority of tumors that occur in man?
Can tumors occur spontaneously, without any external cause?
Is a single gene alteration sufficient to cause a cancer?
What is the smallest dose of a carcinogen sufficient to produce a cancer?
Are animal tumors good models for human tumors?
Are childhood tumors fundamentally different from tumors that occur in adults?
Why aren't the same characteristic genetic changes seen in all tumors of a specific type?
Why do tumors that arise from the ectoderm and endoderm (embryonic layers) account for more than 95% of the tumor burden of humans?
What is a stem cell tumor?
How do germ cell tumors differ from all other tumors of humans?
Why are there different kinds of tumors and how many kinds of tumor are there?
How might a tumor classification be used to guide the development of new chemotherapeutic agents?
Is cancer prevention more important than cancer treatment?
How can our understanding of precancers lead to the eradication of all human cancers?

Twain said, "Everybody is ignorant, only on different subjects." I have written the book with the assumption that for every topic covered, there will be many readers who will be unfamiliar with the subject matter. Within each chapter, only enough information is provided to drive the discussion forward. A large glossary contains the important concepts from different scientific fields that are related to chapter topics. The glossary goes beyond simple definitions and depicts situations in which the precise meaning of a word might sway our conceptual understanding of cancer or might raise additional questions that remain to be answered. I have tried my best to make the glossary a literary work that can stand alone as a teaching text. Throughout the chapters, glossary terms appear in bold font.

Questions intrude relentlessly throughout this book. (The greatest scientific invention was the question mark). A good science book raises more questions than it answers.

As the title of this book would suggest, a set of principles accounts for neoplastic development and diversity. The fields of oncology, tumor biology, and cancer research become easy to understand if you understand these basic principles. The essence of the book is summarized in the final chapter, which lists 10 principles that account for all of the concepts introduced and discussed through the preceding chapters. The list is an oversimplification of reality that can serve as a baseline set of ideas that can be improved or disproved by the readers.

I would like to thank Drs. Andre Kajdacsy-Balla and Donald Henson, who reviewed this book. Their valuable suggestions greatly improved the final manuscript. I thank Drs. Rupert Allan Willis, Robert P. Bolande, Leslie Foulds, and Ernst W. Mayr, whose works had a profound influence on my own understanding of tumor biology and classification. Willis authored *The Borderland of Embryology and Pathology* (1958) (1) and Leslie Foulds wrote *Neoplastic Development* (1969) (2). Through the 1970s and 1980s, Robert Bolande made numerous contributions to the developmental biology of cancer. Ernst Mayr's lifetime contributions to the field of animal classification and speciation, starting with his publications in the early 1940s, were crucial to the creation of the Developmental Lineage Classification of Neoplasms. This book is, in large part, an attempt to provide a few answers to some of the enduring questions raised by these four scientists.

I also thank Dr. Gary M. Williams, who was my thesis advisor during my formative years as a pathologist and who guided my early experiments in the field of carcinogenesis. Of course, I thank my wife and daughters for their encouragement and their patience. It is hard to be a writer, but it is even harder to live with a writer.

Introduction

What Will You Learn by Reading This Book?

You can observe a lot by watching.

—Yogi Berra

In 1971, President Richard M. Nixon signed the National Cancer Act into law, marking the year that the United States launched a war on cancer.

Here are the results (Figure I.1).

There has been a modest decline of the cancer death rate for many tumors this past decade. Nonetheless, we see that after 36 years of cancer wars, the U.S. cancer mortality rate is about the same now as it was when the war began. The greatest drops in mortality have been due to preventive measures. Americans who have stopped smoking have benefited from a greatly reduced lung cancer rate. Women have enjoyed a sustained drop in uterine cancer, exceeding 70%, since the introduction of Pap smear screening (in the 1940s). For reasons that are poorly understood but that may relate to improved refrigeration and preservation of foods, the U.S. stomach cancer rate greatly declined in the twentieth century. By in large, the same cancers that were responsible for the greatest numbers of deaths in 1971 are the same cancers killing us in about the same numbers in 2007 (3).

Why Is Cancer Such a Difficult Disease to Understand?

Medical students are taught to beware the patient with the "positive review of systems." The review of systems is the part of the physical examination of a patient where the clinician asks dozens of questions to determine whether each of the physical systems in the body is working normally. "Do you ever experience ringing of the ears?" "Do you have frequent headaches?" "Do you ever faint or do you fall while standing?" "Do you experience pain while swallowing?" Occasionally, a patient will be encountered who answers "YES!" to every question. When this occurs, the doctor is put in an impossible situation because no known disease can account for a positive review of systems. There

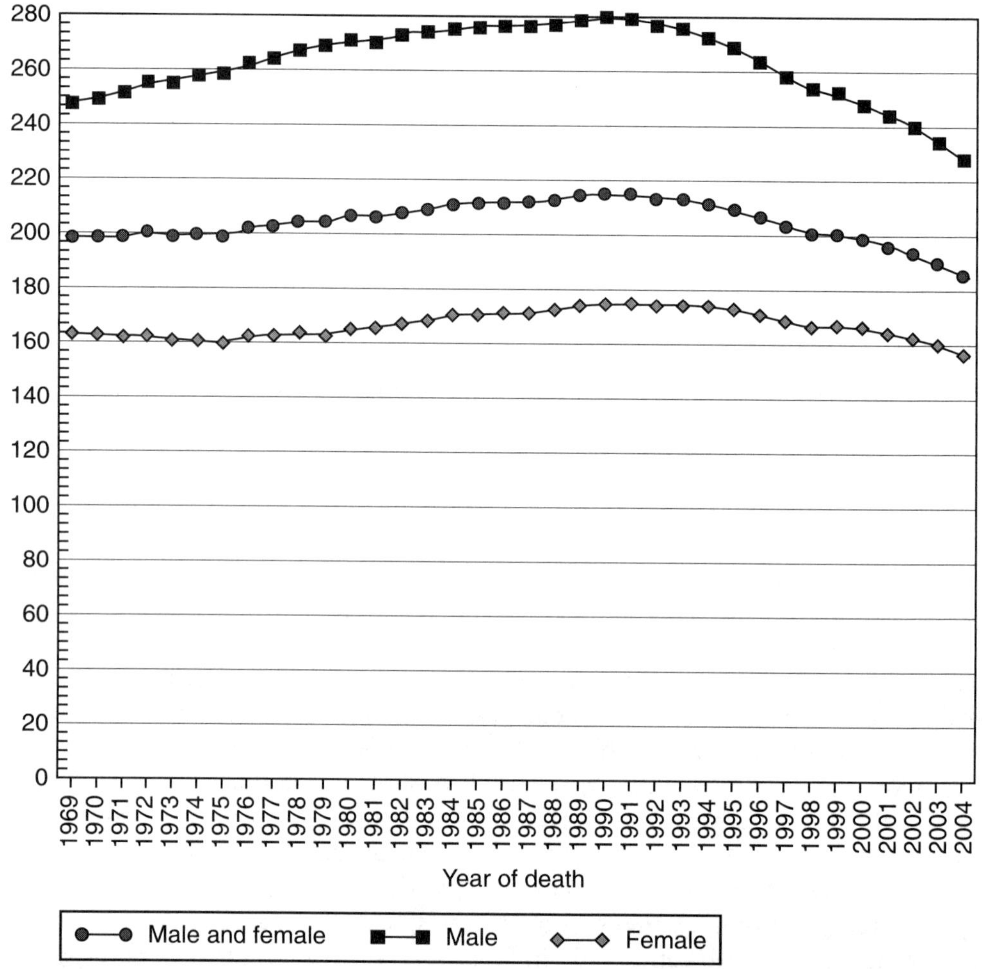

FIGURE I.1 Age-adjusted total U.S. mortality rates for all cancer sites, all ages, all races for 1969–2004 by sex and age—adjusted to the year 2000 U.S. standard population. *Source:* The Surveillance, Epidemiology, and End Results Program.

is no satisfactory approach to the patient who suffers from every possible clinical complaint.

Cancer can be thought of as a cellular disease with a positive review of systems (List I.1).

If the answer to all these questions is YES, our cell has a big problem. Things get even worse when you try to elicit some historical information from the cancer cell (List I.2)

Anything you might want to measure in the cancer cell is different from normal. If everything is abnormal, how can you determine the one abnormality that leads to all

List I.1 Questions we would ask a single cell during a review of systems.

Is your size altered?
Is your shape altered?
Is your nucleus bloated and irregular?
Have you become less aerobic lately?
Have you noticed any changes in your cell membrane?
Have you lost or gained any DNA lately?
Are your methylation levels normal? (4)
Are your histones excessively deacetylated?
Is alternative *RNA splicing* giving you a run-down feeling?
Have your transposons moved much lately?
Have you noticed a sudden increase in the length of your **telomeres**?
Is there a difference in the functionality of your internal and external receptors?

List I.2 Questions we would ask a single cell when we take a medical history.

Have you been exposed to a virus?
Have you been exposed to a carcinogenic chemical?
Have you been exposed to radiation or sunlight?
Do you have any close relatives with cancer?
Can you do the same job that you were doing before you became a cancer cell?
Has your growth rate changed?
Is this growth persistent, or does it sometimes regress?
When was the last time you mitotically divided?

of the other abnormalities? Historically, cancer research has been an endless funding resource for every type of biologist. If your interest is cell membranes, you could study the changes in cell membranes that occur in cancer. If your interest is mitochondria, you could study the mitochondrial changes that occur in cancer. If your interest is energy metabolism, you could study energy metabolism in cancers. There are thousands of different types of cancer, and questions related to any particular interest can be studied in each type of cancer. This could (and does) keep legions of scientists gainfully employed throughout their academic careers. Although we know how cancer cells differ from normal cells, we do not know how to cure a disease that affects every cellular system.

Is Cancer One Disease or Many Diseases?

A recurring existential problem in the field of cancer reduces to this: Is cancer a single disease manifested as many different types of cancers, all following a solitary, mysterious disease process, or is cancer a myriad of different diseases that manifest as abnormal populations of growing cells? If cancer is a single disease, why have we not found a cure? If cancer is thousands of different diseases, how can we hope to find cures for every different type of cancer? A central thesis of this book is that cancer can be sensibly classified into a few dozen categories of disease and that it will be possible to prevent, diagnose, and treat all types of cancer after we understand the biological classification of neoplasms.

Gaining an understanding of cancer does not need to be difficult. Every area of cancer research can be reduced to a few basic principles and a few observations. Connecting all of these diverse areas and synthesizing a simplified view of neoplasms is one of the chief goals of this book.

This book is divided into three major parts.

Part I: Speciation

What determines the types of neoplasms that occur in humans? We are taught that neoplasms are caused by mutations that lead to uncontrolled growth and genetic instability. We are also taught that the genetic instability in cancer cells leads to the progressive accumulation of additional mutations. If this is the fundamental mechanism of carcinogenesis, would you not expect a near-infinite number of cancers resulting from all of the possible different combinations of observed mutations?

The harsh realm of neoplasms is much like the realm of living terrestrial organisms. The world is composed of unique, individual life forms. Somehow, every unique life form belongs to one species from among a finite number of species. The squirrels in your backyard may all look alike, at first glance, but each squirrel has a unique genome and a unique personality. Each squirrel is a life form that has never before appeared on earth and that will never appear on earth again; nevertheless, each squirrel is an unmistakable member of the squirrel species. Every squirrel shares features in common with other squirrels and lacks features that are found in other animal species.

Every neoplasm is unique, with a unique set of genomic and epigenomic variations that can distinguish one tumor from every other tumor. Why do we encounter certain types of tumors with the same morphology, occurring in different people? The shared morphologic and biological features among types of tumors allows us to examine any unique tumor under a microscope and say with complete certainty that the tumor is a testicular seminoma, or a Warthin tumor of salivary gland, or a pilomatrixoma of skin, or any one of the thousands of distinctive species of neoplasms? Why is it that even though humans are different from every other animal, tumors found in fish and rodents may look and behave very much like specific types of tumors found in humans?

Before we can start thinking about how to classify tumors, we need to understand why every unique tumor must belong to one of several thousand species of tumor. This topic is discussed in Part I.

Part II: Classification

After the neoplasms of humans are assigned species, the next step is to divide the species of neoplasms into broad classes. Creating a neoplasm classification is a difficult scientific problem. Lessons learned through the millennia from naturalists such as Aristotle, Discorides, Ray, Linnaeus, Darwin, and Mayr have direct bearing on the classification of neoplasms. Modern classifications serve several purposes (List I.3).

Part III: Cancer Research and the End of Neoplasms

Distinguishing the many different species of neoplasms and grouping them into a classification is a pointless exercise if it cannot lead to a reduction in the number of human deaths caused by cancer. In Part III, we will learn that the agents that cause cancer, the measures we use to prevent cancer, the steps in carcinogenesis, the biological properties of cancers, and the treatment of cancers are all class-dependent. Treatments that may be effective in one class of tumors may not be effective in tumors from other classes. Much more importantly, we can hope that agents that can cure one type of cancer may be effective against all of the other types of cancers that share the same class.

At first glance, cancer cells seem to be characterized by a hopelessly complex set of abnormalities involving every cellular process. If we do not make the intellectual effort to speciate and classify neoplasms, new and effective approaches to eradicating cancer will not be forthcoming. In Part III, we examine a variety of research approaches that can contribute to the eradication of cancer.

List I.3 The purposes of a modern neoplasm classification.

1. To seek universal truths that pertain to all neoplasms
2. To simplify our understanding of an otherwise disconnected list of thousands of neoplasms
3. To serve as a conceptual framework on which hypotheses about the relationships among the different neoplasms, and about the properties of neoplasms in general, can be asked and answered
4. To serve as an information key to which data can be sensibly stored and retrieved
5. To provide a rational basis for developing class-targeted preventions, modifications, and treatments based on biological properties and pathways shared by the members of a class

PART I

Speciation

CHAPTER

What Properties Are Shared by All Cancers?

1

There's nothing new under the sun, but there are lots of old things we don't know.

—Ambrose Bierce

1.1 Background

All malignant tumors share a basic set of properties (List 1.1.1).

Malignant tumors persistently grow and invade their surrounding tissues. **Metastasis**, the seeding of growing tumor cells to locations outside of the vicinity of the original site of tumor growth, is another property that is associated with **malignancy**. Most people who die of cancer succumb to their metastases, not to the damage produced by the primary tumor (the tumor at its original site of growth).

Cancers acquire their properties by altering pre-existing **pathways** found in normal cells. The purpose of this chapter is to review the fundamental features of cancer and to explain how these properties derive from the normal regulatory processes that control the growth, repair, **differentiation**, and death of cells.

List 1.1.1 The minimal properties common to all cancers.

1. Growth
2. Persistence
3. Invasion
4. Metastasis

List 1.2.1 Comparing each cancer property with properties of normal tissues.

Are there normal tissues that grow? Of course.
Are there normal tissues where the growth is persistent? Yes.
Are there normal tissues that invade? Yes.
Are there normal tissues that metastasize? Yes.
Are there normal tissues that have all these properties? No.

1.2 Are There Any Properties of Neoplasms that Are Not Found in Normal Cells?

Cancer cells do not create new biological properties (List 1.2.1). They use the same properties that normal cells use, only they tie them all together in a package that nobody wants to receive. Let us examine the normal tissues that have properties of cancer cells.

1.3 Persistent Growth in Normal Cells

You can think of the body as a collection of two types of cells: those that grow persistently and those that grow intermittently or never. The most numerous persistently growing cells line the outside and the inside surfaces of the body: squamous cells of the skin's epidermis and **enterocytes** of the gastrointestinal tract (the gut). The growing cells of skin and gut exactly balance a second population of cells that are actively dying. The reason that the skin epidermis and the gut **mucosa** maintain a constant thickness is that the dead skin cells are sloughed into the air (as dust specks) and the dead enterocytes are sloughed into the fecal stream. Otherwise, both the epidermis and the gut would double in size every few weeks, a rate of normal growth that easily exceeds the growth rate of neoplasms.

Epithelial tissues (the tissues that line surfaces and populate most solid organs and glands) contain a subpopulation of growing (dividing) cells. Each division of a dividing cell yields two cells: one cell that performs the normal functions of the tissue, and another cell that is capable of dividing. The cell that performs the normal functions of the organ is referred to as a differentiated cell, because it has completed the developmental process and has attained a distinct morphologic appearance. Most differentiated epithelial cells are incapable of dividing and will eventually die. The precise balance of cell growth and cell death yields a zero net growth in normal adult tissues.

Some organs, like the liver, are composed of **epithelial cells** that cease to divide or divide only rarely. The liver of an elderly man who has abstained from alcohol and led a relatively toxin-free life will contain many of the same, hard-working hepatocytes (liver cells) it had when the man was a youth. In the event of a liver injury, the otherwise quiescent hepatocytes quickly divide, enlarging individual acini (i.e., microscopic liver glands), quickly restoring the liver. In the rat, the full weight of the liver is restored in under a week after surgically removing two thirds of the liver. The remarkable regrowth of the liver after partial hepatectomy appears in the myth of Prometheus, whose liver is eaten by eagles each night and restored the next day.

Some nonepithelial tissues lose the ability to divide at about the time that the human body has reached its adult size. Many of the diseases of aging are actually just the consequences of wear and tear in populations of cells that cannot renew themselves. Osteoarthritis is a joint disease that results from damage, over time, to articular cartilage. Cartilage cells (chondrocytes), in contrast to liver cells, have only limited ability to divide. When cartilage is damaged, it usually stays damaged, and this may cause a chronic inflammatory disease involving the damaged joint cartilage and the underlying bone (osteoarthritis). The dementia that accompanies aging is due, in part, to the limited ability of senescent neurons to renew themselves through division. You can think of the degenerative diseases of aging as the medical opposite of neoplasia. Aging results from accumulated damage in cells that cannot divide. Neoplasia is the result of accumulated damage in cells that must divide.

1.4 Invasion by Normal Cells

Invasion is another biological property that is shared by neoplastic and certain normal tissues. An example of an extreme form of normal tissue invasion is found in gestational implantation and placentation. In implantation, the blastocyst (the early embryo) attaches to the endometrium (the lining of the uterus) and invades into the uterine wall. In placentation, the uterine spiral arteries (of the mother) are invaded by **trophoblast** cells of the embryo. Implantation and placentation occur quickly. Normal trophoblastic invasion matches or exceeds the most aggressive invasion seen in neoplasms.

1.5 Metastasis by Normal Cells

In metastasis, tumor cells leave the primary tumor site and travel to some anatomically distant (or at least noncontiguous) location, where they form a growing mass. There are many steps involved in metastasis, and it is no wonder that metastasis is a late event in **carcinogenesis** that follows the acquisition of invasiveness (List 1.5.1).

Each of the steps of metastasis provides opportunities for a fledgling clone of tumor cells to die, and each step may take time (days, months, or years) while the clone acquires the biological prerequisites for the next step. The time between the seeding of a tumor cell into the lymphatic circulation (the small vessels that lead to lymph nodes), and the clinical observation of a metastatic growth, is called **dormancy.** Dormancy is discussed in depth in Chapter 2. The time between tumor seeding and the appearance of a metastatic lesion can depend on a great many factors. If the seeded tumor cells

List 1.5.1 Steps in metastasis.

1. A tumor cell must invade through the wall of a lymph vessel or a blood vessel.
2. Once inside the vessel, the tumor cell must be carried away from the primary tumor to another site, still within the vessel, where it may reside for some indeterminate time.
3. The tumor cells residing inside the vessel must invade through the vascular endothelium into the surrounding tissue.
4. The invading cell or cells must begin to grow.
5. The growing colony must adjust to its new microenvironment and must eventually attract the growth of vessels from which it can derive oxygen and other external nutrients.

have all of the properties necessary to complete the steps of metastasis, then a metastasis may become clinically apparent soon after the primary tumor has begun to invade. For example, if cells were seeded by a metastatic lesion (not from the primary tumor), then the second generation metastatic cells might arrive at their metastatic site pre-equipped for successful growth. In this case, metastases might occur quickly. Alternately, host factors (properties of the non-tumorous cells of the organism) may provide barriers to tumor metastasis. For example, tumor cells that have seeded into the lymphatics might be filtered, trapped, and destroyed in lymph nodes through an immunologic reaction.

Given the complexity of metastasis, you might think that only malignant cells could manage the task. Surprisingly, metastasis is also seen in normal tissues. All blood cells move through the circulation through a process indistinguishable from metastasis. White bloods cells (leukocytes), in response to inflammatory **chemokines**, will attach to a vessel wall, move through the wall, and proceed to invade local tissues until they arrive at the site of inflammation (the site of highest chemokine concentration). Leukocytes are all highly invasive and metastatic. The only feature that distinguishes normal leukocytes from cancer cells is their controlled growth. The growth of the leukocyte population in response to an infectious process is self-limited. Growth stops when the infection ceases. The most significant clinical property that distinguishes physiological neutrophilia (increase in **neutrophils**, the most common type of leukocyte) from chronic myelogenous **leukemia** (the neoplasm caused by an increase in neutrophils) is growth persistence. Chronic myelogenous leukemia gets worse and worse over time. Physiologic neutrophilia eventually plateaus or returns to normal levels. In Chapter 9, we examine some nonmalignant proliferative disorders of myeloid and lymphoid cells, all of which have clinical features that closely resemble leukemias and **lymphomas**.

In **chinchillas**, the normal placenta both invades and metastasizes (5). Trophoblastic tissue invades into uterine vessels, passes into the general circulation, and seeds the lungs. It is normal for a chinchilla mother to harbor pulmonary placental nodules that persist for up to 2 months after delivery. In humans, trophoblastic invasion of the

myometrium must be well controlled. Otherwise, harm can come to fetus and mother. Placenta accreta occurs when the placenta attaches directly to the myometrium with deficient development of decidual (maternally-derived) cells between the invading trophoblasts and the myometrium. If the placenta invades deeply into the myometrium, this is called placenta increta. If the placenta invades through the myometrium, this is called placenta percreta. All three conditions are associated with medical risks to baby and mother. We discuss human **gestational trophoblastic disease** in Chapter 20.

1.6 Is There a Common Temporal Sequence Leading to the Development of Cancer?

The explanation of how cells achieve the observable traits of cancer (the so-called **malignant phenotype**) varies greatly from tumor to tumor and accounts for much of the mystery in the field of cancer research. Despite the observable morphologic and behavioral differences among different neoplasms, they all seem to acquire their phenotype through a common sequence (List 1.6.1).

Exceptions occur when one or more of the properties of the malignant cell happen to be constitutive properties of the normal cells from which the tumor arises. For example, all normal white blood cells are invasive and metastatic. Blood cells use these two properties to travel through the body and to move through tissues. When neutrophils become malignant, they do not need to acquire the properties of invasion and metastasis.

In Chapter 4, we learn that characteristic molecular events precede the development of the malignant phenotype (i.e., growth, persistence, invasion, and metastasis) and that early changes distinguish normal cells from cells undergoing carcinogenesis (the process that leads to cancer).

1.7 Why Is It Important to Treat Cancers Early?

The easiest and most effective way to cure a cancer is by removing it from the body before any of the cancer cells have had a chance to metastasize to another site. No matter how rapidly a tumor may grow, or how invasive it may be, if the primary cancer is removed before it metastasizes, the patient is always cured. If a primary cancer is

List 1.6.1 The defining properties of tumors arise in a prescribed order.

1. First comes growth.
2. Then comes persistence.
3. Then comes invasion.
4. Then comes metastasis.

removed after it has metastasized to distant sites, the patient will often die. This is true even if the primary tumor is a slow-growing tumor with minimal invasion. Left untreated, most cancers will eventually metastasize. Metastases are responsible for the vast majority of deaths caused by cancer. If we knew how to stop tumors from metastasizing, we could cure most of the cancers of humans.

A tumor that is not invasive cannot metastasize. This means that noninvasive tumors can be cured with simple excision. Invasion is a property that develops in tumors over time. The earlier you find a neoplasm, the less likely that the tumor has begun to invade.

After invasion begins, metastasis becomes possible. As time passes, the chance that a metastasis will occur becomes greater and greater.

1.8 Cancer Morphology

In 1858 (1 year before Darwin published *Origin of the Species*), Rudolph Virchow published *Cellular Pathology*, describing the principles of pathology that were known to Virchow and his contemporaries. These same principles have been studied for the past 150 years. Simply put, diseases result from changes that occur in our cells. By examining diseased cells under a microscope, we can identify the disease, and we can often find the cause and cure of the disease. Since the mid-19th century, virtually every advance in medicine has required the careful pathologic examination of diseased cells.

Cancers are diagnosed by biopsy followed by microscopic examination. A biopsy is a sampling of tissue and derives from the Greek *bios* meaning life and *opsy* meaning appearance. The biopsied cancerous tissue is preserved in a chemical fixative that kills the cells in the tissue while maintaining their shape and substance. The fixed tissue can be sectioned into thin slices (less than the thickness of a single cell) and mounted on a **glass slide**. The tissue slice on the glass slide can be stained with chemical dyes that colorize cells. The stained cells can be examined under a microscope by an **anatomic pathologist**. Anatomic pathologists are physicians who specialize in evaluating cells and tissues under a microscope.

In most cases, an anatomic pathologist can examine a stained sample of tumor and determine its precise **diagnosis** in under 1 second (comparable to the time that you would need to identify the *Mona Lisa* or to distinguish an apple from a banana).

1.9 General Rules for Naming Neoplasms

Although there are thousands of different neoplasms, a few tricks will help you understand the **nomenclature** (List 1.9.1).

There are many exceptions to these general rules: eponymous tumors (**Warthin tumor**, **Teilum tumor**, Kaposi sarcoma), lymphoid malignancies (we use malignant

List 1.9.1 Generalizations for tumor nomenclature.

1. There are two major classes of malignant tumors: those that derive from **connective tissue** (the sarcomas) and those that derive from glands and surface epithelium (the carcinomas).
2. Carcinomas that arise from glands are given the more specific name adenocarcinoma (carcinoma of glands). Examples are adenocarcinoma of colon, adenocarcinoma of breast, adenocarcinoma of prostate, and adenocarcinoma of lung.
3. Names of malignant tumors are often created by combining the name of the cell or tissue that most closely resembles the malignancy followed with the major **class** of tumor as a suffix or as a separate word, followed by the anatomic location. Examples are squamous cell carcinoma of skin (carcinoma of squamous surface epithelium), ductal carcinoma of breast (carcinoma of ductal epithelium arising from the ducts of the breast), and rhabdomyosarcoma of thigh (sarcoma of striated muscle cells of the thigh).
4. **Benign tumors** (tumors that do not invade or metastasize) are given the indeterminate suffix "oma" and otherwise retain the same descriptors used by malignant tumors. Examples are colon adenoma (the benign equivalent of colon adenocarcinoma), breast adenoma (the benign equivalent of breast adenocarcinoma), and dermal fibroma (the benign equivalent of fibrosarcoma of skin).

lymphoma, not lymphosarcoma), and inscrutable terms (e.g., **mycosis fungoides** is a neoplasm and not, as you might expect, a type of mushroom).

1.10 What Is a Cytologic Diagnosis?

Tumor cells are different from normal cells because they have genetic alterations. Because genomic DNA is confined to the **nucleus**, you might expect the nucleus of a malignant cell to have a different morphology than the nucleus of a normal cell. This is, in fact, the case. Observations made on cancer cell nuclei form the basis of diagnostic medical cytology (the study of cells).

Cytologists practice their profession by observing collections of stained cells obtained from body fluids (such as pleural or peritoneal fluids or urine) or from scrapings of epithelial surfaces (such as **Pap smears** from cervix) or from fine-needle aspirations of internal or external nodules. A representative cell population of just a few cells from a neoplasm can be sufficient for a cytologist to render a diagnosis.

Malignant cells have a different morphology than their normal tissue counterparts (the normal cells that look most like the malignant cells and from which the malignant cells probably arose). Let us examine a fine needle aspirate of pancreas (Figure 1.1).

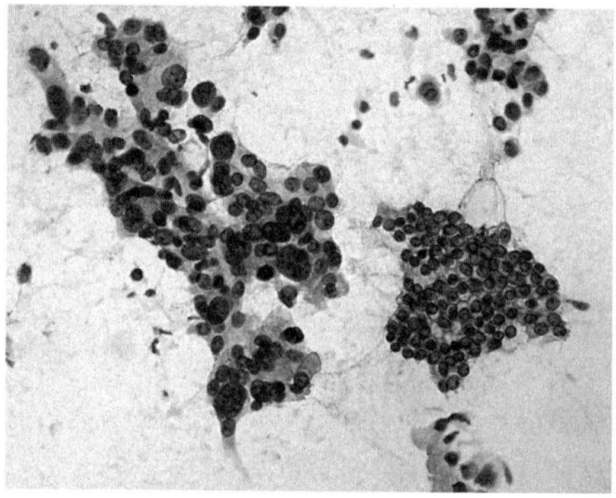

FIGURE 1.1 Fine-needle aspirate of pancreas. The large clump of cells on the left side of the image is cancer. Likewise, the two small clumps of cells seen at the top border of image, middle and right, are also malignant cells. The clump of cells on the right side of the image is a normal clump of epithelial cells seen from on top. The small clump of six cells at the bottom right border of the image is another clump of normal epithelial cells laid out on their side.

Three clumps (the large clump on the left and the two smaller clumps at the top margin of the image) are malignant. Two clumps of cells are benign (the large clump on the right of the image and the small clump of cells on the bottom right border of the image). Without any instruction in cytologic diagnosis, can you look at these different clumps and find differences between the normal clumps and the cancer clumps? In the next section, we discuss the features of malignant cells. For now, we can observe the features of normal cells. A clump of normal cells is an orderly collection of nearly identical cells, each containing a round nucleus. The nuclear diameter of most epithelial cells in humans (and most animals for that matter) is about 7 to 9 microns. The clumped normal cells are geometrically spaced, much like the cells in a honeycomb.

1.11 Morphology of Malignant Cells

We can determine whether a cell is malignant or benign based on morphologic observations of the nuclei. Here are the changes that contribute to the diagnostic nuclear changes found in malignant cells.

1. Nuclear size and volume.

   The atypical nucleus is virtually always larger than the nucleus of a normal, nondividing cell. The normal nucleus of a nondividing cell is about the same size as a mature red blood cell (about 7 to 9 microns in diameter). The nuclei of malignant cells can range from just larger than normal to over 50 microns in diameter. The difference in size between normal and malignant nuclei is one of the most dependable criteria for malignancy. Some tumors are highly cellular, with very little cytoplasm. Examples are small cell carcinoma of lung and the small blue round cell

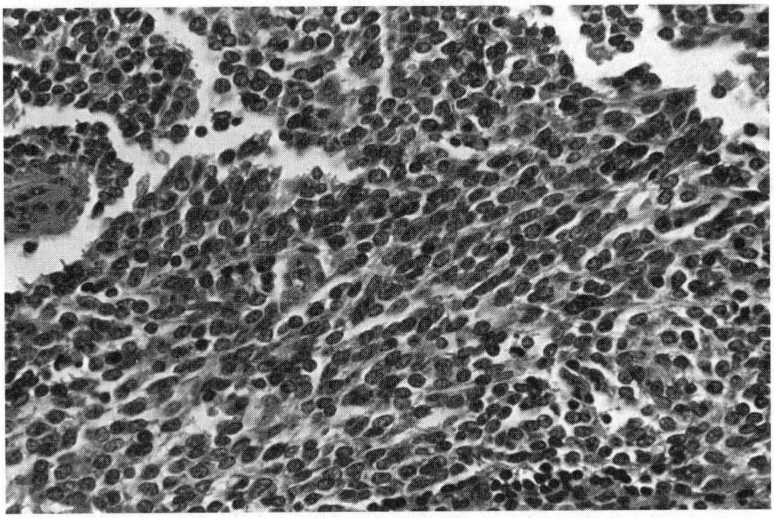

FIGURE 1.2 Small cell carcinoma of lung. The tumor is composed of densely packed cells with hyperchromatic nuclei and scant cytoplasm. In places, the nuclei are pushed together so that the nucleus of a malignant cell is molded by an adjacent cell.

tumors of childhood (Figure 1.2). The cells comprising these tumors are small and closely packed. They provide a large number of cells within a single microscopic field (the area viewed when you look through the binoculars of a microscope). This creates an illusion of smallness; however, by comparing tumor nuclei with red blood cells (which have a diameter equivalent to the diameter of normal cell nuclei) residing within the tumor, we see that the nuclei of most tumor cells have diameters that are larger than the diameter of red blood cells.

In addition to their increased nuclear size, tumor cells have increased nuclear variation. A tumor may have a cell with an enlarged nucleus adjacent to another cell with a nucleus of twice the first cell's size. A nonneoplastic population is composed of cells with very little variation in nuclear size among the members of the population. Careful examination of a group of cells from a cancer will almost always reveal many small variations in nuclear size.

2. Nuclear contour.

   Cancer nuclei are not as smooth or as round as the nuclei of normal cells. They tend to have a more complex contour with jagged bumps and pits. The irregular shape of one cancer cell will tend to be different from the irregular shape of adjacent cancer cells.

3. Nuclear texture.

   Nuclear texture relates to the distribution of chromatin. Chromatin is the nuclear material that stains blue in the standard **hematoxylin and eosin** stain that is used to colorize tissues mounted on glass slides. Chromatin is a mixture of DNA, RNA, and attached proteins that reside within the nucleus. The blue hematoxylin is preferentially bound to nuclear chromatin. Pink eosin is also bound to chromatin, but to a lesser extent. In a normal cell, the chromatin is evenly distributed, and the nucleus has a near-uniform distribution of chromatin. In a malignant cell, the distribution of chromatin will vary greatly from place to place inside the nucleus, with large dense clumps of chromatin admixed with smaller clumps, separated by light areas where there seems to be no chromatin staining.

In cancer cells, the distribution (texture) of chromatin will be different from nucleus to nucleus.

4. Nucleolar variability in size, shape, and number.

   Each chromosome of a normal nucleus contains a **nucleolar organizing region** that may be associated with a **nucleolus**, a circumscribed location within the nucleus in which RNA is synthesized and constructed into functional subunits.

   Though every nucleus is entitled to a nucleolus, the normal, quiescent cell usually contains between zero and six observable nucleoli, and these are typically small, round, discrete structures that stain pink (eosinophilic) with hematoxylin and eosin stain. Malignant cells tend to have more visible nucleoli than normal cells, and the nucleoli tend to vary greatly from cell to cell in size, shape, and number.

1.12 Cancerous Atypia and Reactive Atypia

Atypia simply indicates that a change has occurred from the normal morphology of a cell. Whenever cells are exposed to a toxic agent, the cell responds with a large variety of morphologic changes that alter the morphology of the nucleus and that bear some resemblance to the morphology of cancer cells. Cells with reactive atypia either die or recover without acquiring a genetic change that can confer the malignant phenotype to daughter cells. The job of distinguishing reactively atypical cells from cancer cells often falls to the cytology laboratory.

The field of cytology deals with the study of cells, the basic living unit of all organisms. Medical cytologists are highly trained microscopists who screen the specimens received in cytology departments. **Cytopathologists** are medical pathologists who have received additional, specialized training in cytology. In this book, medical cytologists and cytopathologists are referred to under the generic term "cytologist." Cytologists can diagnose a wide variety of pathological processes, including cancer, by examining individual cells (rather than entire tissues).

Cytology specimens are different from tissue specimens. Tissue sections are thin slices of a tissue that preserve normal tissue architecture and the relationships among all the different cell types that compose the tissue. Cytology specimens consist of loose clumps or single cells that were smeared from a tissue surface (as in Papanicolaou smears of cervix), or scraped from an internal tissue (as in a fine needle aspiration), or floated in a body fluid (such as urine). About 50 million human cytology specimens are collected each year in U.S. cytology laboratories. The field of medical cytology is built on the premise that you can distinguish one pathologic process from another just by careful observation of individual cells.

1.13 How Can You Distinguish Reactive Atypia from Cancerous Atypia?

There are three major categories of atypical cellular morphology (List 1.13.1).

The atypical dying cell is one that has received an irreversible toxic insult that will eventually result in the death of the cell. Programmed cell death, also known as *apoptosis*, is a pattern of morphologic changes that occur when a cell literally kills itself (by activating cytolytic, neucleolytic pathways and by altering the flux of compounds through the cell membrane) after a toxic event. The most prominent morphologic change observed in **apoptosis** is pyknosis, the shrinkage of the cell and nucleus accompanied by nuclear hyperchromasia (increase in coloration). Apoptotic atypia is easily distinguished from the atypia associated with cancer.

Radiation-induced atypia is an exceptional phenomenon in which scattered cells enlarge and become stellate and their nuclei become large and distorted. Cells changed by radiation atypia are sprinkled through a tissue and do not cluster like a growing neoplasm. Radiation atypia will occur in different cell types within one anatomic location. For example, vascular endothelial cells, fibrous cells, and epithelial cells may all show signs of radiation atypia. Cancer is a disease that arises from a single clone of cells, and it will not cause atypia in scattered cells of different histologic types. Although a single cell with radiation atypia may be difficult to distinguish from a single cell with neoplastic atypia, the pathologist who examines many different cells within a specimen will usually distinguish these two conditions.

The term "reactive atypia" contrasts with the atypia of cancer cells that persists in the absence of any external stimulus. Inflammation, repair after injury, and compensatory **hyperplasia** in response to the loss of a subpopulation of cells due to a toxic or degenerative process are all causes of atypia in growing cells (Figure 1.3). In general, reactively atypical nuclei are enlarged but maintain a round or oval contour. The nuclei are marked by hypochromasia more often than hyperchromasia. Nucleoli may be enlarged and increased in number, but are usually round.

List 1.13.1 Types of cellular atypia.

1. Atypical dying cells
2. Reactive atypia (atypia in non-neoplastic cells)
3. Atypical precancerous cells
4. Atypical cancerous cells

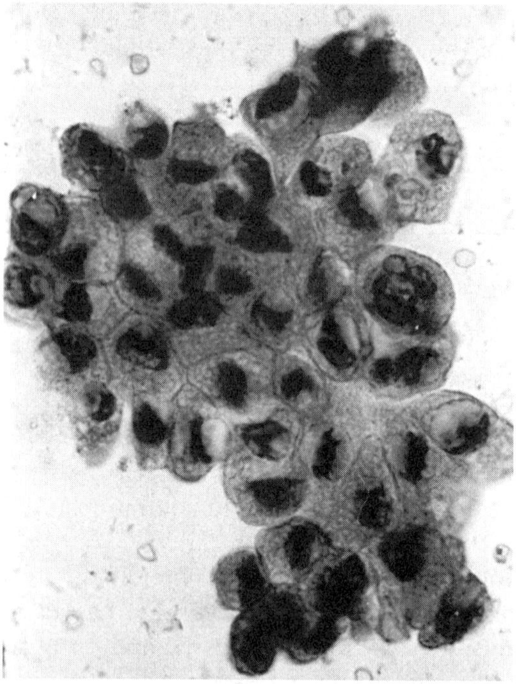

FIGURE 1.3 A flat clump of metaplastic cells with posttracheostomy reactive atypia. One multinucleate cell is present. Some cells have cytoplasmic globules compressing nuclei, a feature suggestive of a cellular morphology that is transitional between squamous and glandular cells (6). These reactively atypical cells can easily be confused with neoplastic cells.

1.14 Dysplastic Cells and How They Differ from Cancer Cells

The term "**dysplasia**" (from the Latin, "wrong growth") has several different medical meanings. Pediatricians use the term dysplasia to indicate a dysmorphic developmental disorder (i.e., a disorder in which the shape or size of the skeleton or some other observable body part is abnormal). This use of "dysplasia" is not related in any way to the neoplastic process. The term dysplasia is also used to describe a growth insufficiency in newborns (bronchopulmonary dysplasia) and a type of growth defect in bone (fibrous dysplasia of bone). These uses of the word dysplasia are likewise unrelated to cancer.

Those of us who work in the cancer field use "dysplasia" to describe a type of **nuclear atypia** that is passed from a cell to its progeny. Pathologists try to reserve the term dysplasia to describe the early changes in nuclei that characterize **precancers** (the lesions that precede the development of invasive neoplasms). These early changes in precancer cells cannot always be distinguished from changes seen in cancerous cells, and thus, both precancers and cancers are technically dysplastic.

1.15 Nuclear Atypia in Cancer Cells

The nuclear atypia found in cancer cells is the morphologic expression of accumulated genetic and epigenetic changes. Epigenetic changes occur in chromosomes and include structural changes, nuclear protein changes, and modifications of DNA adducts that do not involve alterations in genetic code.

Because cytologists examine many different nuclei in a cytologic preparation, they draw inferences about the tumor based on observed differences between many different tumor cells. The ability to make fast and accurate diagnostic judgments about a population of cells is something that no other methodology can rival. For example, a **cytogeneticist** can spend hours unraveling the complex cytogenetic features of a single malignant cell. The cytogeneticist cannot assess all the cytogenetic features of all the cells obtained from a single tumor sample. Likewise, molecular biologists who search for mutations in tumor cells look at a DNA sample pooled from a collection of cells. The molecular biologist cannot assess mutational differences between the individual cells that contributed to the DNA sample.

Only the cytologist can synthesize a diagnostic assessment that combines careful examinations of individual nuclei (like the cytogeneticist) and aggregates of nuclei (like the molecular biologist). Only the pathologist, who has access to tissues, can synthesize a diagnostic assessment based on the cytologic properties of the tumor cells and on relationships between tumor cells and the tissues in which they grow (e.g., depth of invasion, adequacy of excision, and size of tumor).

1.16 Why Are the Nuclei of Malignant Cells Different from Nuclei of Normal Cells?

We often hear that malignancy is caused by a mutation in a single cell, leading to altered growth and the eventual emergence of a clonally-derived neoplasm. We know that malignant cells are characterized by a variety of morphologic abnormalities. You might wonder why, if **malignant** cells are caused by one or two small molecular mutations, there is such a profound change in the appearance of cancer cells. This is one of the most important questions in the field of cancer research and will be the subject of later chapters. For now, it suffices to indicate that cancers are characterized by genetic instability and the progressive accumulation of genetic and epigenetic alterations (7). After a cell has received a carcinogenic mutation, it progresses through a long process during which it collects additional genetic and epigenetic alterations leading to morphologic atypia and cellular hetergeneity.

1.17 Tumor Monoclonality

It is often possible to show that a tumor has a monoclonal origin (i.e., every cell in the tumor is a descendant of a single cell). Patterns of x-chromosome inactivation (also called Lyonization) in tumors occurring in women can establish tumor monoclonality

(8). In women, at an early stage in embryogenesis, one **x chromosome** is randomly inactivated in each cell. Afterwards, all of the progeny of a cell will have the same x chromosome inactivated. Every woman is a genetic mosaic in the sense that half of her cells have one x chromosome inactivated and the other half of her cells have the other x chromosome inactivated. Most normal cells obtained from females have a small, heterochromatic chromosome pressed against the nuclear membrane. This small chromosome is the inactivated x chromosome and is called a Barr body. Men lack Barr bodies. On this basis alone, cells from women are morphologically different from the cells of men. In women, a gene that happens to have a different **allele** (gene form) on each of the two x chromosomes, will be expressed in one of two different protein forms in every cell, depending on which x chromosome happens to be inactivated in the cell. About half the cells in the body express one form. The remaining cells in the body express the other form.

In a series of studies in the 1960s, a variety of different tumors were examined from women who had x chromosome heterozygosity for a protein of interest. If all the cells in a tumor grew from a single ancestral cell, you would expect that the tumor would express one allele or the other, but not both. This was the case. Most tumors expressed one allele only, consistent with a monoclonal origin of neoplasms.

In the past decade, there have great advances in tumor genetics. Many tumors have been found to have highly characteristic cytogenetic markers or gene markers. The occurrence of a characteristic gene aberration occurring in every cell of a tumor and in the tumor cells of virtually every tumor of the same kind from every patient who bears the tumor is very strong evidence of the monoclonality of neoplasms.

1.18 Monoclonal Proliferative Lesions

Not every monoclonal proliferation is neoplastic (List 1.18.1). Although monoclonality is characteristic of cancer, there are instances when a clonal expansion deriving from a single initiating cell may occur in nonneoplastic tissues. Every developmental lineage in an animal results from the clonal expansions of a small number (possibly one) of progenitor cells of the lineage. Antibody specificity is a manifestation of a clonal growth of plasma cells that yields a common protein product. You can even

List 1.18.1 Observations on monoclonality.

There are monoclonal cell populations that are nonproliferative.
There are monoclonal proliferative populations that are nonneoplastic.
There are neoplastic monoclonal populations that are benign.
There are malignant neoplastic populations that have a monoclonal origin but that become polyclonal as a consequence of **tumor progression**.

think of any animal as the nonneoplastic monoclonal expansion of a single cell (the fertilized ovum).

Paroxysmal nocturnal hemoglobinuria (PNH) is an example of an acquired genetic disorder manifested by a clonal expansion of hematopoietic stem cells with a mutation in PIGA (phosphatidylinositol glycan) synthesis. PIGA is a cell membrane constituent that anchors certain types of protein to the cell membrane. Red cells deficient in PIGA have high rates of hemolysis.

In PNH, the bone marrow slowly becomes a mosaic of cells, some of which are normal and some of which are PNH cells. PNH is not a cancerous condition. For the most part, PNH cells behave just like normal cells, although PNH red blood cells are more prone to lysis than normal red blood cells. In PNH, over time, the number of PNH cells in the marrow may increase. In this case, we infer that PNH cells have a growth advantage over normal cells. The greater the number of marrow PNH cells, the greater is the clinical effect of the disease.

1.19 Clonal Expansion in Paroxysmal Nocturnal Hemoglobinuria

The specific cause of the clonal expansion found in PNH is unknown, but we can infer that a progenitor PNH cell results from a mutagenic event and that the mutated PHN progenitor cell yields a clonal population that accounts for the disease.

Approximately 30% of PNH cases follow a bout of **aplastic anemia**. In aplastic anemia, there is a profound depletion of bone marrow elements, which results in a life-threatening pancytopenia (reduction of all types of circulating blood cells). Aplastic anemia is a rare disorder that can be caused by a variety of toxins (benzene, radiation, pesticides, chloramphenicol, medications), viruses (hepatitis, parvovirus B19, HIV, infectious mononucleosis), and autoimmune diseases. Without therapy, severe cases of aplastic anemia are often fatal. Survivors of aplastic anemia experience a gradual repopulation of the bone marrow from a small number of stem cells. Mutations that may be present in any of these surviving stem cells will be passed to all of their progeny cells. It is easy to imagine that during the recovery phase of aplastic anemia any mutation present in surviving stem cells will be prevalent in the recovered population.

The bone marrow is unique among tissues because it poses no structural limitation to the local dissemination of clonal populations. Contrast the bone marrow with the skin. When a skin stem cell grows, the progeny it produces are confined to the epidermis adjacent to the stem cell. If the stem cell proliferates faster than its neighbors, it may produce a small clonal patch, but the clonal patch will be found in close vicinity to the stem cell. The skin has a variety of conditions wherein a small focus of abnormal cells is found at a location (e.g., seborrheic keratoses, actinic keratoses, nevi, hemangiomas, and verrucae). In contrast to the confining growth of clonal skin lesions, the cells of the bone marrow are suspended in a loose mesh of adipose tissue and fibrovascular tissue, and the myeloid precursors are mobile. Clones can expand surreptitiously and can extend throughout the marrow. In PNH, a mutated stem cell with a growth advantage expands through the marrow. This means that a clonal growth within marrow can replace normal marrow cells without producing a focal lesion.

PNH is not the only disorder that can follow a bout of aplastic anemia. Many cases of **myelodysplasia** follow aplastic anemia. We discuss myelodysplasia in detail in Chapter 9. For now, all you need to know is that myelodysplasia is a neoplastic preleukemic condition. Both PNH and myelodysplasia result from the clonal expansion of an abnormal hematopoietic stem cell. In the case of PNH, the clonal expansion leads to a hemolytic disease. In the case of myelodysplasia, the clonal expansion can lead to leukemia.

1.20 Clonal Expansions of Normal Cells that May Not Lead to Cancer

Clonal expansions of cells are physiologic events. In the normal response to an antigenic stimulus, a unique immunocyte capable of producing an antibody that precisely matches the antigen is created and selected. The normal mature immunocyte grows as a clonal expansion.

Clonal aggregates of lymphoid cells are commonly found in a variety of autoimmune disorders, and these clonal aggregates have a clonal genotype (9).

1.21 Polyclonal Expansions that May Lead to Monoclonal Cancer

Although cancer is a clonal disease, not all cancers begin with a single altered cell. Acutely transforming viruses may infect and transform a large number of cells all at once. This can result in many different competing clones of abnormal cells. Early lesions can have polyclonality. Typically, over time, one clone outgrows all the others, and the resulting tumor is monoclonal. We discuss the special case of viral carcinogenesis in Chapter 9.

1.22 Tumor Growth Regulation and Tumor Autonomy

Cancers are sometimes described as unregulated growths of abnormal cells. Actually, the growth of cancer cells is regulated. It is just that cancer growth is not as exquisitely regulated as the growth of normal tissue.

Let us look at a hypothetical model of unregulated cell growth. A cell and all its progeny divide continuously, with no cell death. With each cycle, each cell gives rise to two child cells. A single **hepatoma** cell has a volume of 30,000 cubic microns (10). If the cell cycle time is 1 day, the volume of the hepatoma after a brief 45 days of growth would be 1 cubic meter. In 55 days, the volume would exceed 1,000 cubic meters. If an unregulated tumor were grown for the normal lifetime of a person, it would come to occupy much of the measured universe.

In fact, tumors have lower proliferative rates than some normal tissues. The normal rate of proliferation of gastric epithelium is higher than that of virtually any cancer. The enormous numbers of dying gastric epithelial cells are conveniently sloughed into the gut lumen and eliminated, along with dead epithelials collected from the entire alimentary tract, in feces. High proliferation rates only result in net growth when cell proliferation rates exceed cell death rates.

Terrestrial organisms are adept at homeostasis. In most instances, organs proliferate early in development, reach a certain size, and stay that size for the life of the organism. The homeostatic mechanisms that achieve this remarkable balance must involve multiple pathways whereby growth is balanced by death, and both processes are fine-tuned by feedback inhibitors and **enhancers**. Tumors are controlled by these same pathways, but not as well. Accumulated mutations in growth-regulating pathways take their toll. As we describe in Chapter 2, precancers (lesions that precede fully malignant cancers) often regress. It is likely that as mutations accumulate in small populations of precancerous cells, some mutations will favor cell death over cell growth. When cell death prevails over cell proliferation, the population regresses. Fully developed cancers are characterized by persistent growth. Mathematical models have shown that very small deviations from the normal balance between cell proliferation and cell death can account for the observed growth kinetics of tumors (11–13).

1.23 Limits on Tumor Autonomy

Neoplasms that grow without external stimuli are said to be autonomous; however, even autonomous tumors may respond to modifications of their environment.

In rodents, there are a variety of tumors whose growth is dependent on maintaining a favorable hormonal milieu. Shen and coworkers, using Spontaneous Dwarf Rats, have reported that tumors exceeding 1 cm in diameter (very large for a Spontaneous Dwarf Rat) regressed completely after growth hormone supplementations were stopped (14).

Similarly, in many mammals, growth of tumors stops during hibernation (15). The ability of advanced tumors to respond to normal growth controls (such as hormonal withdrawal and hibernation) indicates that some pathways involved in the control of cell growth still function. Control of these conserved pathways offers another approach to cancer and precancer therapies (16).

Chapter 1 Summary

Cancer cells do not invent new biological properties through which they achieve the malignant phenotype. They simply borrow existing cellular pathways. The principles that determine which pathways are available to which cancer cells is the subject of this book.

All cancers are characterized by persistent growth, invasion, and the potential to metastasize. In almost all cases, cancers attain their properties sequentially, beginning with persistent growth, acquiring invasion, and then acquiring metastasis.

In general, the easiest and most effective way to cure a cancer is by removing it from the body before any of the cancer cells have had a chance to metastasize to another site. No matter how rapidly growing the tumor may be or how invasive it may be, if it is removed before it metastasizes, the patient is always cured.

Cancers have morphologic features that distinguish one cancer from another and that distinguish cancer cells from normal cells. Cytologists can determine whether a cell is malignant by examining the morphology of its nucleus.

CHAPTER

Carcinogenesis

2

Cancer cells don't read textbooks.

—Anonymous

2.1 Background

Carcinogenesis is the process by which cells acquire a series of genetic and epigenetic alterations that result in a malignant phenotype (i.e., continuous growth, invasiveness, and metastasis). We need an entire chapter to discuss carcinogenesis because the specific events in the neoplastic process vary greatly among tumors. An understanding of the different types of neoplastic events and the different ways in which these events unfold will help us understand how we might interfere with the neoplastic process, thus preventing or eradicating cancer.

To provide some perspective to the problem, we can list some of the many options for the neoplastic process (List 2.1.1).

Although carcinogenesis provides many different options for tumor development, the steps of carcinogenesis occur sequentially (List 2.1.2).

The last step, progression, begins after the cancer has formed and continues for the life of the neoplasm. Progression has several components (List 2.1.3).

Two kinds of genes play a particularly important role in carcinogenesis: **oncogenes** and **tumor suppressor genes**. A single mutation in an oncogene or the inactivation of a tumor suppressor gene can precipitate the sequence of events that may lead to the emergence of a cancer. The purpose of this chapter is to describe carcinogenesis and to review a few things that we seem to know about the role of oncogenes and tumor suppressor genes in this process.

List 2.1.1 Properties of carcinogenesis and tumor progression.

1. Neoplasia can occur over a few weeks or over many years.
2. Neoplasia may involve few or many events that alter genes.
3. The events in the neoplastic process can arise under many different circumstances.
4. The neoplastic events leading to one tumor type may be different than the neoplastic events for another tumor type.
5. As a corollary to point 4, genetic events that may cause cancer in one cell type may have no effect at all when they occur in another cell type.
6. The neoplastic events leading to a tumor of a certain type may be different than the neoplastic events leading to another tumor of the same type (e.g., there may be many different genetic paths to colon cancer).
7. Progression is never finished. Neoplasia is an ongoing process that proceeds even after a tumor has achieved a malignant phenotype.

List 2.1.2 Time-ordered steps in carcinogenesis.

1. Initiation
2. Latency
3. Precancer
4. Cancer
5. Progression

List 2.1.3 Components of progression (can occur concurrently).

1. Increasing aneuploidy
2. Increasing **tumor heterogeneity**
3. Increasing aggressiveness
4. Metastasis
5. Dormancy
6. Accelerated tumor growth

2.2 Oncogenes and Tumor Suppressor Genes

Oncogenes and tumor suppressor genes are altered forms of genes that are expressed in normal cells. When an oncogene is overexpressed or altered to a more active form, it can produce a profound effect on a narrow range of cells, conferring a neoplastic phenotype (List 2.2.1). The normal (nonactivated) form of the oncogene is called the proto-oncogene. The more active form of the gene is called the oncogene. Some retroviruses contain oncogenes and can cause tumors by inserting their oncogene into the host genome.

A tumor suppressor is a gene that codes for a normal cellular control mechanism (List 2.2.2). When a tumor suppressor is inactivated, normal cellular controls are diminished or lost. Most genes have two copies (because human chromosomes are **diploid**). To eliminate the expression of a tumor suppressor gene, both copies of the gene must somehow be inactivated, by either a deletion, an inactivating mutation, gene

List 2.2.1 Types of oncogenes.

Receptor tyrosine kinases (e.g., EGFR, PDGFR, VEGFR, HER2/neu)
Cytoplasmic tyrosine kinases (e.g., Src-family, Abl)
Growth factors (e.g., c-Sis)
Cytoplasmic serine/threonine kinases (e.g., Raf kinase, cyclin-dependent kinases)
Regulatory GTPases (Ras protein)
Transcription factors (MYC)

List 2.2.2 Some regulatory activities of tumor suppressor genes.

Cell growth
Cell cycle checkpoint responses
Detection of DNA damage
Repair of DNA damage
Protein **ubiquitination**
Protein degradation
Mitogenic signaling
Cell differentiation
Cell migration
Cell death
Angiogenesis

silencing, and so on. This is usually referred to as biallelic inactivation (referring to the two alleles of the gene). Each of the inactivated alleles can be inactivated by independent mechanisms (e.g., one allele can be deleted, and the other can be mutated).

In nature, well-controlled processes usually have intricate mechanisms that balance stimulatory and inhibitory actions. The simple act of eating a meal is an extraordinary feat of neuromuscular balance and rhythm. The fork is held just tightly enough to sustain an angle and a tension adequate to harpoon a bit of food and draw it back to the mouth, where it deposits the food with just the right speed and pressure for the tongue to cradle and at just the correct time for the mouth and teeth to close without biting on the fork. Then comes chewing and glutition, both of which continue while the fork is propelled toward another morsel of food. The process of eating, like all control process, involves numerous opposing inhibitory and stimulatory signals. Likewise, many different genes contribute, in one way or another, to finely-tuned cellular control pathways. When a tumor suppressor gene is inactivated, some of the fine-tuning is lost. New tumor suppressors will likely be found for every aspect of normal cellular control, including growth, cell death, **DNA repair**, metabolic pathways, protein syntheses, transcription, translation, cellular degradation, transport, and so on.

2.3 The Narrow Range of Tissue-Specificity of Oncogenes

Malignancy requires the expression of abnormal **phenotypic** properties (persistent growth, invasion, and metastasis) in a clonal population. Observations on the occurrence of oncogenes in tumors would suggest that an oncogene has a limited range of **cell types** in which it can orchestrate a malignant phenotype. Several observations support this assertion (List 2.3.1).

There is a good reason why oncogenes restrict their neoplastic expression to a narrow range of cell types. Oncogenes work by altering pre-existing cellular pathways. If the pathways are not already present in the cell, the oncogene has nothing to work with; therefore, oncogenes can only impose a neoplastic phenotype on a narrow range of cells that come pre-equipped with the relevant pathways. In Chapter 14, we learn that

List 2.3.1 Three reasons why we know that oncogenes are cell-type specific.

1. Genetically engineered mice with inserted oncogenes develop tumors in narrowly restricted tissue types (17).
2. Germline mutations of activated oncogenes in humans result in tumors that occur on one or several tissues and in no others (e.g., the activated abr-bcl fusion oncogene specifically produces myeloid leukemias).
3. Even when mutagenesis events occur in many different cell types (as in radiation), the resulting (postradiation) neoplasms develop in a narrow range of cell types.

cellular pathways are inherited through a developmental lineage and that the developmental lineage of cells help us classify neoplasms into groups that share a common set of cellular pathways.

2.4 Are There Tumors that Have Multiple Activated Oncogenes?

Genetic instability is one of the hallmarks of cancer. As tumors grow, they acquire more and more genetic alterations. One would expect that, over time, tumors would acquire multiple active oncogenes. The co-occurrence of multiple oncogenes in a tumor might lead to a tumor with exaggerated properties of the malignant phenotype and a poorer **prognosis.**

To the best of our current knowledge, this is just not the case. Human tumors examined for the presence of multiple activated oncogenes seem to yield only one (18–21). A similar finding applies to genetically engineered mice. Cardiff and coworkers found that when genetically engineered mice had multiple different oncogenic transgenes, a single oncogenic transgene dominated the type of tumor that emerged (17).

If these very preliminary observations hold up against the test of time, we can infer a few fundamental lessons about carcinogenesis and tumor diversity. The absence of multiple activated oncogenes in tumors strongly suggests that additional oncogene activation confers no growth advantage to tumor cells. The increasing aggressiveness sometimes encountered in advanced cancers could be explained by alterations in nononcogenic genes or the epigenetic influence on pathways that enhance the malignant phenotype.

We can also infer that the growth advantage conferred by an activated oncogene is something that works for some cells but not for others. This would explain why people who inherit **germline mutations** that predispose to cancers have, in most cases, an increased likelihood of developing cancer in a narrow range of cell types. Most cells of in the body gain no growth advantage from any particular activated oncogene.

There are tumors that can be caused by any one of several different activated oncogenes but for which only one such oncogene is found in any single tumor. After a set of pathways has produced a malignant phenotype, there would be no benefit to acquiring an additional oncogene that activates the same pathways (21).

2.5 Do Tumor Suppressor Genes Cause Cancer?

Unlike oncogenes, some of which have the power to confer a malignant phenotype on otherwise normal cells, tumor suppressor genes usually play a passive role. In effect, inactivating a tumor suppressor gene increases the likelihood that an oncogene-driven clonal population will emerge.

After the discovery of the tumor suppressor gene p53, it was observed that inactivated p53 is found in the majority of all human tumors. It was certainly tempting to hypothesize that p53 inactivation was the underlying molecular mechanism that caused most human cancers. Sansom and Clarke review this idea in light of studies on p53 null mice (lacking any p53 activity) (22). Despite a profound absence of p53, these mice navigate the vicissitudes of embryonic and fetal development without developing congenital tumors (tumors present at birth). The p53 null mice have a higher-than-normal chance of developing cancer after birth, but if p53 in inactivated in the earliest embryonic cells, should we not expect p53 null mice to develop tumors during fetal growth?

We know that cancers contain many different mutations, an indication of genetic instability (23, 24). We also know that cancers can contain multiple mutations in tumor suppressor genes that normally regulate genetic stability (25). The loss of a tumor suppressor gene, even when it occurs in embryonic cells, does not confer a neoplastic phenotype. It simply reduces the cell's stability and increases the likelihood that a **proto-oncogene** will activate at some later time.

There are examples in nature where defects in genetic stability precede defects to proto-oncogenes (e.g., the **Li-Fraumeni syndrome**, the chromosome breakage syndromes, and the DNA repair deficiency syndromes) and examples where oncogene activation precedes genetic instability (e.g., RET oncogene mutation in familial **medullary** carcinoma syndrome). Mathematical analysis indicates that both types of mutations (those leading to proto-oncogene activation and those leading to tumor suppressor gene inactivation) occur most often when tumor suppressor inactivation precedes oncogene activation (26).

2.6 Can Genetic Alterations in Cancer Cells be Mediated (and Reversed) Through Epigenetic Modifications?

We can think of two distinct mechanisms by which the phenotype of a somatic cell can be heritably changed: genetic and epigenetic. All genetic mutations in a cell are passed to the progeny of the cell after normal **mitotic** division. Similarly, epigenomic modifications to a cell are preserved in the progeny after cell division. This is the basis of cellular differentiation. All cells in the body have the same genome (i.e., complete genetic sequence). Epigenomic modifications distinguish a liver cell from a muscle cell. During embryonic development, epigenomic modifications are preserved among the cells of a common lineage, and these modifications determine the differentiated phenotype of the different cells in the body.

Genetic mutations in cancer cells are recognized as physical changes in DNA molecules. Epigenetic changes are changes in nucleotide methylation or in the proteins that modify the structure of DNA without modifying the sequence of nucleotides in DNA.

Leukemias are neoplasms of blood cells. Normally, promyelocytes differentiate to become nondividing myelocytes (including neutrophils). Neutrophils are the major circulating nucleated cell and play a crucial role in inflammation and the body's defenses against infections. In acute promyelocytic leukemia, a gene **translocation** produces the PML/RAR(alpha) fusion protein (27). This fusion protein causes a progenitor promyelocyte to divide, producing a clone of mutated promyelocytes. Eventually, the population of clonal promyelocytes arising from the neoplastic progenitor cell will attain a sufficiently large number to be recognized clinically as a promyelocytic leukemia.

Acute promyelocytic leukemia is one of the few cancers that can achieve clinical remission without treatment with cytotoxic agents. Complete remission is achieved with all-trans retinoic acid, which somehow causes the neoplastic promyelocytes to differentiate and become nondividing mature myelocytes (28).

The mechanism by which the neoplastic fusion protein, PML/RAR(alpha), induces a neoplastic phenotype and the mechanism whereby all-trans retinoic acid reverses the neoplastic phenotype seems to be mediated through the epigenome. It is hypothesized that PML/RAR(alpha) modifies **histone** deacetylase complexes, resulting in the inappropriate transcriptional repression of genes would normally inhibit promyelocyte proliferation. All-trans retinoic acid reverses this effect (27). If this hypothesis is true, promyelocytic leukemia would serve as an example of a neoplasm that uses epigenomic alterations to sustain the neoplastic phenotype.

2.7 Tumor Initiation

Tumor **initiation** occurs at the moment when a normal cell is exposed to a **carcinogen**, producing a cellular lesion that, over time, may result in additional changes leading to the emergence of a neoplasm from a clonal descendant of the initiated cell.

Most encounters between cell populations and carcinogens do not result in the emergence of a tumor. Cells exposed to a carcinogen suffer reversible damage (i.e., they recover), or they die. Carcinogenesis is a rare event in experimental systems (2, 29). In the absence of initiation, cancer is an even rarer event. Some tumors have identifiable initiating events. For example, some cases of lung cancer are initiated by cigarettes, **oral cancer** by chewing tobacco, and angiosarcoma of the liver by polyvinyl chloride exposure.

As far as anyone knows, after a dividing cell has become initiated, it looks like any other cell. We can infer this assertion several ways. First, after tissues are exposed to a carcinogen (hence initiated), microscopic evaluation of the tissue does not reveal a morphologically altered subset of cells.

Even if you could produce a morphologic change using a carcinogen, there would be no reason to suspect that the altered cells were initiated cells. Following radiation treatment for cancer, cytopathic effects occur in connective tissue cells (fibroblasts, muscle cells, and vascular endothelial cells) within the radiation field. Scattered large cells, often with bizarre shapes, are seen. The nuclei of these cells are atypical, often multinucleated. These are the cells that survived a dose of ionizing radiation, and somehow continued to live. These cells are not initiated cells. They are scattered relics, incapable of sustained cell division. These postradiation effects signify that the connective tissue is unhealthy and cannot be used by surgeons to reconnect parts of tissues in future surgical procedures.

2.8 Tumor Latency

The **latency** period is the time between exposure to a carcinogen (the initiating event) and the emergence of a detectable neoplasm. The latency period is one of the least understood aspects of carcinogenesis.

We know that in some experimental systems, after initiation, the initiated cells look like normal cells and can participate in normal differentiation. Gestl and coworkers used a transgenic mouse strain containing the Wnt oncogene, but in which Wnt expression is doxycycline dependent (30). These transgenic mice, when provided with doxycycline in their diet, will almost always develop Wnt-initiated mammary adenocarcinomas. In this system, the process of postinitiation carcinogenesis can be interrupted by simply withdrawing doxycycline from the diet of the mice. Initiated mammary tissue can be transplanted into cleared fat pads of uninitiated transgenic mice (that have not had doxycycline in their diets) and then observed. The initiated mammary tissue, now growing without the influence of the Wnt oncogene, grows as normal mammary explants, producing ducts and lobules that are indistinguishable from the ducts and lobules of normal mammary tissue. If doxycycline is added to the diet of the mice receiving the transplants, the initiated tissue resumes its carcinogenic pathway to quickly yield multiple synchronous carcinomas. This tells us that initiated cells can look and behave like normal cells, under experimental conditions. Because tumors arose rapidly after doxycycline-dependent Wnt expression was restored, it suggests that the growth of initiated (but latent) cancer foci can be manipulated. This means that we may be able to develop new therapeutic interventions that stop the growth of latent cancers.

Currently, nobody knows all the conditions that lengthen or shorten the latency period. Because carcinogenesis is a multistep process, we can guess that every condition that modifies early steps in carcinogenesis will have some effect on the latency period.

In carcinogenesis experiments, we can expose a large number of rodents of a single strain, gender, and age to a single carcinogen at the same dose. In the 1970s, cancer researchers used this approach to investigate whether the initiating dose of a carcinogen would modify the latency period. Specifically, they wanted to determine whether low doses of a carcinogen would result in a latency period that exceeded the life expectancy of the animal. If this were the case, there would be no practical problem with exposing animals (or people) to doses of carcinogen that produce a longer-than-lifetime latency.

Many rodent trials ensued. Latency was measured as the time between initiation and the time that the first tumor occurred. It is not surprising that high doses of carcinogen produced short latency periods. Low doses of carcinogen resulted in long latency periods. Very low doses of carcinogen produced very long latency periods. Extremely low doses of carcinogen usually produced no more tumors than no carcinogen at all.

In a mathematical analysis conducted in 1977, Guess and Hoel showed that these prior measurements of tumor latency were simply manifestations of dose-dependent changes in tumor incidence. If a high-dose drug produces many more tumors than a low-dose drug, and if the tumor latency period is random (i.e., not determined by the dose of drug), you would see that tumors occurred earlier in the high-dose subjects. Why? The more tumors that occur, the likelier it would be that some tumors will occur early rather than late. If a low-dose carcinogen produces very few tumors, the time before the first tumor develops will likely be longer, simply because there are fewer tumors (31). Today, low-dose risk extrapolations take into account dose-dependent changes in tumor incidence. Despite decades of research, tumor latency remains one of the murkiest areas of carcinogenesis. We do not know whether there is a safe dose for carcinogen exposure.

2.9 Tumor Dormancy

Dormancy occurs when an established cancer (one that has passed through the stages of initiation, latency, and precancer) stops growing for a period of time. When people use the word dormancy, they are usually referring to one of three situations (List 2.9.1).

In each of these cases, the tumor undergoes a period of dormancy. Biologically, each instance of dormancy is a distinctly different process. In the first instance, an entire tumor ceased growth. In the second instance, occult metastases (i.e., metastases that were not evident at the time of diagnosis) persisted without noticeable growth for some period of time before becoming clinically obvious. In third instance, a patient with known metastatic cancer was treated with systemic chemotherapy. The metastatic lesions survived chemotherapy and eventually grew to become clinically detectable masses.

List 2.9.1 Examples of tumor dormancy.

1. A patient has an invasive **brain tumor** and has refused any form of therapy. The tumor is followed by radiologic imaging at regular intervals. The tumor that had been growing suddenly ceases to grow, for several years, before resuming growth.
2. An invasive primary tumor is excised. There is no evidence at the time of excision that the tumor has metastasized. Several years later, the patient develops clinically obvious metastatic lesions in multiple organs.
3. A patient with carcinoma is diagnosed with cancer. The cancer is staged. Several regional lymph nodes (near the primary cancer) are found to have small deposits of metastatic cancer. The patient is treated with excision of the primary tumor followed by intensive chemotherapy. Several years later, the patient develops clinically obvious metastatic lesions in multiple organs.

The phenomenon known as tumor dormancy probably represents several different processes that have the same observed outcome: growth cessation and growth renewal of a formerly growing neoplasm. In the case of the second instance (occult metastases growing after a period of time), we can speculate that the process of metastasis selected for a cell that escaped from a primary tumor and settled in a distant anatomic site. The selection process for a metastatic cell may not have favored a cell that was particularly well equipped to grow into a thriving clone. It is likely that most metastatic cells die without producing clinically detectable metastatic lesions. The emergence of a growing metastatic subclone may have required multiple successful generations of very slow growth, during which a fast-growing descendant cell was selected.

2.10 Spontaneous Regression of Latent Tumors

Although we know very little about the cellular kinetics of latent neoplasms, we may infer a few principles based on models of clonal growth and general observations made on normal and abnormal tissue growth.

We have already noted that tumors do not grow like bacteria, through an unrestricted exponential expansion of a clonal progenitor. If tumors grew like bacteria, they would achieve the size of refrigerators after a few dozen cell generations. Tumors, like every tissue, balance proliferation rates against cell death rates. Homeostasis is achieved in adult tissues when the cell growth rate equals the cell death rate. When you think about the constantly proliferating tissues of the body (e.g., the epidermis and the gut mucosa), it is remarkable that we virtually never encounter diffuse overgrowths of these tissues.

Mathematical models of clonal cell growth include growth restraints that mimic the growth restraints observed in normal tissues. After a cell divides, the offspring may consist of two dividing cells, or the offspring may consist of two nondividing cells (that live for a while and then die). The offspring may also consist of a dividing cell plus a nondividing cell. The probabilities for these events become part of the mathematical model. The model computes the possible outcomes, over many generations, of clonal growth, using different initializing conditions.

In a series of articles that I published in the early 1990s with G. William Moore, simulations showed that clones, with growth properties observed in human neoplasms, will, as a rule, reach extinction (11–13). This means that if you start with a single cell imbued with a higher-than-normal cell growth rate and a lower-than-normal cell death rate, the clone of cells that grows from the progenitor cell will, in most cases, die out!

The same growth kinetics that support the relentless growth of established tumors will lead to the extinction of early clones of tumor cells. Why? In a computer simulation, when a clone is small (one cell or maybe a few dozen cells), there is a reasonable chance that all of the cells will be unlucky and will reach a nondividing state. When that happens, the simulated clone "dies." In a tumor, composed of many millions of cells, there is essentially no chance that all of the cells will attain a nondividing state.

Basically, large populations grow differently than small populations, even when the individual cells in the small and large populations obey the same growth kinetics.

Genes mutate spontaneously. Oller and coworkers have estimated that the spontaneous mutation rate is approximately 0.2 per million genes per cell division (32). This estimate is in the same ballpark as estimates from other laboratories, all of which are somewhat speculative. This may seem like a small number, but there are billions of dividing cells in the body, and each cell contains over 25,000 genes. Thus, you might expect cancer-related mutations to occur all of the time. The predicted low success rate of clonal growth during the latent period of carcinogenesis (after an initiating mutation and before the appearance of a detectable precancer) may explain why cancer has not wiped out all multicellular life on earth.

2.11 Are Tumor Latency and Tumor Dormancy Related Phenomena?

It might appear that dormant tumor cells are genetically very different from latent tumor cells, as they derive from very different points in tumor progression. Still, there are striking similarities (List 2.11.1).

A dormant cell may have simply lost use of a pathway that conferred net growth. A latent cell may be "waiting" to gain the pathway that confers net growth. The observation that both latency and dormancy are benign states that precede and accompany the malignant phenotype might indicate that there is a biological process through which malignant cells may evade the malignant phenotype.

2.12 Chemical Carcinogenesis

Nobody knows, with any certainty, the proportion of human cancers that are caused by chemicals found in our environment; however, scientists seem to agree on some approximate numbers, and these numbers appear countless times in the popular literature. Nonbiological agents (including chemicals and radiation) account for about 85% of human cancers. Biological agents (viruses, bacteria, parasites) account for 15% of human cancer. Inherited conditions account for 15% of human cancer. The total comes to 115%.

List 2.11.1 Similarities between latency and dormancy.

Latent and dormant cells sit for long periods without net growth.
Latent and dormant cells can regress.
Latent and dormant cells can suddenly acquire the malignant phenotype (**promotion** of initiated cells, stimulation of dormant cells).

The reason that the total exceeds 100% is that carcinogenesis is a multievent process, and different events may have different causes. One person my have an inherited predisposition to a virally induced cancer. In that case, a virus and an inherited gene take credit for the same tumor. This may not be mathematically rigorous, but it makes a point. When we study cancers, we tend to look at part of a very big and complex story that takes place over decades of time. The concept of a "cause" becomes very fuzzy when stretched over a lifetime of events.

Most of the chemicals that cause cancers do so following metabolic transformation to an active form that interacts with DNA and produces mutations by damaging the DNA molecule (breaking strands, cross-linking strands, and so on) or by forming adducts (added chemical moieties) on the DNA molecule. Damage to DNA is passed to succeeding generations of cells unless the damaged portion of the DNA can be quickly repaired.

2.13 Why Do Chemical Carcinogens Need to be Activated by Cells in our Bodies Before They Can Cause Cancer?

The active moieties of most chemical carcinogens will react with virtually any molecule in their vicinity: air, water, cell membrane, proteins, cytoplasm, or anything. For this reason, chemicals that are chemically reactive in their natural state are almost never carcinogens: they expend all their reactivity on nongenetic molecules before reaching the nucleus! A potent carcinogen is a molecule that can pass through membranes and through the interior of a cell without reacting with other molecules. An effective carcinogen waits until it gets near to its target molecule (DNA) before it activates. In most cases, activation is achieved within the smooth endoplasmic reticulum. In the endoplasmic reticulum are a variety of enzymes that can activate some molecules and deactivate others. Carcinogens are activated to a highly reactive **species**. Reactive species formed deep within a cell may find their way to DNA. Some of the DNA alterations may lead to mutations that are passed to progeny cells. Some of those mutations may lead to cancer.

There are exceptions. Some unstable molecules spontaneously achieve an active state (without the participation of cellular enzymes), but the rate at which the reactive species are generated is sufficiently slow to give the nonreactive form enough time to gain proximity to the nucleus. This is the case for several carcinogenic alkylating agents.

Other agents produce mutations in DNA without chemically reacting with bases. Intercalating agents (such as ethidium bromide) are examples of carcinogens that are not activated within cells. Silicates and so-called foreign body carcinogens and a range of inert fibers (including asbestos) do not require enzymatic activation. Still, most of the carcinogens delivered in food, air, and water are chemical carcinogens that require intracellular activation.

2.14 Do We Know Which Chemicals Are Likely to Cause Cancer in Humans?

There are about 100,000 different chemicals that are registered for commercial use in the United States. A good portion of these chemicals are produced in only small quantities, and many of the chemicals simply never come into contact with humans. In 1984, the U.S. National Academy of Sciences found 12,860 chemicals that were produced in volumes exceeding 1 million pounds per year. For the majority of these compounds, we have no toxicity data. Nearly half of food additives have no toxicity data, and about a third of registered pesticides have no toxicity data (33, 34). The number of industrial chemicals far exceeds our ability to test all these substances for carcinogenicity.

2.15 Naturally Occurring Carcinogens

It is easy to think of any organism as a small chemical plant. For chemical diversity and metabolic activity, the plant, fungi, and bacteria kingdoms easily surpass the animal kingdom. A simple mustard seed contains over 1,000 different chemical compounds, and we know almost nothing about the carcinogenic potential of these ingredients. When we add herbs and spices and plant extracts to our foods, we know nothing about the long-term consequences of our food choices.

Basically, humans and mammals have never been much good at metabolizing chemicals. Humans cannot metabolize cellulose, the most abundant structural protein of the plant kingdom. Animals that derive nutritional benefit from cellulose do so by using bacteria or fungi as metabolic servants.

The reason that plants and insects excel at chemical synthesis and metabolism relates to their methods of defense. Plants defend themselves from bacteria, fungi, insects, and animals through chemical warfare. Plants produce a wide variety of chemicals designed to incapacitate and discourage animals. Whenever we eat plants, we are eating toxic chemicals. A few examples of the kinds of carcinogens ingested with plants and fungi are cycasin, aflatoxin, and gyromitrin.

Gyromitrin is a hydrazine compound present in most members of the False Morel genus (*Gyomitra* spp.) of mushrooms. Many hydrazine compounds are carcinogenic, and it is quite possible that gyromitrin is a human carcinogen. Although not as well known to the public as alpha-amanitin, a liver toxin found in the *Amanita phalloides* mushroom, gyromitrin is an example of a widely distributed compound with an unknown effect on the incidence of tumors in humans.

Cycasin is a toxin found in every genus of Cycad and is found in the seed and the pollen. It produces a variety of tumors when fed to animals (35).

Aflatoxin is a strong hepatocarcinogen produced by *Aspergillus flavus,* a fungus that grows as a common contaminant on peanuts and other crop plants.

Aflatoxin is not removed from peanuts and peanut butter purchased in stores. Food manufacturers are forbidden to add known human carcinogens to foods, but they are not required to remove naturally occurring carcinogens that happen to be in the foods that we eat. Peanut butter manufacturers take great pains to insure that peanuts are harvested under conditions that minimize their contamination with aflatoxin, and they measure the amounts of aflatoxin in manufactured peanut butter to ensure that batches that exceed an allowed level will never reach the market.

There is no way of knowing all of the carcinogens that we eat and drink. Because a plant may have many different marketed varieties and may contain thousands of chemical constituents, and because the chemicals synthesized by a plant may vary based on growth and storage conditions, it is impossible to know exactly what we are eating when we munch on a fruit or vegetable.

A recent study found no evidence of a beneficial effect of a diet high in fruits and vegetables for women who had received treatment for breast cancer (36). This expensive randomized controlled trial involved multiple institutions, over 3,000 women and extended over a decade.

Despite these considerations, experts at the American Cancer Society and the U.S. National Cancer Institute have urged us to consume five servings of fruits and vegetables each day. When you think about it, the concept of a diet "high" in fruits and vegetables is vague and immeasurable. Which fruits and vegetables, consumed in what particular combination, will lower the incidence of what types of cancer for which people?

These comments are only intended to indicate that cancer prevention is complex. At this point, the most we can say with any certainty is that it is wise to lower your exposure to carcinogens when feasible. Cooked meats are known to contain mutagenic pyrolysis products (chemicals formed by the action of high temperatures on meat products) (37). Eating fruits and vegetables to reduce exposure to the known carcinogens present in cooked meats might be a good idea. Sometimes our only choice in life is to pick our favorite poison.

2.16 Do Viruses Cause Cancer?

Some tumors of man and other animals are caused by viruses (List 2.16.1). We discuss the biology of viral carcinogenesis in Chapter 9.

Some of these viruses profoundly influence the number of cancer deaths worldwide. These include hepatitis B virus (associated with an increased incidence of **hepatocellular carcinoma**) and human papillomavirus (which causes cervical cancer). **Liver cancer** is the third leading cause of cancer deaths worldwide, accounting for 611,000

List 2.16.1 Some viral causes of human cancer.

Epstein-Barr virus (B-cell lymphomas, Burkitt's lymphoma, nasopharyngeal cancer, Hodgkin's disease, and T-cell lymphomas)
Hepatitis B virus (hepatocellular carcinoma)
Human papillomavirus types 5, 8, 14, 17, 20, and 47 (**skin cancer**)
Human papillomavirus types 16, 18, 31, 33, 35, 39, 45, 52, 56, and 58 (cervical cancer, anogenital cancer)
Human papillomavirus types 6 and 11 (verrucous carcinoma)
Human papillomavirus types 16, 18, 33, 57, and 73 (cancers of oral cavity, tongue, larynx, nasal cavity, and esophagus)
Human T-lymphotropic virus type 1 (adult T-cell leukemia)
Human herpesvirus 8 (Kaposi's sarcoma)
Hepatitis C virus (hepatocellular carcinoma and low-grade lymphomas)
JC, BK, and SV40-like polyomaviruses (tumors of brain, pancreatic islet tumors, and mesotheliomas)
Merkel cell polyomavirus (Merkel cell carcinoma of skin)
Human endogenous retrovirus K (**seminomas** and germ cell tumors)

deaths in 2000 (38). Cervical cancer is the 11th leading cause of death worldwide, accounting for 254,000 deaths in 2000 (38).

2.17 Are There Tumors that Arise Spontaneously, Without Initiation, and Without Any Contribution from an External Carcinogen?

We can think of cells as tiny chemical factories. Other than size, the major differences between chemical factories like Dow or Monsanto and chemical factories like the hepatocyte or the proximal tubular epithelial cell is one of chemical diversity. The cells handle many more chemicals, produce many more intermediate and final chemical species, and do it all simultaneously and without resorting to physical extremes of temperature or pressure.

During the normal course of events, our cells produce potent endogenous **mutagens** and carcinogens (ethanol, formalin, free radicals, and so on). Based solely on our production of endogenous carcinogens, you might be tempted to predict that humans must expect an irreducible number of spontaneously occurring neoplasms.

Despite these self-imposed attacks on our genetic material, our cells are built to cope with endogenous and exogenous carcinogens through the deployment of efficient and dependable DNA repair mechanisms.

Some scientists suspect that cancer never results from normal cellular activities. They believe that cancer only occurs when there is an influx of exogenous carcinogens that overwhelm the intrinsic repair mechanisms in our cells, or if there is an acquired or inherited flaw in the DNA repair mechanisms.

This idea of "perfect" DNA repair becomes important when we try determine the "safe" exposure levels from low-dose exposure assays in animal models. If you believe that normal individuals can repair the damage produced by a low level of carcinogen, then exposure to the carcinogen is safe so long as the levels of the carcinogen do not exceed the maximal repair rates in our cells. Exposures above that level would incur some cancer risk. If you believe that normal cellular repair is imperfect and cannot reliably repair every DNA mutation, then you might wish to eliminate all exposure to known carcinogens. Currently, we simply do not know which belief is correct.

2.18 Cancer Regression

The general rule for natural progression of cancer growth is unrelenting growth followed by invasion, metastasis, and death. This is not always the case. Sometimes, neoplasms simply stop growing and even disappear (regress).

Regression is a poorly understood phenomenon that occurs commonly in precancers. Precancers are non-invasive, usually small, lesions that precede the development of invasive cancers. We know that regression occurs because we can observe animal models in which large numbers of induced precancers simply disappear, leaving a small number of precancers that persist (2, 29, 39). The most easily studied human precancers are actinic keratoses (precancers for squamous carcinoma of skin) and dysplastic nevi (precursers for **malignant melanoma**). Both these lesions are very common, and their progression can be followed over time. Countless observations of these lesions in humans indicate that the likelihood for any single lesion (either actinic keratosis or dysplastic nevus) to progress to cancer is quite small (40). Although rare, we occasionally encounter fully formed cancers that regress. There are many possible mechanisms for tumor regression, and different mechanisms may apply to different steps in the process of carcinogenesis (List 2.18.1).

Obviously, if we could induce tumors to follow naturally occurring **regression pathways**, we could eliminate tumors. The mechanisms of tumor regression are a vastly underexamined area of cancer research.

2.19 The Inherited Syndromes that Cause Cancers

Inherited mutations are germline mutations. A mutation in the germline is present in every diploid cell of the developed organism. There are many inherited diseases that are associated with a heightened risk of developing cancer. Often, the inherited syn-

List 2.18.1 Possible mechanisms of tumor regression.

1. The population of neoplastic cells may not have acquired all of the biological properties needed for sustained growth. These would include angiogenesis, heightened **glycolysis** (to survive a hypoxic microenvironment), invasiveness, and so on.
2. The population may be unlucky. As noted in a prior section, Monte Carlo simulations of tumor growth have shown that small populations often die out when the cell growth rate and the cell death rate of a neoplastic population are of similar magnitudes (11, 12).
3. The population of tumor cells is sufficiently immunogenic to stimulate a host response that eliminates the primary cancer (sometimes seen in small primary **melanomas** and benign nevi and in immunogenically stimulated transitional cell tumors of bladder) (41, 42).
4. The tumor differentiates to a nondividing population of cells (the putative mechanism for regression in **keratoacanthoma**) or to a benign neoplasm (as occurs in some cases of neuroblastoma that develop into **ganglioneuromas**) (43, 44). Alternatively, the tumor cells may fail to immortalize (i.e., cannot attain unlimited cell replication) (45).
5. The growth advantage of the tumor, compared to normal cells, is lost when a growth stimulus is removed from the host environment (e.g., hormone-responsive tumors, some cases of *Helicobacter*-induced **MALTomas**, possibly immuno-stimulated lymphomas).
6. Virally induced tumors for which viral replication ceases (e.g., tumors occurring and regressing in transiently immunosuppressed patients) (46).

drome is associated with a specific type of cancer or with several different cancers that share a common property. We discuss the inherited cancer syndromes in Chapter 8.

2.20 Minimal Number of Events Necessary for Tumor Development

As discussed, carcinogenesis is a slow process. Following initiation, there is a long latent period before a precancer emerges. The precancer often grows for years before progressing into an invasive cancer. During the latency and precancer stages, the fledgling tumor is much more likely to regress than to progress to an invasive cancer.

Sometimes, however, we encounter tumors that seem to arise **ab initio**, from a single biological event. In the case of acutely transforming retroviruses, the usual multistep process of carcinogenesis is bypassed. The Rous sarcoma virus (RSV) is a transmissible agent that produces sarcomas in chickens. The acutely transforming variant of RSV contains a mutated version of the src oncogene (v-src), which it must have captured from an animal host genome at some time in the past. When the virus is injected into chicks, the v-src gene incorporates into host DNA of connective tissue cells. The v-src gene transforms infected cells, establishing the full malignant phenotype in a single generation. Within a few days, multiple tumors emerge, killing the chick (47). So powerful is the integrated src oncogene that injection of DNA extracted from Rous sarcoma virus is sufficient to produce sarcomas in chicks (48).

2.21 Fast and Slow Carcinogenesis

It is useful to think of two types of carcinogenesis: slow and fast. Slow carcinogenesis accounts for most tumors that are caused by chemical and physical carcinogens and for most inherited cancers (i.e., over 99% of the cancers that are encountered in man and animal). Fast carcinogenesis occurs very rarely in nature, but it can be replicated and studied under laboratory conditions.

Most of the retroviruses that induce cancer in humans and animals are not acutely transforming and do not contain viral oncogenes. Retroviruses that lack a viral oncogene are called proviruses. They produce cancer by inserting themselves randomly into host DNA. Sometimes, the inserted viral gene occurs near one of the host's cellular oncogenes. The provirus alters the expression of the host cell's proto-oncogene, and this seems to be biologically equivalent to alterations in cellular oncogenes produced by chemical carcinogens.

Following cellular oncogene activation by a retrovirus, the initiated cells pass through a latency phase and a precancer phase, just as they would if activated by a chemical carcinogen.

Fast carcinogenesis follows infection with an acutely transforming retrovirus containing a viral oncogene. Under some conditions, a suitably altered oncogene, delivered to a susceptible tissue in a susceptible host and inserted into the correct location in the genome, will result in the immediate neoplastic transformation of a cell.

2.22 Iatrogenic Insertional Carcinogenesis

Iatrogenic means "caused by a physician" and comes from the Greek "iatros," physician, and "genic," the source. In a disappointing miscarriage of the Law of unintended consequences, the field of gene therapy was dealt a stunning setback that has had repercussions within the field of cancer research. One of the few instances in which gene transfer experiments have yielded positive results has been in the correction of x-linked severe combined immunodeficiency. This disease is caused by an inherited deficiency of adenosine deaminase and results in profound immunodeficiency. As therapy for this condition, a retrovirus is used to carry the missing gene into the genome of autologous lymphocytes of affected children. In some children who received gene transfer therapy, a peculiar clonal proliferative process of T-cell lymphocytes develops about 3 years after therapy. Examination of the proliferating T lymphocytes showed that the retrovirus integrated in proximity to a promotor of the LMO2 oncogene, leading to overexpression of LMO2 (49). The overexpressed oncogene resulted in the uncontrolled growth of cells, which eventually manifested as a T-cell lymphoma. This is another example of a peculiar property of hematopoietic cells: that sustained hyperproliferation in the absence of any additional acquired phenotypic properties is sufficient to produce a leukemia or lymphoma.

2.23 Can Cancer Be Reversed by Modifying a Single Oncogene?

In a remarkable paper by Shachaf and coworkers, a malignant tumor was cured by inactivating its oncogene (50). The authors used a transgenic mouse that conditionally overexpressed the MYC proto-oncogene in liver cells. All mice that overexpressed MYC in liver cells developed liver tumors in about 12 weeks. The tumors had features of hepatocellular carcinoma or of hepatoblastoma. The tumors were locally invasive, metastatic, and transplantable.

In the transgenic mice used, overexpression of MYC occurs unless the MYC promoter is inactivated by doxycycline. Liver-derived cells (normal and neoplastic) from these mice, when treated with doxycycline, do not express MYC. When transgenic mice carrying liver cancers were treated with doxycycline, overexpression of MYC ceased, and the liver tumors regressed. In fact, every liver tumor in every mouse regressed, even when the mice were moribund from their tumor burdens. Furthermore, the liver underwent a restorative process, recruiting normal hepatocytes and bile duct cells from the formerly cancerous cell population. Even nontransgenic mice, carrying transplanted liver tumors from the transgenic strain, enjoyed complete, sustained tumor regressions after doxycycline treatment.

Do these observations herald a cure for all human cancers? Probably not. The liver tumors occurring in the MYC-activation murine model were driven by a pleiotypic response (a response in which multiple cellular processes are altered by a single gene) that quickly yielded tumors (in approximately 12 weeks). In this model, carcinogenesis did not involve a latency period, an in situ growth period, or a postinvasive tumor progression period during which multiple mutations accumulated and heterogeneous subclonal populations emerged. The nuclear cytologic features in the liver tumor cells were characterized by enlargement and the appearance of prominent nucleoli, changes that can be observed in nonneoplastic growing cell populations (i.e., not indicative of **aneuploidy**). Comparative genomic hybridization, a technique that detects DNA copy number anomalies (e.g., amplifications, losses, and duplications of DNA sequences), was performed to compare normal hepatocytes and tumor cells. With the exception of the amplification of c-MYC in tumor cells, the gene copy numbers appeared similar in normal cells, tumor cells, and tumor transplants. This suggests that changes produced by c-MYC overexpression in the transgenic mouse model were limited and reversible. In contrast, profound genetic abnormalities are seen in all of the commonly occurring tumors of humans.

Would the outcome have been different if the mouse tumors had developed additional mutations contributing to the malignant phenotype? Giurato and coworkers used a similar MYC inactivation model (51). In their model, aggressive lymphomas occurred in transgenic mice with MYC overexpression in lymphoid precursor cells. Treatment with doxycycline resulted in a regression of all primary tumors; however, treatment with doxycycline did not result in the regression of all transplanted tumors in **syngeneic** hosts. One out of four transplanted tumors persisted. Why? The authors demonstrated that the persistently growing tumors had acquired another abnormality: p53 inactivation. After this second genetic abnormality was acquired, MYC inactivation alone could no longer regress the malignant phenotype.

Chapter 2 Summary

Carcinogenesis is the collection of biological processes that precede the development of a clonal lesion with persistent growth that can invade, and metastasize.

The earliest event in carcinogenesis is initiation. Initiation occurs at the moment when a normal cell is exposed to a **carcinogen**. Following initiation is a period of latency, in which events proceed without demonstrable changes in cell morphology.

During latency, there is an expansion of the population of initiated but morphologically unaltered cells. Some time later (days, weeks, years, or decades), a morphologically altered population of cells appears, marking the end of latency and the beginning of the precancer phase. In experimental systems, the latency period can be drastically shortened with promoting agents, which induce the appearance of many subclonal populations of growing cells that can be distinguished from the normal cells. Most of these promoted subclones of initiated cells will regress without forming cancers.

The altered population of cells that emerges after the latency period is called a precancer. Precancers lack the ability to invade or metastasize, and most precancers lack the property of persistent growth. Therefore, most precancers regress without ever leading to cancer. A subset of precancers continues to grow, and subclones of precancerous cells emerge. Eventually (days, weeks, years, or decades later), a population emerges with the properties of persistent growth and invasion. At this point, the precancer has become a cancer (Figure 2.1).

Cancers contain mutations in oncogenes and in tumor suppressor genes. Oncogenes provide a growth advantage to a narrow range of cell types. People who inherit germline mutations that predispose to cancers have, in most cases, an increased likelihood of developing cancer in a narrow range of cell types. Other cell types in the body gain no growth advantage from the oncogene.

Most tumor suppressor genes work by interfering with processes that would normally regulate cell homeostasis and integrity. Tumor suppressor genes do not have the same narrow range of cell specificity observed with oncogenes. One tumor suppressor, p53, is found to be inactivated in the majority of the different cancers occurring in humans.

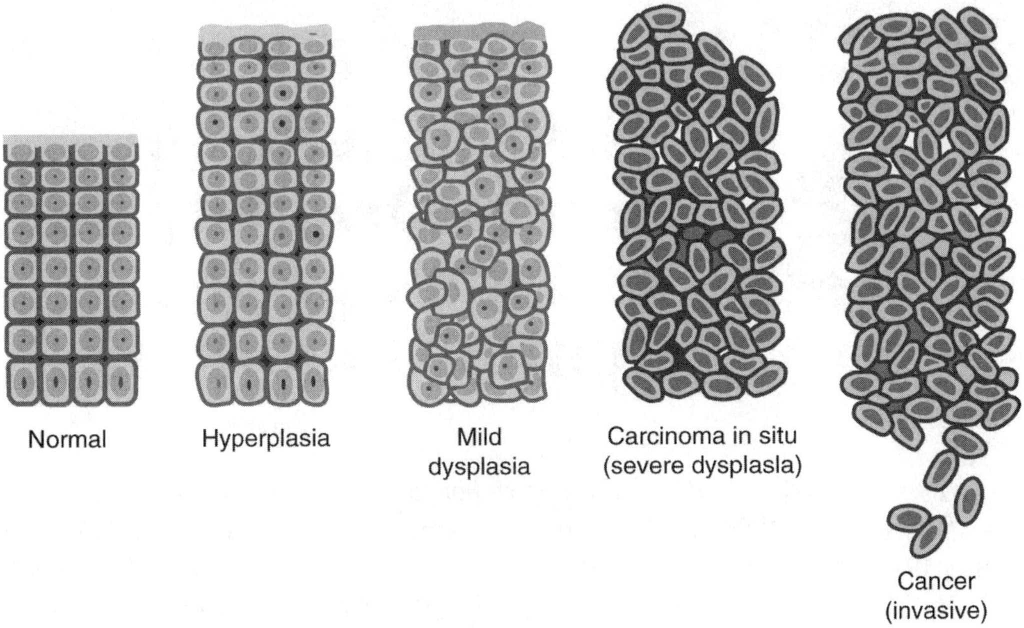

FIGURE 2.1 Schematic showing the cellular progression from precancer to invasive cancer. Cytologic atypia and tissue disorganization increase over time within the dysplastic epithelium.

Following the emergence of a fully malignant cancer, the tumor undergoes progression. Progression is the process by which subclones appear in the tumor, and these subclones have diverse properties. Over time, the biology of the tumor may change, often to a more aggressive and metastasis-prone phenotype, due to the emergence of a dominant subclone. Progression occurs as the result of genomic and **epigenomic instability** within the cells of the cancer.

When humans die from cancer, they usually succumb to their metastases, not their primary tumor. Sometimes, following excision or chemotherapeutic treatment of a primary tumor, the metastatic cells that have seeded the body cease to demonstrate net growth for some indeterminate period of time—the dormancy period.

It is not unusual for a person with cancer to enter a remission after their treatment. During remission, the person has no signs of cancer. Sometimes the cancer never recurs. Occasionally, the remission period is followed by the sudden reappearance of cancer in one or more anatomic locations. We can infer that the metastatic cells were dormant during the remission (disease-free) period. If we knew how to premanently arrest or kill dormant cells, we would be able to save many lives.

CHAPTER

3 The Mysteries of Tumor Diversity and of Type-Specific Tumor Uniformity

But when I studied more detailed properties of cellular automata, what I found was that most of these properties were closely correlated with the classes that I had already identified. Indeed, in trying to predict detailed properties of a particular cellular automaton, it was often enough just to know what class *the cellular automaton was in.*

—Wolfram S. *A New Kind of Science.* Champaign, IL: Wolfram Media, Inc., 2002, p 235.

3.1 Background

In Chapter 1, we described neoplasms and discussed the properties that define all cancers: persistent growth, invasion, and metastasis. In Chapter 2, we discussed the process of carcinogenesis and how this process leads to continuous tumor progression in cancers. As the result of progression, cancers usually become more atypical, more aneuploid, more heterogeneous, and more aggressive over time.

If there are thousands of different carcinogens, and if carcinogens cause many nonspecific mutations in DNA, and if the resulting tumors are constantly accumulating more and more mutations over time, why do we see a distinct set of tumors that occur in humans? Why is there just a limited number of types of primary tumors that occur in any organ of the body? Why does a tumor of a given type (e.g., a Warthin tumor of salivary gland, or a pilimatrixoma of skin, or a clear cell carcinoma of kidney) look almost identical to every other specimen of the same type, occurring in different individuals, even though we know that each of these tumors is genotypically unique? How do **pathologists** define and reliably diagnose the few thousand tumors that encompass all of the tumors that occur in humans? Why is it that even if we encounter new carcinogenic chemicals and viruses, we can be virtually certain that these new agents will produce one or more of the known types of cancers and will not produce a type of cancer that has never been previously encountered? Why do animals, with different genomes than humans, develop tumors of the same cell types as the tumors that develop in humans, often with the same oncogenes as those that occur in human tumors?

The paradox of cancer is the same paradox that we see in animal speciation. Each individual member of a species is unique; nevertheless, every individual creature on earth fits neatly into a class of animals with a distinguishing set features common to every member of the class and absent from every animal that is not in the class. The same is true of cancers. Each particular tumor of a named type is different from every other tumor of the same named type, yet every tumor of a named type shares features that permit pathologists to name the tumor and to distinguish the tumors from all other named tumors.

The purpose of this chapter is to explain how tumors manage to be unique and still fall into distinct biological species.

3.2 Tumor Heterogeneity

Tumor heterogeneity can be perceived as an unintended consequence of tumor clonality. For a cancer clone to arise, some event must have occurred that permitted a single cell to grow out from a population of pre-existing cells. Tumor heterogeneity is the presence of multiple, distinguishable subclones arising within a clonal tumor. Presumably, the same neoplastic process that led to the emergence of the original progenitor cell has also led to the appearance of subclones arising in the clonal tumor. In Chapter 9, we will see that tumor heterogeneity is a consequence of tumor genetic and epigenetic instability.

Tumor heterogeneity poses an enormous challenge to the successful treatment of certain classes of tumors. In essence, the existence of tumor heterogeneity is equivalent to the existence of multiple tumors within a single tumor mass. A treatment that is active against one of the subclones within the tumor may be ineffective against other subclones of the same tumor. The persistence of a single subclone of tumor cells after treatment may result in the rapid emergence of a new, aggressive, and treatment-resistant population of tumor cells. In Chapter 23, we discuss strategies for treating heterogeneous tumor populations (List 3.2.1).

List 3.2.1 Inferences from heterogenous tumors.

1. They have genomic and epigenomic instability (they continue to progress and change over time).
2. They have nuclear atypia.
3. They are more likely than homogeneous tumors to develop multidrug resistance.
4. They are more likely to occur in adults.
5. They are more likely to continue enlarging over time.
6. Heterogeneity is observed less in metastatic lesions than in the primary tumor.

Cancer progression is the acquisition of additional properties of the malignant phenotype over time. Progression is achieved through a variety of mechanisms (e.g., genetic instability, epigenetic instability, and aberrant cell death regulation).

The direct result of tumor progression is tumor heterogeneity. A cell that has acquired a new genetic aberration will produce a new, growing subclone within a clonal neoplasm if the new aberration confers a survival advantage. Tumor heterogeneity is almost always a bad prognostic indicator because the new subclones will tend to have phenotypic properties that permit them to grow under conditions that prevail in an enlarging tumor. These will likely consist of enhanced anaerobic energy metabolism (a lessened dependence on respiration), enhanced invasiveness (providing growing cells with a mechanism to avoid being crushed by neighbor cells), increased growth rate (the ability to grow when other clones have slowed or stopped), a greater incidence of metastasis (the ability to seed to distant sites after invading through the walls of lymphatics or blood vessels), and shortened dormancy (the ability for cells seeded in distant anatomic sites to grow as metastatic subclones).

3.3 How Are Clonality, Progression, and Heterogeneity Related?

Life on earth is basically clonal *and* heterogeneous. There are millions and millions of animal species on the earth, but every organism is descended from the same primordial life form that appeared on earth three or four billion years ago. This life form acquired the skill of self-replication, and after that process began, alternate forms of life appeared on earth. What we have today are a myriad of subclonal descendants.

Neoplasms, too, are both clonal and heterogeneous. Studying a cancer is much like walking into a room of identical quadruplets and finding that one of them is tall, another is short, one is fair skinned, and another is swarthy. A clonal population of neoplastic cells tends to produce subclones, and each subclone produces cells that are different from the original clone. After a few generations of subclones spawning new subclones, the resulting population may consist of cells with diverse phenotypes. Every diverse cell is a clonal descendant of the same progenitor cell.

Because members of a clone are identified by a genetic alteration that is shared among all of the cells in the clone, cancer is often regarded as an acquired genetic clonal disease. In the next two sections, we examine the Janus-like phenomena of neoplastic clonality and heterogeneity.

3.4 The Complexity of Cancers

Most classes of malignant tumors are much more complex than the normal tissues from which they derive.

Complexity results from genetic instability, yielding a complex and chaotic phenotype.

The idea that cancer is a highly complex cellular system is contrary to beliefs held for many decades, that cancer represents a stripped-down and undifferentiated version of cellular existence. This view was popularized by Otto Heinrich Warburg in 1924. He correctly noted that cancer cells reverted to a largely anaerobic state, in which energy creation (ATP) is created through nonoxidative glycolysis. Normal, well-differentiated cells create most of their ATP through oxidative (oxygen-dependent) pathways. Advanced cancers of every type often converge toward a similar morphology, lacking differentiated cellular structures, such as cilia, mucin, and glands, found in the normal cells from which they derive. The reversion to primitive metabolic pathways and the loss of advanced differentiated features led tumor biologists to assume that cancer cells were simplified variants of normal cells.

The difference between current and past concepts of tumor cell complexity might be explained by a simple analogy selected from a common childhood experience. If you have ever dabbled in finger painting, you know that a few strokes using several different paints can produce a variegated pattern. If you take your fingers and scramble them around the paper, the colors will begin to run into themselves, and you end up with a somewhat monochromatic, dull-green product. Nonetheless, the final product is complex. There are hundreds of strokes, and many thousands of colors represented in a scrambled fingerpainting. The painting appears monochromatic and without pattern to you because your eyes cannot distinguish all of the strokes and color combinations that you have clumsily created.

At a distance, advanced cancer cells may all look alike. At a molecular level, cancer has a much greater complexity than normal cells. This explains the observation, made in the Introduction, that cancer cells differ in virtually every measurable way from normal cells.

3.5 Tumor Diversity and the Disconnect Between Genotype and Phenotype

Every tumor is unique, and every cell in every tumor is unique. An individual cancer cell is not necessarily characterized by persistent growth. As the tumor grows, many of the individual cells within the tumor are dying. Persistent growth of the tumor requires only a net surplus of proliferating cells over dying cells.

Likewise, invasion is a feature that does not characterize all of the cells in a tumor population. Cells deep within a tumor do not have the opportunity to invade adjacent tissues. Only cells at the edge of a tumor have the opportunity to invade adjacent tissue.

In disorders that are inherited through the germline, every cell in the body harbors the same altered gene, but only a few types of cells will display the inherited phenotype (i.e., only a few types of cells will be diseased). Why aren't all cells in the body affected by the germline disorder? The answer to this question is found in the miracle of cell lineage. During development, cell lineages achieve a differentiated phenotype through a genetically determined epigenetic modification that is permanent in normal circumstances. It is the lineage of a cell that determines the expression of an inherited gene. The lineage of a cell determines the cellular pathways used by the cell. If the cellular pathways in a cell do not include the expressed inherited gene, then the cell will not be affected by the inherited disorder. Despite the genetic heterogeneity that develops within tumors, there will always be a lineage-determined set of pathways that determine how an oncogene exerts its effect on the cell.

3.6 Animal Speciation and Tumor Speciation

This section is probably the most important section in the book. In this section, we directly confront the biological phenomenon of speciation and how it applies to every organism on earth and every tumor that occurs in every organism on earth.

Animal speciation is one of those concepts that everyone thinks that he or she understands; however, most people confuse speciation with mutation, or they incorrectly equate it with "survival of the fittest." Speciation is a process that occurs constantly in nature and results from the individuality of terrestrial organisms. Because every member of a species is a genetically unique individual, it is possible to breed new organisms that have unique genotypic features inherited from their unique parents and that also share many constitutive genetic features that were present in every member of the ancestral species. Over time, if any group of animals is constrained to mate within a closed group, a new species may emerge that can no longer mate with its ancestral group. Members of the new species will share a distinctive set of features that separate the new species from its direct ancestor species and from all other species. But every member of the new species will retain a genome that is, with only a few exceptions, identical to the genome of the ancestor species. The phenomenon of speciation is a direct consequence of the much more fascinating phenomenon of the genetic uniqueness of every animal on earth. The greater the pre-existing genetic diversity in the population of a species, the easier it is to speciate.

The same principles of animal speciation apply to neoplasms: Distinct types of tumors arise from a genetically diverse population of cells. Although every tumor that occurs in humans is unique and genetically distinct from every other tumor that has ever occurred in any human, tumors speciate into one of several thousand species that share common morphologic and biologic features. Speciation involves the imposition of pathways, specific for the cell lineage from which the tumor species derived, upon a cell that has been altered by an oncogenic process. Oncogenes influence pre-existing cellular pathways that are available to cells within a narrow range of developmental lineages. These oncogene-influenced pathways produce stereotypical morphologic and biological changes in unique cells, and this process accounts for the distinctive types of tumors in animals. During tumor progression, tumors may acquire additional mutations, and these mutations modify the biology of individual tumors; however, they do not change the species of the tumor (e.g., a melanoma will always be a melanoma and cannot transform into any other type of tumor).

No simple calculation allows us to predict the number of tumors species in humans, but there are approximately 200 differentiated cell types in humans. There may be several hundred additional types of stem cells that precede the development of fully differentiated cells. Oncogenes influence a narrow range of cell types with lineage-dependent pathways conferring a limited number of neoplastic templates. This process accounts for the several thousand different species of human neoplasms that occur in humans.

3.7 Benign Tumors

If the acquisition of genetic aberrations is a feature of cancer progression in all tumors, how might tumors behave in an alternate universe, where genetic instability is absent from neoplasms (List 3.7.1)?

Every listed feature that one might expect to find in a genetically stable tumor applies to benign neoplasms. Most benign neoplasms are diploid or near diploid. Those gene alterations found are often limited to one or two alterations that are characteristic of the tumor (List 3.7.2).

List 3.7.1 Life without tumor heterogeneity.

1. There would be no tumor latency period and no precancer lesions that were separable from fully developed cancer lesions.
2. Early tumors would behave much like late tumors.
3. Tumors would be diploid.
4. All tumors of a particular type would have about the same clinical course.
5. There would be only minimal differences between tumor cells and the normal cells; cytopathologists would be hard pressed to earn an honest living.
6. Tumors would be diagnosed by molecular biologists by finding the stable genetic alteration responsible for the malignant phenotype.
7. Tumors could be cured with a single nontoxic drug targeted to the tumor oncogene or to a critical pathway controlled by the oncogene.

List 3.7.2 Some simple gene alterations encountered in benign neoplasms and nonmalignant proliferative lesions.

Cylindroma, cyld1 gene mutation (52)
Polycythemia vera, JAK2 mutation (53)
Pilomatrixoma, beta-catenin (54)
Pleomorphic adenoma, PLAG1 (55)
Ovarian granulosa-stromal cell tumors, diploid or trisomy 12 (56)
Clear cell hidradenoma of the skin, TORC1-MAML2 gene fusion (57).
Lipoblastoma, rearrangements of chromosome bands 8q11-13, possibly involving PLAG1 (58)
Angiolipoma, diploid (58)
Lipomas, HMGIC
Hemangioma, HMGIC
Chondroid hamartoma, HMGIC

Benign tumors can be thought of as impaired malignant tumors. Most of the features of malignant tumors are present in benign tumors to a lesser degree. They grow persistently, but not as fast as most malignant tumors. They have some genetic instability, but not as much as that of most malignant tumors. They show some ability to progress (to a more malignant phenotype), but this occurs only rarely.

3.8 Are There Benign Oncogenes?

Lipoma, hemangioma, and chondroid hamartoma of lung, all benign mesenchymal tumors, are sometimes associated with the gene that codes for HMGIC, a member of the high mobility group protein family of transcription factors. An HMGIC gene mutation is occasionally found in many different types of benign mesenchymal tumors, but HMGIC mutations are not specific for benign mesenchymal tumors. The mutation is found in some cases of malignant tumors, such as sarcomas and acute leukemias. When we discuss **Cowden syndrome** in Chapter 4, we learn that an inherited syndrome characterized by the occurrence of a variety of benign hamartomatous neoplasms will also increase the occurrence rate of several types of malignant tumors. There seems to be no example of a purely "benign" oncogene.

Nonetheless, a few generalizations hold (List 3.8.1).

Benign tumors should be studied as extensively as are the malignant tumors, but they are not. In fact, virtually all funding in the field of cancer research goes toward malignant tumors. This is unfortunate because when we understand the benign tumors, we are really understanding the essence of neoplasia without the confounding factors of tumor progression and genetic instability.

List 3.8.1 A few generalizations about benign and malignant tumors.

1. Some benign tumors almost never lead to malignant tumors, and these tumors tend to be diploid or near diploid and genetically stable.
2. Benign tumors that occasionally progress to become malignant tumors tend to have fewer genetic aberrations than their malignant counterparts (59).
3. Some benign tumors, such as colon adenomas, have a significant likelihood of progressing to cancer (Figure 3.8.1). Progression to cancer usually occurs in large, aneuploid adenomas.
4. Any benign tumor, without exception, if grown to enormous size, may develop malignant foci within the tumor.
5. Clonal growths characterized by genetic stability are usually benign neoplasms. Clonal growths characterized by progressively worsening aneuploidy are most often cancers or precancers.

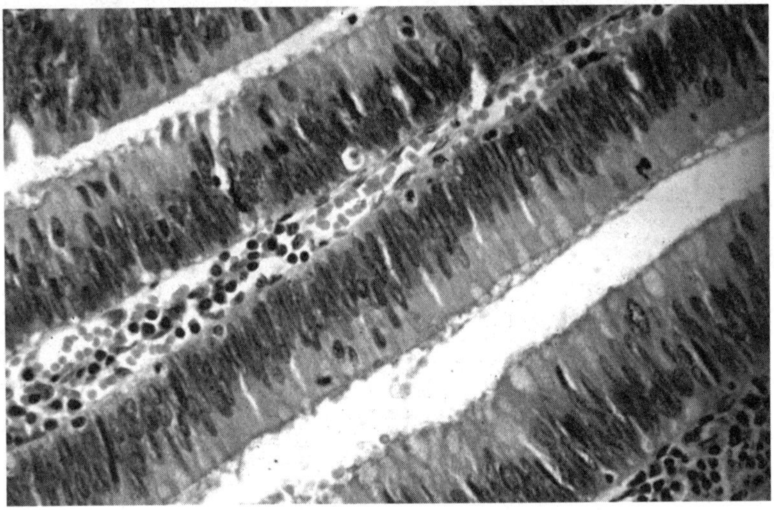

FIGURE 3.1 Histology of colon adenoma. Well-formed glands are lined by uniform glandular cells. Cells contain mucin droplets (white area) at apex—the side of the cell facing the lumen (empty part) of the gland. Notice the atypia of each cell. Cells have nuclei that are several times the size of nuclei of cells in **lamina propria** (e.g., lymphocytes). The enlarged nuclei are hyperchromatic, with variation in nuclear shape and chromatin distribution, from cell to cell.

Chapter 3 Summary

Tumor biology is a contradictory amalgam of genetic heterogeneity and biological uniformity. A finite number of different tumors occur in humans and animals. A named tumor in one person will often have the exact same morphology and biological behavior as the tumor of the same name occurring in another person.

Although every tumor that occurs in humans is unique and genetically distinct from every other tumor that has ever occurred in any human, tumors fall into one of several thousand tumor species. Each tumor of a given species shares common morphologic and biologic features with every other tumor of the same species. Tumor speciation is the process whereby oncogenes influence a distinctive set of cellular pathways that are available to a narrow range of cells, determined by developmental lineage. These oncogene-influenced pathways produce stereotypical morphologic and biologic changes in cells that account for the tumor species.

Tumor progression (the accumulation of genomic alterations in cancers over time) modifies the malignant phenotype but does not eradicate the identity of tumors. In most cases, increasing aneuploidy and tumor heterogeneity advance in a predictable manner. Because speciation is determined by developmental lineage, and because the number of lineage-specific progenitor cells is limited, the number of different types of tumors in animals is limited. Because the types of tumors of man are limited, we can create a tumor **taxonomy** (a listing of every species of tumor). Without an understanding of tumor speciation, the field of cancer research remains hopelessly complex. If you understand speciation, you can start to classify tumors into biological groups whose growth and development can be understood based on shared pathways and common vulnerabilities. It may be possible to design cancer treatments that are effective against classes of neoplasms that share a common developmental lineage. The discussion of class-based cancer therapies will be revisited in later chapters.

CHAPTER

The Range of Neoplastic Phenotypes

4

When a scientist doesn't know the answer to a problem, he is ignorant. When he has a hunch as to what the result is, he is uncertain. And when he is pretty darn sure of what the result is going to be, he is in some doubt.

—Lee A. Dubridge

4.1 Background

There are four types of neoplasms: cancers, benign tumors, hamartomas, and precancers. The purpose of this chapter is to provide examples of each type of neoplastic condition and to discuss what these different kinds of neoplasms reveal about carcinogenesis.

4.2 Neoplastic Lesions

Cancers are the most important type of neoplasms because they account for most (not all) of the deaths that are caused by tumors. A neoplasm is a focal growth composed of genetically abnormal cells (List 4.2.1). The term "neoplasm" excludes tissues that grow reactively as part of a normal physiologic process, such as hyperplasias (the overgrowth of otherwise normal tissues) and inflammatory swellings.

We can provide a definition for each term (List 4.2.2).

Although there are precise definitions for the different types of neoplasms, it can be very difficult to assign lesions to one or another category with any certainty. If I were to say that a seborrheic keratosis is a hamartoma, some people might agree with me, and others would argue that it is a clonal growth, a benign tumor, or none of the above. If I say that ductal carcinoma in situ (DCIS) of breast is a type of cancer, some might

List 4.2.1 General classes of neoplasms.

Progressively expanding tumors
- Cancers
- Benign tumors

Hamartomas and choristomas
Precancers

List 4.2.2 Definitions for the classes of neoplasms.

1. Cancer. A progressively expanding and invasive clonal lesion that will, over time, metastasize to anatomic sites distant from the primary growth location of the tumor.
2. Benign tumor. A expanding clonal lesion that neither invades nor metastasizes.
3. Hamartoma. A malformative overgrowth of tissues normally found in the location of growth that grows to a certain size and stops growing. It sometimes regresses. Some pathologists group these tumors with malformations, not neoplasms. We include the hamartomas with neoplasms because they occur in some of the same inherited syndromes that are associated with cancers and because cancers are known to occasionally develop from hamartomas. The common mole (also known as nevus) and the common seborrheic keratoses of skin are examples of hamartomas.
4. Choristoma. A hamartoma composed of a tissue not normally found in the location of the growth.
5. Precancer. A clonal growth of cells that does not always grow progressively, never invades, does not metastasize, but sometimes develops into a cancer.

argue that it is only a precancer because it lacks the constitutive cancer properties of invasion and metastasis. If I labeled DCIS a precancer, others would say that, as its name implies, DCIS is a variant of breast cancer.

The basis for disagreements among cancer biologists and pathologists often reduces to a subtle existential rift in the way that different people assess reality. Some people will look at a boy sitting quietly in class and say, "There is a good boy!" because the boy is being good at the moment. Others will look at the same scene and say, "There is a bad boy" because they caught him bullying another boy last week, and they expect that such behavior will lead to something worse in the future. For some people, a fertilized human ovum is just another cell. It may have interesting observable properties, but so does every other human cell. For other people, a fertilized ovum is a human because it is an early stage in the life of a human being.

Some pathologists restrict the use of a neoplasm term to the current observable properties of the lesion. For these people, a precancer is not a type of cancer. It is a lesion that occasionally precedes the development of cancer. Other people use words to encompass the biological stages through which the lesion may progress. To such people, a precancer is equivalent to a cancer because a precancer is a stage in the development of the cancer.

When we discuss benign tumors, we run up against recurring, and unresolved, conceptual problems in pathology. Should we evaluate a lesion based on its observable biological properties, or should we expand our evaluation to include properties that a lesion acquires over time? Fundamental differences in the way we conceptualize neoplasms have led to profound differences in our understanding of neoplasms.

4.3 Do Benign Tumor Cells Develop into Malignant Tumor Cells over Time?

If a tumor has the potential of becoming malignant but it is not currently malignant, should we say that the tumor is benign (based on its observed biology) or malignant (based on future expectations)? If we say it is benign and it later metastasizes, might a lawyer argue that we were wrong in our original diagnosis? If we say that the tumor is malignant and the patient undergoes disfiguring surgery to remove the tumor, might a lawyer argue that we had misdiagnosed a benign tumor, causing the patient unwarranted pain and suffering?

What is the process whereby malignant tumors arise from benign tumors? Often, the overtly malignant portion of a benign tumor can be distinguished from the benign tumor from which it arises. The malignant part of the tumor contains cells that are morphologically different from the benign part of the tumor. We recognize these malignant cells by their visualized behavior and by the cytologic features of cancer. The cells from the malignant portion of the tumor are found invading surrounding tissues. They may grow as streaming cords. They may show distinctive nuclear atypia that is absent from the adjacent benign tumor cells. The malignant focus may show necrosis, hemorrhage, cysts, and foci of hypercellularity. The pathologist usually assumes that the transformation of a benign tumor into a malignant tumor is one of clonal selection of a malignant progenitor cell that develops from a population of benign cells. This means that the population of benign tumor cells does not actually become malignant. They merely serve as the source of cells from which a malignant subclone might arise.

Follicular adenoma of the thyroid is a benign tumor arising from the glandular epithelium of thyroid cells. It is almost always encapsulated (surrounded by a rim of fibrous tissue). Over time, follicular adenomas may develop into follicular adenocarcinomas. The diagnostic feature that distinguishes a follicular adenoma from a follicular adenocarcinoma is invasion. Pathologists will carefully scrutinize the capsule of a thyroid follicular neoplasm, searching for invasion of the tumor through the fibrous capsule. If invasion through the capsule is found, the lesion is called a follicular adenocarcinoma. If no invasion (of capsule or vessels) is found, the lesion is called a follicular adenoma.

The problem with this simple approach to diagnosis becomes evident when we try to find markers that distinguish follicular adenoma from follicular adenocarcinoma. Cytopathologists have grudgingly accepted that in most cases follicular adenoma cannot be distinguished from follicular adenocarcinoma based on cellular examination. No one has found a genetic or histochemical marker that can consistently distinguish follicular adenoma from follicular carcinoma. If the two lesions are biologically different (noninvasive adenoma and invasive adenocarcinoma), why can they not be distinguished on a cytologic basis? The answer is that the invasive focus of a follicular carcinoma is a small number of cells that have subcloned from a population of benign "adenoma" cells. The two populations, benign adenoma cells and invading carcinoma cells, live together in a single tumor mass, with the adenoma cells accounting for most of the mass. We cannot distinguish the properties of a follicular adenoma from a follicular carcinoma when the lesion that pathologists call a follicular carcinoma is a mixture of adenoma cells and carcinoma cells.

4.4 How Do We Know When a Lesion Is a Neoplasm?

We do not always know when a mass-occupying lesion is a neoplasm. This is especially true if we cannot observe growth in the lesion, or if the lesion contains several different cell types, or the cells in the lesion have the morphology of normal cells, or we can think of an etiology for the lesion other than neoplastic growth.

Over the decades, a number of lesions have been described whose neoplastic origins were the topic of great controversy (List 4.4.1).

Molecular genetics can determine whether a lesion is neoplastic. When a lesion contains a genetic alteration that is present in all of the cells of a lesion and in all occurrences of a lesion in multiple individuals and that is absent from normal tissues, then

List 4.4.1 A few lesions whose neoplastic nature has been doubted.

Aneurysmal bone cyst (60)
Atrial myxoma (61, 62)
Inflammatory pseudotumor (63)
Nodular fasciitis (64)
Idiopathic hypereosinophilic syndrome (65)

the lesion must be a neoplasm. If the lesion is composed of exclusively aneuploid cells, then the lesion must be a neoplasm.

Aneurysmal bone cyst is a bone lesion that was thought to be a reactive (nonneoplastic) process, resulting from bone damage. Oliveira and coworkers have described a recurring chromosomal translocation t(16;17)(q22;p13) in **aneurysmal bone cysts**. The translocation produces a CDH11-USP6 fusion gene. CDH11 is expressed strongly in bone, and these findings suggested that the highly active CDH11 promoter deregulated USP6 transcription (60). The finding of a characteristic, clonal genetic alteration in aneurysmal bone cysts is evidence that the lesion is a neoplasm.

Nodular fasciitis is a benign proliferation of connective tissue cells (fibroblasts and myofibroblasts) arising from the fascia that separates the (fibrous) dermis from the deeper, fatty subcutis. These lesions are generally small (1 or 2 centimeters) with self-limited growth. They can occur in adults and in children. Whether they are reactive lesions or true neoplasms has been a point of debate among pathologists. Recently, Donner and coworkers have found a clonal gene rearrangement in a nodular fasciitis, suggesting a neoplastic origin (64).

Atrial myxoma occurs as a mass attached to the walls of the atria of the heart. Histologically, atrial myxoma has many of the features of an organizing thrombus (blood clot). Like a thrombus, it sometimes dislodges from the atrium and travels through the arterial system to cause an abrupt ischemic event in an extremity. It occurs in families with Carney's syndrome. Mutations in PRKAR1A have been found in some atrial myxomas (61), as well as aneuploidy (62), both suggesting a neoplastic origin for this lesion.

Inflammatory pseudotumor (now called **inflammatory myofibroblastic tumor**) can arise from multiple locations. It is composed of spindle cells and inflammatory cells, suggesting a chronic inflammatory condition. The name inflammatory pseudotumor would seem to imply that the lesion is not a neoplasm. Today, the lesions that were once considered nonneoplastic are now considered true neoplasms of myofibroblasts. The inflammatory component, which masks its neoplastic identity, is preserved in the current terminology, "inflammatory myofibroblastic tumor." Recent work has identified a rearrangement of the ALK (Anaplastic Lymphoma Kinase) gene in these tumors (63).

A variety of conditions, such as allergies and parasitic infections, cause a reactive (physiologic) increase in blood **eosinophils**. For a long time, all cases of hypereosinophilic syndrome were considered nonneoplastic. We now know that when the circulating eosinophils increase with no apparent cause and the increase is high (more than 1,500 eosinophils per cubic millimeter of blood) and sustained (more than 6 months), the patient is likely to have a clonal, neoplastic expansion of eosinophils. In some patients with neoplastic hypereosinophilic syndrome, a characteristic fusion gene, PDGFRA-FIP1L1, is found (65). These cases are responsive to imatinib (Gleevec).

Sometimes molecular studies indicate the nonneoplastic origin of a suspected neoplastic process. Myositis ossificans is a localized lesion composed of reactive hypercellular fibrous tissue and bone. Genetic studies have shown that it lacks a monoclonal origin, suggesting a nonneoplastic lesion (66).

4.5 Borderline Tumors

Borderline ovarian tumors (BOTs) are uncommon, enigmatic neoplasms that have inspired a level of confusion, apprehension, and altercation disproportionate to their incidence. Pathologists have not been able to decide whether these tumors should be considered "benign" or "cancerous." They cut their differences by creating a separate classification just for these entities: "borderline tumors." Cancer Registrars are the people who collect clinical data on all malignant neoplasms received in a hospital. If a tumor is borderline, it is not malignant and is not captured by the Cancer Registrar. In California, Registrars are given special instructions for handling borderlines.

> Borderline ovarian tumors changed from malignant to borderline in ICD-O-3 (International Classification of Diseases, *Oncology*, version 3), a standard nomenclature used by hospitals to collect information on new cases of cancer. The California Cancer Registry will require the continued collection of borderline ovarian tumors due to their potential to become malignant and because there are a number of studies underway in the state that include these tumors. These same tumors, categorized as cancers under one set of guidelines, are recorded as benign tumors under guidelines issued by the American College of Surgeons. Under these guidelines, a special subcode is used to indicate "borderline behavior" (67).

The plot thickens. In 2003, Dr. Steven Silverberg and I organized a workshop on borderline ovarian tumors; it was sponsored by the U.S. National Cancer Institute. In my workshop summary, I described the basic problem with the borderline category of tumors (68).

After a tumor is given a special category, such as borderline, all of the pathologic descriptors that apply specifically to benign or to malignant tumors must be abandoned and translated into a new terminology created just for borderline tumors. BOTs obviously cannot invade (that is the kind of thing that malignant tumors do). When a tumor that resembles a BOT is found to have invasion, pathologists withdraw the diagnosis of BOT and label the tumor a malignant ovarian carcinoma. If any focus within a malignant ovarian carcinoma has morphological features of a BOT, pathologists will typically omit any reference to the BOT component to spare the surgeons any confusion. This act of kindness eliminates documentation of those cases where BOT and cancer coincide, obscures our clinical understanding of tumors with mixed BOT/carcinoma morphology, and leaves us clueless regarding the relationship between BOT and cancer. Although invasive BOTs do not exist (a borderline tumor cannot be invasive), anatomic pathologists reserve a category of microinvasive BOT. It seems that a BOT can have microinvasion, but a BOT cannot have invasion.

Because BOTs are neither benign nor malignant, the extraovarian spread of BOT cannot proceed through metastasis (a property exclusively associated with malignant tumors). BOTs spread through the pelvis via "implantation." Unlike the situation of metastatic spread of tumors (where a metastasis is always bad), the clinical relevance of implants is determined by the morphologic features of the implanted tumor. If the implant is invasive, the prognosis changes. Interestingly, a patient with the *diagnosis* of BOT who develops a highly invasive implant (e.g., an implant attaching to the colon and invading deeply into the colon wall) does not have cancer. In this case, the patient has a BOT with invasive implant (i.e., the implant is invasive but the BOT is not)! Remember now that a BOT with invasion limited to the ovary is called a carcinoma. To the best of my knowledge, BOT is the only tumor where the diagnosis is changed *not* by the presence of invasion but by the location where the invasion occurs (ovary versus implantation site). Local invasion changes the diagnosis to cancer, erases the BOT, and turns any implants into metastases! A BOT with distant invasion (even if we someday learn that implantation occurs through a mechanism equivalent to metastasis) is still a BOT.

The borderline tumors provide a cautionary tale. The words we use to categorize types of tumors may have enormous clinical consequences to individual patients and to cancer researchers. Outside the field of cancer, the same caveats apply. Richard Gordon describes an interesting development in England, where the simple act of counting the dead changed the perception of life (69). Beachy Head is a cliff in Southern England facing the English Channel. Here the flat grassy earth abruptly gives way to a 532-foot drop. The vertical, chalky-white cliff face is a popular last stop for suicides. In 1975, the Beachy Head suicide rate dropped by half. What public health measures may have accounted for this dramatic improvement? A new coroner had decided that bodies found at the foot of the cliff would be assayed for blood alcohol content. A body with alcohol in the blood would henceforth be labeled an accidental death. To be a suicide, the body would need to be dead and ethanol-free. The notion that someone bent on suicide might take a drink to fortify his or her resolve was not to be entertained. In essence, how you categorize things determines medical reality.

4.6 Precancers

Precancers are neoplastic growths that precede the development of cancers. The precancers are extremely common lesions that can often be biopsied (excised as a tissue sample) and diagnosed by pathologists. In most cases, clear-cut morphological features distinguish precancers from cancers.

Suppose this were not the case. Suppose we lived in a world wherein precancers could not be distinguished from cancers, and suppose that we had no means by which we could study the biology of these important lesions. Even if we had no physical examples of precancers, we could infer the existence of these lesions from simple observations of tumor biology.

Here is an example. In the *New York Times* of December 15, 2006, there was a provocative article by Gina Kolata (70). Ms. Kolata wrote:

> Rates of the most common form of breast cancer dropped a startling 15 percent from August 2002 to December 2003, researchers reported yesterday. . . . The reason, they believe, may be because during that time, millions of women abandoned hormone treatment for the symptoms of menopause after a large national study concluded that the hormones slightly increased breast cancer risk.

Here is the problem with this news. The 15% drop in breast cancer rates occurred in the interval between August 2002 and December 2003, immediately after the large drop of hormone use among the population at risk for breast cancer. For most other diseases, this might lead you to suspect causality, but carcinogenesis is a multistep process that extends over many years. Based on our understanding of carcinogenesis, there is no reason to think that a decision to discontinue a drug could possibly result in a measurable drop in cancer incidence observed the following month!

It is commonly accepted that breast carcinogenesis takes about 15 years (from initiating event to tumor detection). Let us pretend that the 15-year number is good. Then if there is a large drop of cancers beginning in August 2002, would not it make sense to look for some change in carcinogen exposure that occurred starting in August 1987?

The only way around this objection is to assume that the women who opted out of hormone treatment in August 2002 (and who may have accounted for the drop in breast cancer incidence starting the following month) harbored a lesion that was not quite cancer but that could develop into cancer. The "not quite cancerous" lesions are called precancers. The observations raise the question, "Has the reduction in hormone treatment in menopausal women resulted in a decrease in the progression of precancerous breast lesions?"

Decades ago, epidemiologists closely studied the effects of smoking cessation on the incidence of **lung cancer**. The mathematical models for smoking-induced lung cancer did not fit the observed rates in lung cancer incidence in people who had ceased smoking. Epidemiologists inferred the there must be some phase of cancer development, independent of the continuing presence of the carcinogen, that transitions to cancer when smoking stops.

These lesions that occur after initiation and that preceded the development of cancers are called precancers. Almost every cancer occurring in humans is preceded by an obligatory precancer. Understanding and controlling these lesions are among the most important goals in cancer research.

4.7 Consensus Definition of Precancers

In 2004, I helped organize a workshop, along with Dr. Donald Henson and Dr. Cheryl Marks, on the precancers. This workshop, held November 8–9, 2004, was co-sponsored by the National Cancer Institute and the George Washington University Medical Cen-

List 4.7.1 The five defining properties of precancers (all must apply).

1. Evidence must exist that the precancer is associated with an increased risk of cancer.
2. A precancer must have some chance of progressing to cancer, and the resulting cancer must arise from cells within the precancer.
3. A precancer must be different from the normal tissue from which it arises.
4. A precancer must be different from the cancer into which it develops. A precancer may have some, but not all, of the molecular and phenotypic properties that characterize the cancer.
5. There must be a method by which the precancer can be diagnosed.

ter. A definition of precancers was developed over 2 days of workshop discussions (List 4.7.1) (71).

4.8 The Difference Between a Precancerous Condition and a Precancer

There is an important distinction to be made between a precancer and a **precancerous condition.** A precancer is a collection of abnormal cells from which a cancer arises. A precancerous condition is a disease or syndrome or physiologic status that predisposes a person to the development of precancers and cancers.

An example of a precancerous condition is **ulcerative colitis**, a chronic inflammatory condition of the colon. Over time, patients who develop ulcerative colitis will have a high likelihood of developing colon cancer. Sometimes, the inflammation in the colon can be abated with medication. Frequent biopsies of multiple sites within the colon are obtained to follow the response of the disease to treatment. If dysplastic (precancerous) lesions are noted, the patient will usually be given the option of having a prophylactic total colectomy. In this example, ulcerative colitis is the precancerous condition. Focal colonic dysplasia is the precancer.

Many conditions are associated with an increased risk of developing precancers and cancers (List 4.8.1).

4.9 What Is the Difference Between a Benign Tumor and a Precancer?

If benign tumors, left untreated, typically progress to become malignant tumors, should we think of benign tumors as types of precancers? It seems as though all benign tumors are sources of proliferating cells from which a malignant clone may arise. Technically, all benign tumors are types of precancers.

Usually, when we think of precancers, we think of intraepithelial lesions that are small, morphologically altered areas of a mucosal surface. They are not tumors because

List 4.8.1 Some precancerous medical conditions.

Ulcerative colitis
Stewart-Treves syndrome
Cirrhosis and chronic hepatitis
Tylosis
GERD (gastroesophageal reflux disease)
Cigarette addiction
AIDS

they do not grow as a mass of measurable size. When a benign tumor yields a malignant tumor, it usually follows a long period of benign growth during which the benign tumor may reach an enormous mass sometimes measurable as kilograms. It seems likely that the development of cancer from benign tumors occurs through a very different process than the development of cancers from intraepithelial lesions. This would imply that that precancers are a biologically diverse category of lesions.

4.10 Is There a Difference Between Precancer Genes and Cancer Genes?

Cancers have phenotypic properties that precancers lack (chiefly, invasion and the potential for metastasis). It seems self-evident that a cancer would also have genotypic properties that precancers lack. At this time, however, we do not know of any characteristic genetic alteration that distinguishes a precancer from the cancer into which it develops.

The reason that we cannot currently distinguish precancers from cancers using genetic markers is that the precancers that have been analyzed are **intraepithelial neoplasms** (i.e., the precancers that precede the common epithelial tumors). Epithelial tumors have a multitude of genetic alterations, as do their precancers. With only a few exceptions (described in Chapter 15), the key genetic changes that produce the malignant phenotype in epithelial cancers are unknown. Without this knowledge, it is virtually impossible to separate the genetic changes that are sufficient to produce the precancer phenotype from those that are necessary to produce the cancer phenotype.

Based on the definition of the precancers, we can predict that a precancer would occur, on average, at an earlier age than the cancer into which it develops. Also, because some precancers regress, we can be certain that the incidence of precancers is higher than the incidence of cancers. In those cases for which precancers are persistent lesions, we might expect that the prevalence of precancers (i.e., the number of people carrying the lesion) will exceed the prevalence of the developed cancer.

It is important to distinguish precancer from early cancers or small cancers. An early cancer is a fully developed cancer that happens to have existed for a shorter time than a late cancer. Currently, there is great emphasis on diagnosing and treating cancers that are "early." In many studies, collections of early invasive cancers may have inadvertently included precancers, and the successes attributed to early cancer treatment are likely due, in part, to the ease with which the included precancers were cured.

4.11 Is There Value in Treating Precancers?

Many more people are alive today as the result of precancer treatment than are alive due to the treatment of cancers.

The successful reduction in deaths from cervical cancer provides a good example of the effectiveness of precancer treatment. Cervical cancer is a type of squamous cell carcinoma that develops at the junction between the ectocervix (the squamous lined epithelium) and the endocervix (the glandular lined epithelium) in the os of the uterine cervix of women. Before the introduction of cervical precancer treatment, cervical carcinoma was one of the leading causes of cancer deaths in women. Today, in many countries that have not deployed precancer treatment, cervical cancer is the leading cause of cancer deaths in women (72, 73). The relatively low number of cervical cancer deaths in the United States is the result of a 70% reduction in age-adjusted mortality after the introduction of Pap smear screening (74–76). No effort aimed at treating invasive cancers has provided an equivalent reduction in the number of cancer deaths as this simple procedure for treating precancers.

Today, we know that almost all cervical cancer is due to infection by one of several strains of human papillomavirus. The strains of human papillomavirus that cause cervical cancer are transmitted during sexual intercourse by men infected with the virus. In the late 1940s (and really up until the early 1980s), the viral etiology of cervical cancer was unknown. We did know that morphologic changes in cervical epithelial characterized the early steps in cervical cancer development. By sampling and examining cervical specimens from women, it was possible to accurately determine whether precancerous changes were present. If precancerous changes were present, a gynecologist could remove a superficial portion of the affected epithelium, and this would, in the vast majority of cases, stop the cancer from ever developing.

Thanks largely to the persistence of Dr. Papanicolaou and his coworkers, a **screening test** was developed to detect cervical precancers. The Pap smear is obtained by scraping or brushing the junction between the endocervix and the ectocervix and spreading the detached cells onto a glass slide. The cells are then stained with a histologic reagent (the Papanicolaou stain) that allows the cytologist to visualize subtle alterations in cell morphology. A typical Pap smear contains about 20,000 cells, and every cell must be inspected to rule out dysplasia and other pathologic abnormalities. A large cytology laboratory can handle hundreds of thousands of Pap smears in a year. In the last half of the 20th century, the Pap smear cytologic evaluation led to at least a

70% drop in the number of deaths from cervical cancer in every country that fully deployed the test.

Morphologic and epidemiologic observations on Pap smears provided clues that led to the identification of several strains of human papillomavirus as the major causes of cervical cancer. Recently, a vaccine against carcinogenic strains of HPV has been developed. Gardasil and Cervarix are two HPV vaccines that are currently available. If all goes well, most cases of cervical cancer will be prevented by a cancer vaccination. The Pap smear industry will shrink as fewer and fewer women test positive for cervical dysplasia. Examining Pap smears is just one of the many activities of a cytology laboratory. Cytologists routinely diagnose cells obtained from virtually any anatomic site and from any body fluid.

4.12 Hamartomas and Choristomas

Not all neoplasms have continuous growth. Examples are the common mole and seborrheic keratoses, both of which are new growths (i.e., neoplasms). Moles are the small brown or black, round, flat, or slightly raised bumps that occur on skin. Typically, they grow to the size of a few millimeters and stop. Moles arise from a specialized cell, closely related to the melanocyte (pigment producing skin cell) found in the skin dermis and in the lowest part of the epidermis. Most moles, after they stop growing, maintain a constant appearance throughout the life of an individual. Most people can be identified by the unique distribution and morphology of their moles. Sometimes a mole will disappear (regress). Rarely, a malignant melanoma will arise from a mole.

Seborrheic keratoses are slightly raised plaques occurring on the skin. They typically occur on sun-exposed areas and appear with increasing frequency as we age. They are often brown or tan and may vary greatly in size, reaching several centimeters in some instances. Patients often liken their seborrheic keratosis to stuck-on flattened drops of tar (that will not come off). Seborrheic keratoses arise from a small, round epithelial cell in the epidermis. Typically, seborrheic keratoses grow for some months and then stop growing.

Moles and seborrheic keratoses are new growths, consisting of a collection of dividing cells that produce a small mass that we can easily see and feel. When viewed under a microscope, they have characteristic morphologies, and anatomic pathologists can diagnose each lesion with high confidence. Moles and seborrheic keratoses are examples of hamartomas.

4.13 Are Hamartomas Types of Neoplasms?

Hamartomas are benign focal overgrowths of tissues and are often congenital (present at birth). Although they sometimes grow and can become quite large, they do not grow

List 4.13.1 Features shared by neoplasms and hamartomas.

1. Both hamartomas and carcinomas can be caused by carcinogens.
2. Hamartomas occur in the same inherited conditions that produce carcinomas.
3. Hamartomas and carcinomas are composed of cells that have genetic abnormalities.
4. Hamartomas and carcinomas are focal growths.

progressively. In almost all cases, hamartomas will eventually stop growing. Many types of hamartomas will regress over time.

Hamartomas are among the most common neoplasms of man. These would include common nevi (moles), seborrheic keratoses, hemangiomas, and bile duct hamartomas (of liver). Seborrheic keratoses occur late in life, and presumably environmental factors (e.g., sunlight) trigger a growth response in latent hamartomatous progenitors. The **Leser-Trelat sign** is a **paraneoplastic syndrome** characterized by the appearance of new seborrheic keratoses or an increase in the size of existing seborrheic keratoses as a consequence of a cancer somewhere in the body. Ellis and coworkers have speculated that the sign of Leser-Trelat is caused by increased levels of transforming growth factor alpha (TGF-α) produced by cancer cells and stimulating the focal growth of non-malignant skin epithelial cells (77).

A closely related entity is "**choristoma.**" Choristomas are **malformations** composed of an ectopic tissue (a tissue not normally found where the choristoma appears).

Many types of benign tumors have been called hamartomas, and there is no doubt that there are some hamartomas that should be called benign neoplasms and some benign neoplasms that should be called hamartomas. The inclusion of a variety of biologically distinct entities in the hamartoma category has greatly obscured our current understanding of this type of lesion.

Some nomenclatures do not categorize hamartomas among the neoplasms. For instance, the NCI *Thesaurus* places all hamartomas under class "Diseases and Disorders," separate from class "Neoplasm" (78). By assigning "Hamartoma" to a separate class of diseases and not as a subclass of neoplasms, the NCI *Thesaurus* loses the close relationship between hamartomas and other neoplasms. In the Developmental Classification of Neoplasms, to be discussed in later chapters, hamartomas are included as neoplasms because several important properties of neoplasms also characterize the hamartomas (List 4.13.1).

Hamartomas are easy to study because they lack the complexities of cancers, such as tumor heterogeneity, tumor progression, and genetic instability. By studying hamartomas, we may have our best chance to find agents that modify or control the fundamental neoplastic processes of tissue growth.

4.14 Hamartoma Syndromes

Cowden syndrome is an inherited disorder caused by inactivating mutations in the phosphatase and tensin **homolog** gene (PTEN) tumor suppressor. It includes the closely related syndromes, **Lhermitte-Duclose** Disease and Bannayan-Ruvalcaba-Riley syndrome, both of which are also caused by PTEN mutations. PTEN mutations are also found in sporadically occurring brain, breast, endometrial, and prostate cancers.

Cowden syndrome is characterized by hamartomatous neoplasms of the face, breast, colon, and other organs. The skin lesions of the face are primarily trichilemmomas and fibromas (79). Palmar and plantar hyperkeratotic pits and papillomas of the oropharynx are also found.

Women with Cowden disease have a high incidence of breast lesions, including ductal hyperplasia, intraductal papillomatosis, adenosis, fibroadenomas, and breast hamartoma. Fibrotic hyalinized nodules are commonly found. Invasive and noninvasive ductal carcinomas are found in a high percentage of women with Cowden disease (80).

Hamartomatous polyps of the intestines, particularly the colon, characterize the Bannayan-Ruvalcaba-Riley syndrome. This syndrome can be mistaken for multiple polyposis occurring in unrelated inherited syndromes.

Lhermitte-Duclos Disease seems to be a variant of Cowden Disease and is characterized by a hamartomatous overgrowth of hypertrophic ganglion cells which replace the granular cell layer and Purkinje cells of the cerebellum. This results in global hypertrophy of the cerebellum, with enlarged, coarse gyri.

The Online Mendelian Inheritance in Man (OMIM) database has collected additional hamartomas and carcinomas known to have occurred in people with Cowden disease (81, 82). Cancers found in some Cowden disease patients have included breast cancer, ovarian carcinoma, cervical carcinoma, uterine adenocarcinoma, thyroid cancer (follicular cell), transitional cell carcinoma of the bladder, colon adenocarcinoma, and meningioma.

In Cowden syndrome, we see a pattern common to other hamartoma syndromes: a tumor suppressor gene that is associated with different kinds of hamartomas as well as different kinds of cancers.

4.15 Is There a Pure Hamartoma Syndrome?

What can we infer from observations of Cowden syndrome. First, we must wonder whether tumor suppressor genes can ever cause a syndrome that exclusively produces one type of neoplastic process. In Cowden syndrome, we see a predominance of hamartomas, but we also see hyperplasias (e.g., global hypertrophy of the cerebellum and ductal hyperplasia of the breast), benign tumors (e.g., meningioma), and cancers (e.g., ductal carcinoma of the breast).

Tumor suppressors control the normal regulatory processes in cells (such as the control of cell growth or cell death or DNA repair). When tumor suppressors are inactivated, a sustained loss of control may heighten susceptibility to additional changes that cause various types of neoplastic lesions to emerge.

At this time, there seems to be no genetic lesion that causes one type of neoplastic process (cancer, benign neoplasm, hamartoma, or precancer) to the exclusion of the other neoplastic processes. Neoplasms of all types have these properties in common: They are all focal growths; they are all caused by genetic alterations; and their phenotypes can be modified by additional mutations or epigenomic alterations.

Chapter 4 Summary

There are four types of neoplasms: cancers, benign tumors, precancers, and hamartomas. A single inherited mutation or a single carcinogen may produce all of these types of lesions. Three of these lesions (benign tumors, precancers, and hamartomas) may evolve into cancers.

Nonetheless, all of these types of neoplasms have characteristic and distinguishing properties. By studying the properties that distinguish one type of neoplasm from another, we learn valuable lessons pertaining to the steps in carcinogenesis.

Precancers are lesions that precede the development of cancers. A precancer will either regress, stabilize, or continue to become a cancer. Every cancer occurring in adults seems to derive from a precancer. A single type of precancer may give rise to more than one type of cancer. For example, dysplastic bronchial epithelium may give rise to bronchogenic squamous carcinoma, adenocarcinoma, or small cell carcinoma of the lung. At this time, we know virtually nothing about the biological processes that lead to **precancer regression**, but we have experience with a variety of pharmaceuticals that increase precancer regression in experimental and clinical conditions. Developing effective drugs that can safely be provided to human populations at risk for precancers should become a major focus for cancer research.

Benign tumors are progressively growing neoplasms that do not invade or metastasize. Most benign tumors in humans do not progress to become cancerous. However, virtually every benign tumor has at least a small chance of becoming malignant. Unlike precancers, which tend to be very small and which tend to show marked dysplasia, benign tumors tend to have a morphologic appearance that deviates only slightly from normal cells. Most benign tumors are characterized by hypercellularity and persistent growth. Benign tumors, unlike cancers, tend to be near diploid and tend to have a simple genetic defect characterized by a single cytogenetic marker or by a single characteristic fusion protein.

The indolence of benign tumors may be the result of relatively low levels of genomic and epigenomic instability (compared with cancers). When a benign tumor becomes malignant, it often results from the selection of a subclone that has malignant properties (high levels of genetic and epigenetic instability, aneuploidy, heterogeneity, dysplasia, invasiveness, and metastases).

Hamartomas are localized overgrowths of tissues. Choristomas are hamartomas composed of tissues not normally present at the site of the lesion. Hamartomas are most commonly found at birth, but some types of hamartomas can occur throughout life. Hamartomas, unlike the other neoplasms, grow to a certain size and then stop growing.

CHAPTER 5

What Can We Learn About Human Neoplasms by Studying Animals, Plants, Fish, and Insects?

There are no good animal models, but some are useful.

—Herbert Slade

5.1 Background

Bacteria, protoctists, fungi, plants, and animals all share constitutive molecules and pathways, and they all derive from the same primordial organism.

Unicellular organisms, bacteria and protoctists, do not develop cancer. There are bacterial mutants that grow more rapidly than their wild-type counterparts, but there are no instances of cancerous bacteria that dominate the bacterial world, replacing all bacterial flora in a biological system (such as the gut) with an unregulated mutant clone. Presumably this is because unregulated growth is not a particularly useful feature for a single-cell organism. Organisms need to know when to become quiescent. Unregulated growth, occurring when food supply is limited or when growth conditions are suboptimal, does not provide a survival advantage.

The kingdoms of plants (Plantae) and animals (Animalia) share a susceptibility to cancer. In both of these kingdoms, organisms consist of growing amalgams of different kinds of cells. The cells in plants and animals become specialized, with particular cells differentiating along particular developmental lineages due to epigenomic programming. It seems that these features (growth, multicellularity, and differentiated cell types) provide the necessary setting for the occurrence of neoplasms.

The question posed in the chapter title, "What can we learn about human neoplasms by studying animals, plants, fish, and insects?" is often misconstrued to mean, "What are the animal models for human cancer?" Sadly, much of the history of cancer research has involved the search for cancers in rodents and other animals that closely mimic some human cancer. Traditionally, if an animal tumor has a similar morphology and biological behavior to a human cancer, it is thought to be suitable for human cancer studies (testing for potential human carcinogens, developing human tumor markers, or developing new therapeutic agents). Although progress has been achieved through the "animal model" approach of comparative pathology, many detractors believe that animal models are wasteful, time consuming, and cruel. A large community of animal activists has become repulsed by large-scale testing studies on rodents and other animals and has called for the cessation of many experiments that subject animals to toxic agents and surgical procedures.

The purpose of this chapter is to show that we can learn a great deal about human cancer simply by observing the natural occurrences of neoplasms in plants and animals. A naturalist's approach to comparative pathology will show that we learn more by observing differences between one organism and another than by observing similarities. We see in Chapters 6 and 14 that the methods of animal classification serve as the basis for the classification of human neoplasms.

5.2 Neoplasia in Plants

Crown gall is a tumorous disease that infects many different kinds of woody and herbaceous plants (Figure 5.1). It is caused by a bacterium, *Agrobacterium tumefaciens*, that is transmitted from plant to plant. Bacteria from the periphery of the tumor infect the soil and spread to nearby plants. The tumors (called "galls") grow near the soil line, where the bacteria produce a hyperplastic reaction in adjacent plant cells. Tumor cell proliferation continues in response to extracellular bacterial colonies. Mature galls consist of mitotically active cells that can be multinucleate and that are less differentiated than the cells in the adjacent normal plant tissue. Vessels can grow within the gall, but tumor vessels are disorganized and not well connected to the normal vasculature of the plant. Consequently, galls tend to die, sloughing off the live plant. Aside from bacterial-induced gall tumors, plants are known to develop neoplasms much like humans, as inherited genetic traits (83).

The occurrence of tumors in plants as well as animals would suggest that neoplasia, much like aging, is a constitutive property of multicellular organisms.

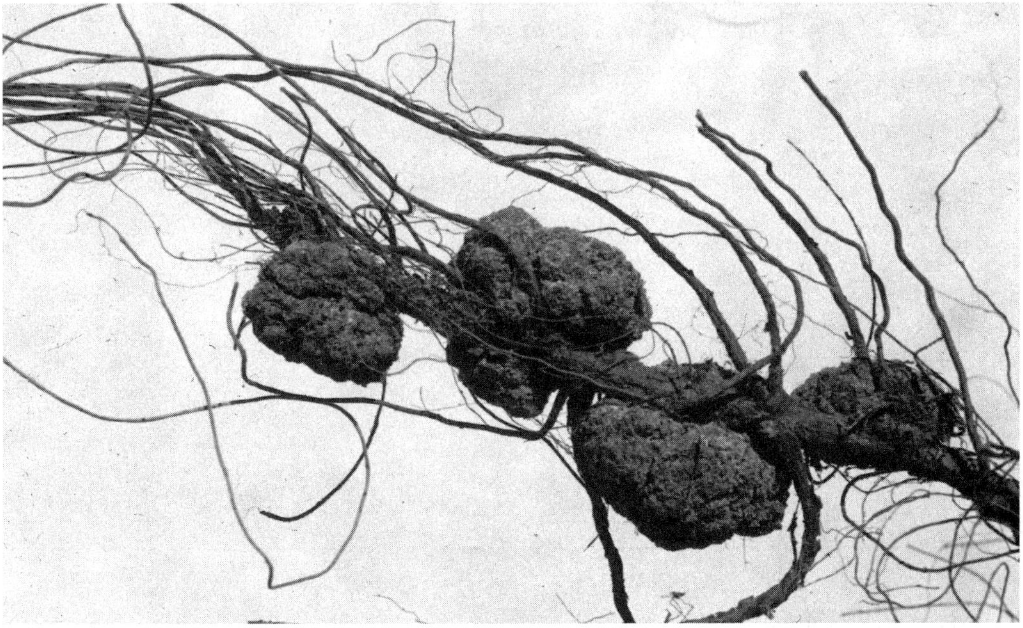

FIGURE 5.1 Crown gall tumors. Multiple tumors grow on plant roots.

5.3 Neoplasia in Insects

Insects are highly resistant to carcinogenesis because most cells in an adult insect are postmitotic (incapable of division). The target cells of carcinogens are dividing cells; however, insects have an Achilles heel: **imaginal discs.** Imaginal discs are foci of stem cells that live in insect larvae and that grow to become the various parts of adult insects. Flying insects have an imaginal disc to create a wing, another imaginal disc to create a leg, another for an antenna, and so forth. Imaginal discs have been the fundamental tool that developmental biologists have used for decades to study the control of organogenesis (84).

Much of the current insect work relating to carcinogenesis is done in *Drosophila melanogaster* (List 5.3.1).

When germline loss-of-function mutations are produced in insects, the imaginal discs may exhibit a variety of growth abnormalities. Tumors arise from a very small

List 5.3.1 Phylogeny of the fruit fly *Drosophila melanogaster* (taxonomy ID: 7227).

Eukaryota
Fungi/metazoa group
Metazoa
Eumetazoa
Bilateria
Coelomata
Protostomia
Panarthropoda
Arthropoda
Mandibulata
Pancrustacea
Hexapoda
Insecta
Dicondylia
Pterygota
Neoptera
Endopterygota
Diptera
Brachycera
Muscomorpha
Schizophora
Acalyptratae
Ephydroidea
Drosophilidae

population of imaginal disc cells, and these insect neoplasms are invasive, metastatic, and transplantable (85). Some loss-of-function genes studied in *Drosophila* are homologous to tumor suppressor genes in humans (86).

In Chapter 18, we describe **tuberous sclerosis**, an inherited syndrome of humans that results in multiple benign hamartomas. The brain tubers (hamartomas) that characterize tuberous sclerosis contain large, sometimes multinucleate neurons. *Drosophila* have the genes that are homologous to the genes known to cause tuberous sclerosis. Loss of function of these genes in *Drosophila* produces enlarged cells with many times the normal amount of DNA (87). For this reason, *Drosophila* is being studied so that we can understand how cell growth is controlled in humans.

5.4 Neoplasia in the Amphibia

In the 1950s, cancer researchers developed an interesting theory of amphibian carcinogenesis based on studies with newts. It seems that newts injected with carcinogens develop local malformations, or the local growth of regenerated limbs, at the injection site. They only rarely develop cancer after the administration of carcinogens (88, 89).

Urodele organisms (newts and salamanders) and anuran organisms (frogs and toads) are members of class Amphibia. Humans and amphibians are members of class Vertebrata. Unlike humans, amphibia have a remarkable ability to regenerate tissues (and limbs) after injury. Because carcinogens caused newts to produce local malformations or to regenerate limbs at the site of carcinogen injection, and because these growths occurred instead of cancers, a new theory of carcinogenesis was proposed. The theory held that cancer is a defect in regeneration occurring in higher vertebrates, such as humans. Humans are incapable of regenerating limbs, so they respond to carcinogens with an uncontrolled growth of stem cells. Newts are capable of regenerating limbs, so they respond to carcinogens with new limbs. It was a simple as that (90)!

Actually, urodele and anuran amphibians develop true neoplasms, both spontaneous and cancer-induced (91). In a large study reported by Khudoley in 1977, tumors were produced in grass frogs at high frequency (up to 50% of exposed frogs) with several different carcinogens. The induced tumors developed in several months (15.6–31.9 weeks) and included tumors of the liver (hepatocellular cancer and hepatic adenomas) and of the hematopoietic system (hemocytoblastosis) (92). This study showed that carcinogenesis can occur in amphibians, much like it occurs in humans. Amphibians develop cancers, benign tumors, and malformations (including accessory limbs) in response to carcinogens. Humans develop cancers, benign tumors, and malformations (hamartomas and choristomas) in response to carcinogens. These findings raise serious doubts about the "failed regeneration" theory of carcinogenesis.

5.5 Carcinogenesis in Fish

In the early 1960s, epizootics of liver cancer occurred in Pacific Northwest trout hatcheries. Aflatoxin B1, produced by the fungus *Aspergillus flavus*, was identified as the cause of liver cancer in fish and led to a greater awareness of the threat of naturally occurring aflatoxins to humans (93).

Soon thereafter, rainbow trout (*Oncorhynchus mykiss*) were used experimentally to study hepatic carcinogenesis. Wales and coworkers found that they could induce liver tumors by exposing trout embryos to a solution of 0.5 parts per million aflatoxin B1 for 30 minutes (94)! To the best of my knowledge this is the smallest dose delivered, over the smallest interval of time, that can produce a high yield of cancers.

Trout possess abundant cytochrome P450 enzymes in liver cells that activate aflatoxin. Trout lack high levels of the glutathione-mediated enzymes that deactivate aflatoxin. Furthermore, trout exhibit poor DNA repair (95). Combine these factors with the observed high sensitivity of young organisms to carcinogens, and you see that the rainbow trout has little chance against even a small dose of aflatoxin.

5.6 Zebrafish

More recently, the zebrafish (*Danio rerio*) has become the fish of choice for a wide variety of biological studies, including carcinogenesis.

The zebrafish is small, easily farmed, almost transparent when young, and has a small and relatively simple set of chromosomes. Most importantly, the zebrafish,

List 5.6.1 The zebrafish (*Danio rerio*) family tree.

Cellular organisms	Actinopteri
Eukaryota	Neopterygii
Fungi/metazoa group	Teleostei
Metazoa	Elopocephala
Eumetazoa	Clupeocephala
Bilateria	Otocephala
Coelomata	Ostariophysi
Deuterostomia	Otophysi
Chordata	Cypriniphysi
Craniata	Cypriniformes
Vertebrata	Cyprinoidea
Gnathostomata	Cyprinidae
Teleostomi	Rasborinae
Euteleostomi	Danio
Actinopterygii	Danio rerio (zebrafish)

descending from Class **Craniata**, has cells of the same developmental lineages that are found in humans (List 5.6.1) and develops tumors that are histogenetically similar to tumors of humans (96).

All members of class Craniata have **neural crests**, and all animals with neural crests can develop tumors of neural crest origin. The zebrafish develops neural crest tumors and has been used as a model for human **neurocristopathies** (97) (Chapter 18).

5.7 Neoplasia in Mammals

Mammals are hairy. Other classes of animals may have stringy appendages, but only mammals are equipped with the complex anatomic structures known as "hair." Keeping this in mind, let us consider the strange but common skin tumor known as **basal cell carcinoma.**

The two most common cancers of man are basal cell carcinoma and squamous cell carcinoma of skin, which occur on the skin. Each accounts for about 600,000 new cases each year in the United States. Together, these two tumors account for 1.2 million new cases each year, just about the total of every other tumor of humans combined!

Neither of these two common tumors account for many deaths from cancer. These tumors occur on sun-exposed skin, and people tend to be self-conscious of any skin lesion that other people can see. Consequently, most squamous cell carcinomas of skin and basal cell carcinomas of skin are called to the attention of a dermatologist when the tumors are small and relatively easy to excise. In addition, neither of these tumors is likely to metastasize, particularly basal cell carcinomas of skin. A metastasis from a basal cell carcinoma of skin is so unlikely that only a handful have been reported in the scientific literature.

Despite their lack of distant metastases, all basal cell carcinomas are invasive. In fact, there is no recognized noninvasive or precancerous stage of basal cell carcinoma. The growth of basal cell carcinoma is always characterized by the invasion of nests or strands of tumor cells in the dermis (the connective tissue underlying the epithelium of the epidermis (Figure 5.2).

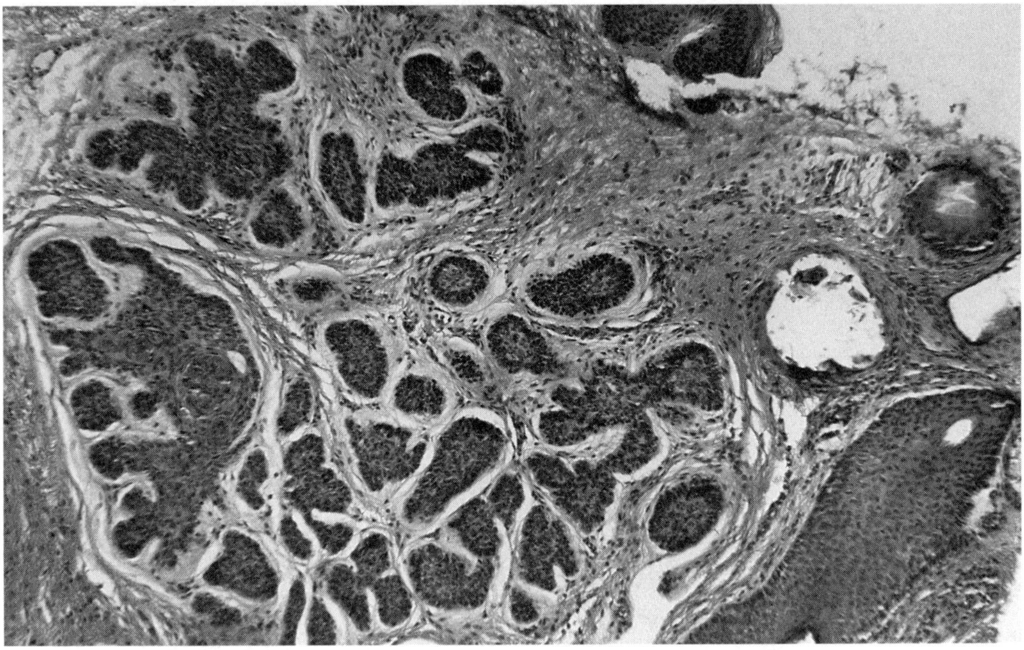

FIGURE 5.2 A typical nodular basal cell carcinoma. Tight nests composed of small, round blue cells are found sprinkled through the dermis.

In Chapter 1, we stipulated that carcinogenesis progresses through sequential events. A noninvasive phase precedes an invasive phase. Once invasion occurs, the likelihood of metastasis increases over time. If that is the case, how is it possible that basal cell carcinomas lack a noninvasive or precancerous phase? How is it possible that basal cell carcinomas, a tumor that grows by invading the dermis, cannot metastasize?

The peculiar behavior of basal cell carcinomas is explained by their histogenetic origin. Unlike squamous cell carcinomas, which arise from the basal layer of the epidermis, basal cell carcinomas arise from the hair follicle epithelium (Figure 5.3). Hairs grow by a cyclic process whereby the follicle of the hair invades into the dermis, providing a growing hair shaft that eventually sheds, leading to another round of invasion. The cycle consists of telogen (follicle resting phase), catagen (follicle involution phase), and anagen (follicle growth and invasion phase). Invasion is a property of normal hair epithelial cells. A tumor composed of hair-derived epithelium invades because invasion occurs in the normal growth of hair. Hair follicles invade, but they do not metastasize. The same applies to basal cell carcinoma, which invades without metastasizing.

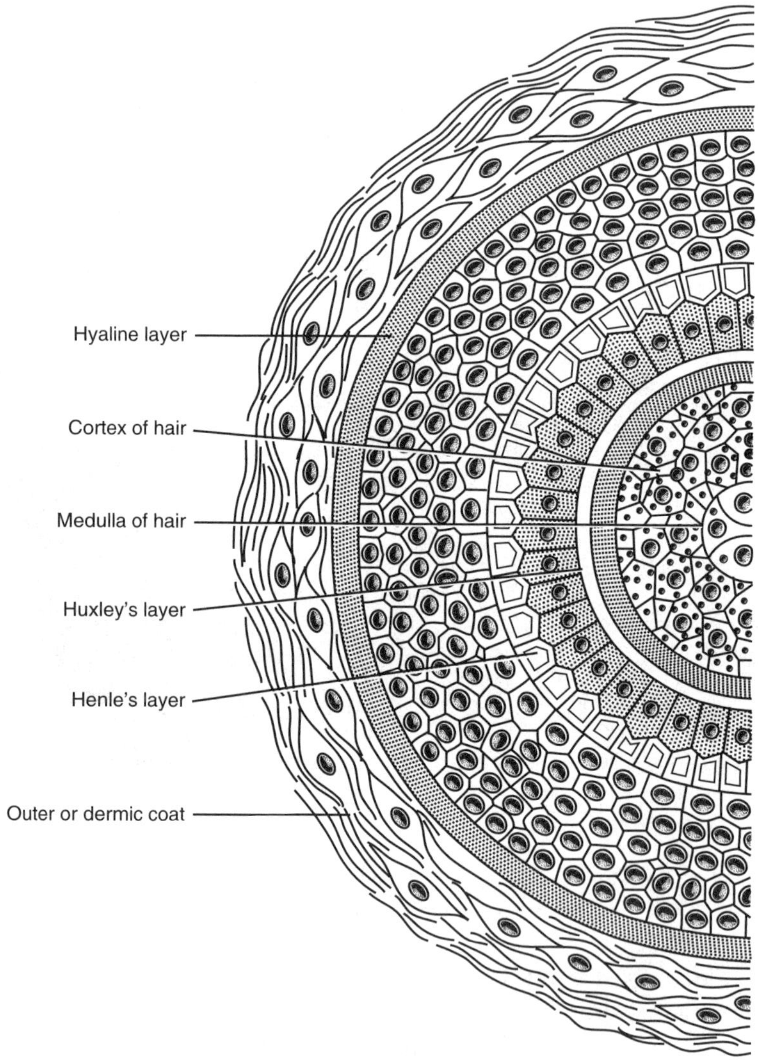

FIGURE 5.3 Diagram of hair shaft. Tightly growing small blue cells, reminiscent of the small blue cells that compose basal cell carcinomas, grow in the hair matrix.

List 5.7.1 Reasons why we think that basal cell carcinomas arise from the hair follicle and not from the epithelium of the epidermis.

Basal cell carcinomas grow in the dermis (like hair roots) and do not arise from the overlying epidermis.

Squamous cell carcinomas can arise from many locations in the body, wherever there is a squamous, pavemented epithelium. Basal cell carcinomas only arise from locations where hair can grow.

Basal cell carcinomas are found in mammals, the only hair-bearing class of animals. Nonmammalian species do not develop basal cell carcinomas.

Why do we think that basal cell carcinomas arise from the hair follicle and not from the epidermal layer of the skin? There are three reasons (List 5.7.1).

The third item in the list illustrates a general point. The types of tumors that we encounter in humans and other animals correspond to the limited biological phenotypes available to the species. We expand this concept in Chapter 14.

5.8 Oncogenes in the Animal Kingdom

All of the studied human oncogenes have homologs in ancient, conserved genes from other organisms (98). Not surprisingly, bone cancers have been found in dinosaur fossils. This means that oncogenes and cancer have been around for a very long time. There is no evidence of newly emerging cancer oncogenes.

Occasionally, a new tumor entity is reported in the literature. Typically, after the first case report appears, additional cases are noticed worldwide. Reviews of old cases usually uncover previous cases that had been diagnosed as some other tumor. We never see newly reported types of tumors confined to a contemporary human population.

The variations of today's human population were inherited from past populations. It seems reasonable to infer that ancient humans suffered from the same tumors encountered today and that "new" tumors are simply newly discovered "old" tumors.

5.9 Tumor Suppressor Genes in the Animal Kingdom

Tumor suppressor genes have a very different function than tumor oncogenes. They act to create an environment in which the acquisition of an active oncogene is likely to occur, persist as a stable variant, or cause growth. In many cases, tumor suppressor genes increase the incidence of a wide range of tumors derived from many different anatomic sites. Unlike oncogenes, which tend to cause tumors in a narrow range of cell types, tumor suppressor genes in mouse models are associated with neoplasms arising from many different cell types (17). Cardiff and co-workers, in their review of **genetically engineered mouse** models of neoplasms, noted that "tumor suppressor genes generally produce a mixture of tumor phenotypes when they are silenced or knocked-out" (17).

In an attempt to develop an animal model for human retinoblastoma, Nikitin and co-workers produced a strain of mouse that carries a single copy of the wild-type RB tumor suppressor gene; just as humans who have an inherited susceptibility to retinoblastoma carry a single copy the wild-type RB gene (99). In both humans and mice, the loss of both copies of the RB gene is required for tumor development. Unlike humans, mice do not develop retinoblastomas. They develop melanotroph tumors of the pituitary intermediate lobe. Because the mouse tumor is a melanotroph tumor and the human tumor is a retinoblastoma, the transgenic RB-deficient mouse model is considered, by some, a failure.

There is actually much to be learned from the RB-deficient mouse. Why is the mouse retinoblast not susceptible to tumor formation after a loss of function in the mouse RB gene? Although loss of function in a single tumor suppressor gene does not result in retinoblastomas in transgenic mice, retinoblastomas can be induced in the transgenic mouse model by knocking out an additional tumor suppressor that is closely related to RB, p107 (99a). The authors of this 1998 article found that when both RB and p107 are inactivated, five of seven adult pRb/p107-deficient chimeric mice developed retinoblastomas.

In children, the occurrence of bilateral retinoblastomas is sometimes followed by the occurrence of a pineoblastoma, which is referred to as **trilateral retinoblastoma**. Pineoblastomas are tumors that arise from primitive neuroepithelium of the pineal gland. The pineal gland has an evolutionary anlage identical to that of the eye. The difference between the development of the pineal gland and the development of the eye is that **neuroectoderm**-derived photoreceptors of the pineal gland took up residence deep within a midline recess in the brain, whereas the photoreceptors of the eyes developed as paired external structures. The evolved pineal gland, like the evolved eyes, reacts to filtered sunlight, although any light reaching the pineal has had a lot more filtering than the light reaching the retinas. In response to light cessation, the pineal gland releases melatonin, a hormone that influences circadian rhythms (e.g., sleep). The same germline mutation that leads to bilateral retinoblastomas may occasionally cause a pineoblastoma, tumors that share a common morphology and homology with retinoblastomas.

Williams and coworkers induced pineoblastomas in mice with germline mutations of both Rb and p53. This showed that all of the tumors associated with trilateral retinoblastoma in humans can be induced as homologous tumors in mice that harbor the RB mutation (100).

The emergence of retinoblastomas and pineoblastomas in mice following inactivation of two tumor suppressors, including the RB gene, teaches us several lessons (List 5.9.1).

List 5.9.1 Lessons learned from the induction of trilateral tumors in mice.

A specific progenitor cell may require multiple tumor suppressor genes for carcinogenesis to proceed. This observation contrasts with studies of activated oncogenes, which seem to follow a "one oncogene—one tumor" rule (101).

Some tumor suppressor genes produce tumors many years after inactivation (e.g., tumor suppressor genes involved in DNA repair). Tumor suppressor genes that yield retinoblastomas in infant humans or retinal dysplastic lesions in embryonic mice are very fast acting. This means that different tumor suppressor genes may have widely varying mechanisms.

Tumors produced by homologous suppressor genes active in mice and humans may produce homologous tumors.

5.10 Why Do Whales Not Develop More Tumors Than Mice?

In 1968, a comparative pathologist, Dr. Clyde Dawe, posed this question, "Why don't extremely large animals develop neoplasms with a much higher incidence than very small ones since the cell population at risk is greater by orders of magnitude?" (102).

A mouse weighs about 30 grams and lives about 1.5 to 3 years. Mice have a gestation period of about 20 days. An elephant weighs about 4 tons. They can live up to about 60 years and have a gestation period of about 22 months. Otherwise, mice and elephants are just about the same. They both have roughly the same set of organs, and the organs are organized similarly in mouse and elephant. Most mouse cells are the same size as most elephant cells. Most of the differentiated cells of the mouse have equivalent cell types in the elephant. The biggest difference between the elephant and the mouse is the number of cells contained in each animal.

The target of mammalian carcinogenesis is the dividing cell. Therefore, one would expect that the more dividing cells there are in an animal, the more tumors might occur. Tumors take a long time to develop after initiation. Long-lived animal provide initiated cells with a much longer development time. Thus, it stands to reason that large, long-lived animals with many dividing cells will develop many more tumors than small animals with relatively few dividing cells and a short lifespan in which tumors can develop.

The facts do not support this hypothesis. Mice have a very high rate of cancer. Whales and elephants, with hundreds of thousands more dividing cells than mice and with many more years in which to develop tumors, seem to have a lower tumor burden than mice.

What could account for this discrepancy between theory and fact? Nobody really knows, but a simple explanation may lie in the differences in genetic stability between cells of small mammals (like rodents) and large animals (like humans, whales, and elephants).

Normal mouse fibroblasts cultured in plastic dishes (so-called tissue cultures or in vitro cultures) tend to degenerate, after a few cell divisions, into aneuploid continuous cell lines. Cells from large mammals remain diploid in culture and have a limited number of cell divisions before the culture dies out. Rodent cells in culture have less ability to repair damaged DNA than similarly cultured cells sampled from large animals.

Genetic instability is a hallmark of the neoplastic process. Genetically unstable cells can rapidly progress to a malignant phenotype (i.e., the latency and precancer phases are shortened). Although laboratory rodents live only 1 to 2 years, they can easily develop multiple tumors in their lifetime. Large animals, with a superior DNA repair mechanism, usually require many years to progress through the carcinogenic process.

5.11 Evolutionary Mimicry and Evolutionary Convergence

In nature, there are many instances when two different organisms may look alike. Sometimes, morphologic similarity among species is coincidental. Often, the species that mimics another species gains a survival advantage. A predator may ignore a butterfly if the butterfly is mistaken for a leaf.

Sometimes two organisms may come to resemble one another because physical and biological circumstances favor the same adaptive features of size, form, color, and so on.

Evolutionary convergence is different from evolutionary mimicry. In convergence, organisms of different classes evolve the same traits independently. Convergence is more than a morphologic resemblance. It involves the evolutionary invention of similar or identical solutions through unrelated evolutionary pathways. One example is found in new world vultures and old world vultures, which look similar to one another and both of which feed on carrion, but the old world vultures are related to eagles and hawks, whereas the new world vultures are related to storks. An example of functional convergence is found in syntheses by elephants and butterflies of a compound ((Z)-7-dodecen-1-yl acetate), which serves as a pheromone for both species!

Striking morphologic and functional convergence is found in two plant genera, *Euphorbia* and *Cactacea*, neither of which is closely related to the other. Both organisms have similar adaptations to desert life and both organisms have similar morphologies.

The endocrine system in animals may be conceived as a group of diverse tissues that have attained a similar morphology and functionality through convergent evolution. The adrenal gland cortex derives from **mesoderm**, the adrenal medulla derives from neural crest, and the thyroid follicles derive from **endoderm**. These different lineages have produced epithelial cells that secrete hormones into the blood system. Although all three lineages have converged to produce endocrine cells, the tumors that derive from endocrine glands have genetic markers that are specific for their lineage (not for their functionality). In Chapter 12, we discuss endocrine neoplasms. In the next section, we see how convergence informs us about human tumor biology.

5.12 Goodbye to the APUDoma

In the early 1970s, a new concept developed within the field of endocrine neoplasia: the APUD (amine precursor uptake and decarboxylation) tumor. The theory was that many types of endocrine cells have the ability to metabolize amines to produce small, hormonally active amine-based molecules such as serotonin, adrenaline, noradrenaline, melatonin, and calcitonin. All of these cells were characterized ultrastructurally by dense core granules (small, opaque round cytoplasmic objects observed in transmission electron micrographs of tumor cells). The APUD cells included the so-called neuroendocrine cells of the small intestine (progenitors for carcinoid tumors), the neuroendocrine cells of the bronchus (progenitors of bronchial carcinoids and small cell carcinomas of the lung), the islet cells of the pancreas, the endocrine cells lining the gut mucosa, and the neuroendocrine paraganglia, including the chromaffin cells of the adrenal medulla. Most significantly, all of these cells were presumed to have a single embryologic origin in the neural crest! The term "APUD cell" became synonymous with the term "neuroendocrine cell."

As we discuss in Chapter 14, every cell in the body descends from an embryonic lineage. Because tumors arise from cells, every tumor can be assigned a tumor class based on developmental lineage.

In the 1970s (and later), it was believed that cells from the embryonic neural crest migrate throughout the body, including the surface epithelium of the skin (**ectoderm**) and gut (endoderm). The endocrine cells of skin and gut were thought to be specialized derivatives of the neural crest and were called "neuroendocrine cells."

Over the years, a number of observations have cast doubt on the unifying concept of the **APUDoma**. It is observed that bronchial carcinoma of the lung has a tripartite cell population (squamous cells, glandular cells, and so-called neuroendocrine cells). The squamous and glandular cells of the bronchus arise from endoderm and presumably represent different morphologic variants of a single progenitor endodermal cell. If the neuroendocrine cells of lung have a neural crest origin, how could bronchial carcinoma, which contains neuroendocrine cells, have a monoclonal origin? Is it not simpler to assume that the so-called neuroendocrine cells of the lung arise from endoderm?

Slowly, pathologists began to accept that the so-called neuroendocrine cells of mucosal origin are just another type of epithelial cell with no embryologic relation to neural crest. Carcinoid tumors are epithelial neoplasms, and the tumor cells contain dense core granules. At one time, carcinoid tumors were considered a type of APUDoma. It is not unusual for a carcinoid to grow as a mixed carcinoid-adenocarcinoma. When a monoclonal cell can differentiated to both endocrine and glandular epithelium, is it not likely that both morphologic cell types would have the same developmental origin? Today, most pathologists would accept that carcinoid tumors have an endodermal origin, not a neural crest origin (Figure 5.4). Neuroendocrine cells of

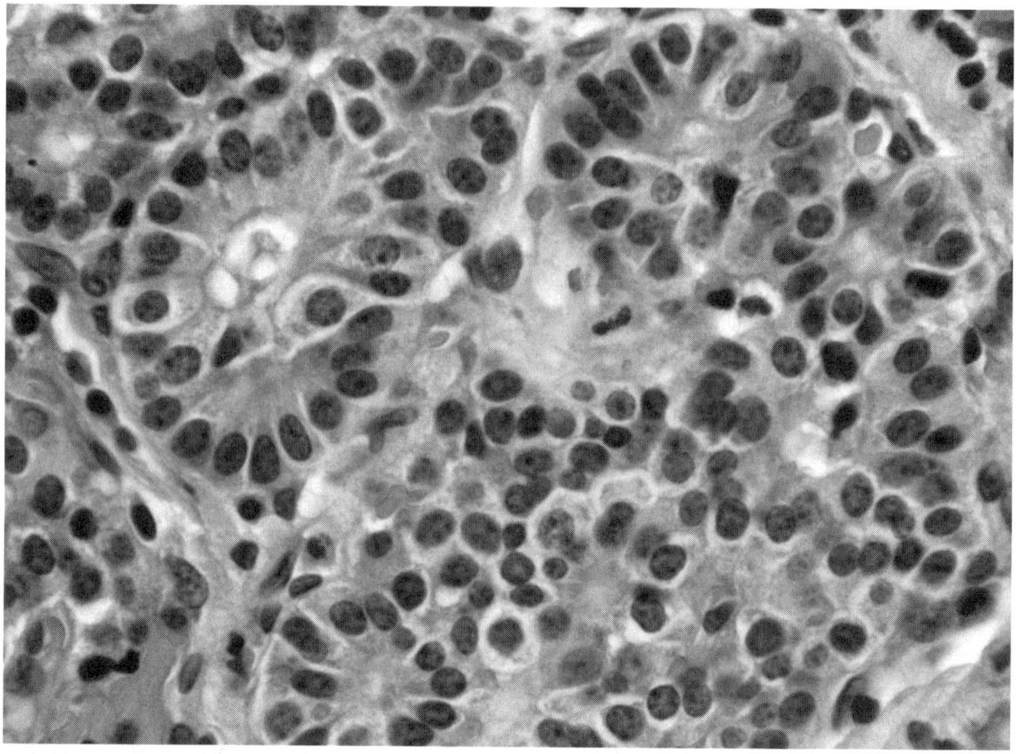

FIGURE 5.4 Pulmonary carcinoid. Small epithelial cells with uniform round or oval nuclei grow as clusters and glands.

endodermal origin share properties with neuroendocrine cells of neural crest origin, through an evolutionary process akin to convergence!

For me, one of the strongest arguments against the neural crest origin of "neuroendocrine cells" comes from a simple observation made on crabs. Crabs, like humans, have a gut, and the crab gut is lined by endocrine cells that can secrete neuropeptides (103). Crabs, unlike humans, are not descended from order Craniata. The neural crest is found only in animals that descend from class Craniata. Because crabs have no neural crest cells, the neuroendocrine cells of the crab gut cannot have a neural crest origin; therefore, the neuroendocrine cells of the gut are not an evolutionary invention of the neural crest.

Although the term *APUDoma* has lost its popularity, the term *neuroendocrine cell* has endured as a popular descriptor for endocrine cells of neural crest origin (a correct use of the term) and for any endocrine cell arising from a mucosal surface (an incorrect use of the term). In a recent paper, Syder and coworkers describe the **transdifferentiation** of a gastric epithelial lineage progenitor to a neuroendocrine phenotype

(104). The authors believe that the gastric epithelial lineage and gastric neuroendocrine cells have different lineages. I believe that the neuroendocrine cells of the gut and the gut epithelium both derive from gut endoderm and that their transitive differentiation is consistent with normal differentiation within one lineage. One of the nice things about science is that it is okay to disagree.

5.13 Are Animal Tumors Good Models for Human Tumors?

Animal tumors are very different from human tumors (List 5.13.1). Does this mean that animals cannot be used in cancer research? We can learn a great deal about human cancer by studying nonhuman organisms. The mistake we often make is to assume that a tumor of a given name in a nonhuman organism is biologically equivalent to the tumor of the same name occurring in humans. Again and again, we are proven wrong when we think that the same carcinogens that yield high rates of tumors in the mouse will yield a similar response in humans or that a cancer treatment that is effective in the rat will work equally well in humans.

Observations that apply in a general way among all organisms are always valuable. That is why we must study cancer in animals; however, a conclusion that is true for a general process will often fail to predict an outcome determined by many different variables applying to a unique situation.

5.14 What Kinds of Questions Can Best Be Answered in Animal Systems?

Humans have killed millions of rodents in the name of cancer research. Both mouse and rat have a short gestation period and are exceedingly easy to breed for laboratory studies.

There are those who believe that mice and rats are excellent models for human disease, and others who believe that humans are so different from rodents that the benefits of mouse and rodent models are negligible, or worse, misleading.

List 5.13.1 Animal tumors differ from human tumors.

Unlike human tumors, rodent tumors develop over a very short period (limited by the short life expectancy of the mouse or rat).
Rodent tumors often lack genetic features that are commonly found in human tumors (e.g., p53).
Animals metabolize carcinogens very differently compared with humans.
Animals are infected by different viruses than humans.
Animal tumors are often morphologically different from human tumors of the same organ.
Animal genes are located at different chromosomal positions than their human homologs.
Animal diets are different from human diets.
Animal host factors (including immune status) are different from those of humans.

After many years of rodent experimentation, we should have a pretty good idea of where these models might be of some value and where these models have conclusively failed. In my opinion, the best uses of animals are to find answers to fundamental questions that are relevant to all animals and from which all animals will benefit from the answer.

All life should be respected, and scientists should avoid inflicting discomfort, deprivation, pain, and death on animals. The term "deprivation" would include anything that denies animals joy, such as social interactions with other members of the species, freedom of movement, and freedom of species-specific behavior.

Scientists have elected to use animals to study a wide range of medical questions and will almost certainly continue to do so in the future. Scientists should obey animal welfare guidelines. These include the four Rs of animal research: reduction (fewer animals), refinement (better tests), replacement (developing alternate tests that do not use animals), and responsibility (for the welfare of laboratory animals).

5.15 Is Cancer Directly Transmissible from Organism to Organism?

Cancer is not, generally, transmitted from one animal to another.

There are exceptions. In the case of gestational choriocarcinoma, the choriocarcinoma is a tumor of the trophoblastic (placental) tissue of the conceptus. The tumor that occurs in the mother was transmitted from the conceptus to the mother. We discuss gestational choriocarcinoma and related molar lesions in detail in Chapter 20.

In laboratories, it is common to transplant tumor cells from individual to individual. These transplantations usually involve the injection of millions of cells from one rodent to the next and employ highly inbred strains of rodents that have immunological tolerance among individuals of the same strain. Transplantation of tumor cells from individual to individual does not occur in nature, with one exception: canine transmissible venereal tumor (CTVT). In CTVT, tumor cells are transmitted from animal to animal through venereal exposure. A 2006 article in *Cell* showed that CTVT cells are genetically distinct from the cells of the host. The authors also showed that tumor cells collected from 40 dogs in five continents all descended from a single neoplastic clone. Phylogenetic analysis suggested that the progenitor tumor cell came from a wolf or an East Asian dog and may have first appeared as early as 2,500 years ago (105).

5.16 Transmission of Virally-Induced Cancers

The transmissibility of oncogenic viruses is a great concern. In humans, there are a few instances wherein virally induced cancers appear as a contagion. This might include epidemics of common warts, caused by noncarcinogenic strains of human papillomavirus. Epidemics may occur among visitors at swimming pools or community bathing facilities in which a person or persons with active warts contaminate a shared source.

Carcinogenic strains of human papillomavirus can be transmitted sexually, causing anal condylomas and cervical condylomas, which can progress to anal carcinoma and cervical carcinoma.

The AIDS virus, human immunodeficiency virus (HIV), is transmissible and is associated with a variety of malignancies, none of which are directly caused by HIV. The AIDS virus causes immunodeficiency, allowing endogenous virus infections (such as Epstein-Barr virus, human herpesvirus 8, and human papillomavirus) to produce malignancies quickly.

In animals, we occasionally see epidemics or endemics of viruses that incapacitate or kill large numbers of animals. The Shope papillomavirus in rabbits can produce crops of mutilating skin tumors. Jaagsiekta in sheep produces epidemics of bronchioalveolar carcinomatosis.

5.17 What Can We Learn from the Differences between Animal and Human Tumors?

Enormous effort is spent by experimental pathologists to prove that their various mouse tumors are valid models for human tumors. The perceived issue is whether the animal model is sufficiently close to the human tumor that experiments in the animal will allow us to draw conclusions that apply directly to humans. Often, the specific question asked is whether a certain mouse carcinogen will cause tumors of the same type in humans, with the same carcinogen, with roughly similar exposures.

This question may never be answered to everyone's satisfaction and distracts us from a closely related question: "How can we answer our questions about human cancer by observing animals?" In the next section, we see how we can better understand human tumors by observing cellular relationships in animals whose organs are fundamentally different from ours.

5.18 The Hepatopancreas of Arthropods, Gastropods, and Fish

Mammals have a liver and a pancreas, and the two organs are anatomically and functionally distinct. In some animals, including arthropods, gastropods, and fish, the pancreas and the liver are a single organ. Fish, like humans, belong to class Craniata, indicating that the hepatopancreas is preserved in animals with a close lineage to mammals.

The observation that the hepatopancreas is a single organ in fish helps explain the close similarity between embryonic tumors of liver (hepatoblastomas) and pancreas (pancreatoblastomas) in humans. Both tumors occur primarily in infants. Both tumors have a somewhat similar morphology, characterized by fetal-type acini (glands) and embryonic organ stem cells. Both tumors produce alpha-fetoprotein, a protein secreted in great quantities into the bloodstream by the normal fetal liver, and both tumors are associated with the inherited growth disorder, Beckwith-Wiedemann syndrome. Both tumors have loss of **homozygosity** for markers on the 11p chromosome (106).

The similarity between pancreas and liver seen in tumors of infants also extends to tumors of adults. Hepatoid adenomas occasionally occur in the pancreas (107), and liver differentiation can be induced in hypoplastic mouse pancreas (108). These observations only make sense in light of the evolutionary relationship between liver and pancreas, observed in distantly related animals.

Chapter 5 Summary

Virtually all multicellular organisms develop tumors, and carcinogenesis in nonhuman animals is the same process as carcinogenesis in humans. Virtually every known human oncogene has a homolog in nonhuman animals. Virtually every cell type in humans can be a progenitor cell for one or more tumors, and virtually every mammal (and many nonmammals) have tumors derived from the same cell types as humans.

Because animals are different from humans (different size, metabolism, life-span, genes, etc.), it is unlikely that any tumor in an animal will behave like any human tumor or that any tumor in an animal will follow the same exact carcinogenic process as any tumor in a human. Progress in cancer research may be accelerated if we stop asking whether we have an animal model for this or that particular human tumor. A better question is this: what can we learn from observing the animal tumor that we cannot learn by observing the human tumor?

The differences among all the tumors of all animals are generalizable and relate to the fundamental properties that distinguish any tumor from any other tumor (List 5.18.1).

List 5.18.1 Some factors that modify the behavior of tumors that arise from the same cell types found in humans and other mammals.

1. Differences in available tumor suppressor genes
2. Differences in host factors that may modify the susceptibility of cells to carcinogens or the speed of tumor progression
3. Differences in the kinds of carcinogens and the levels of exposure.

PART

II

Classification

CHAPTER 6

What Does Classification of All Animals Tell Us About the Classification of All Neoplasms?

We are just "a volume of diseases bound together."

—John Donne (1572–1631)

6.1 Background

The Materia Medica was written in 60 AD by Dioscorides, providing one of the earliest and most influential taxonomies of botanical medicine (109). This scholarly work also contained thoughtful discussion on the topic of taxonomic organization. Dioscorides asked why his predecessors had organized their listings of herbs alphabetically. Doing so separated herbs that were similar to each other, making it difficult to appreciate the relationships and shared properties of related plants. These early musings are just as valid today as they were for Dioscorides.

Over the next 2 millennia, the development of a scientific classification for the organisms of earth became one of the most important activities of biologists. This classification, which is still undergoing modification, is one of the greatest achievements of mankind. It was developed by legions of naturalists through a multigenerational effort. It is impossible to think intelligently about the many organisms on earth without knowing their position in the classification of life.

When we identify that a plant is an angiosperm or an animal is a mammal, we can assume that the general properties of the class of the organism will extend to every species of the class and to all its subclasses.The classification of organisms greatly simplifies the field of biology and provides a method for testing new assertions about the essential features of classes and subclasses of organisms.

The enduring principles of animal classification are instructive. Each organism on earth belongs to a species. There are millions and millions of species of organisms on the planet. Hundreds of thousands of species have been catalogued. Species are grouped based on similarity of features and common ancestry (**phylogeny**). The features that determine membership in a class are chosen to contain the essential properties of a species (i.e., properties shared by every member of a class and absent from members of classes that do not share the same ancestor). Each class of organisms is the subclass of an ancestral organism. The **hierarchy** ascends to a common root organism that is the presumed progenitor of all life on earth.

The list of all known species of organisms, including the name of the class to which each species is assigned, is called a taxonomy. The taxonomy of organisms contains hundreds of thousands of records. Its classification (the list of classes and subclasses that contain the taxonomy) is a connected hierarchy of a few hundred items, sometimes called the tree of life. Any child could look at a schematic diagram of the tree of life and obtain, almost at a glance, a deep understanding of the organization of terrestrial life.

A few observations pertaining to the classification of animals have immediate relevance to the classification of neoplasms. First, once you rise above the level of taxonomy (the listing of species), the classification consists of nonexistent entities. The ancestor common to man and ape no longer exists. Basically, the classes that lie above the taxonomy are inferred from observations of existing species, from molecular studies of genomes, and from paleologic discoveries. Second, the postgenomic classification of organisms is basically the same classification known to Darwin and his predecessors. Third, a classification is a hypothesis about the order of life. Hypotheses must yield to scientific evaluation. One of the most striking discordances between taxonomy and genetic phylogeny is found in the water lotus (genus *Nelumbo*), which morphologically resembles another aquatic plant, the water lily (genus *Nymphaea*). Genetic studies have shown that the delicate and petite water lotus is related to the sycamore tree (Genus *Platanus*). The water lily and the water lotus do not share a recent common ancestor. All classifications are subject to change.

The purpose of this chapter is to provide a conceptual understanding of biological classification, and to explain how the process of classifying organisms applies to the classification of neoplasms. This discussion will prepare the reader for later chapters, wherein we develop several different ways to construct a neoplasm classification.

6.2 What Is Speciation and How Does It Relate to Evolution, Taxonomy, and Classification?

Every terrestrial organism is descended from an ancient, primordial organism. Why do we have different species? Why did the primordial organism not simply evolve over time into a single "best" species? This question points to the difference between the concepts of evolution and speciation. Every educated person understands the meaning of evolution (although many seem to doubt its validity), but most people do not really think much about speciation. As discussed in Chapter 3, speciation is a continual, natural process that results from the individuality of every organism on the planet.

Speciation does not require the acquisition of new mutations. It simply uses the existing genomic variation within populations to produce a unique organism with features shared among its close relatives. In the animal kingdom, the test for speciation is based on reproductive exclusivity. Members of a true species can mate with each other to produce additional members of the species. When members of the new species attempt to mate with members of its ancestral species, reproduction will fail, or it will produce offspring that are not members of the species. In most cases, these offspring will be infertile.

Every member of a species is unique and measurably different from all other members of the species. At the same time, every member shares a set of basic features that no member of any other species can duplicate.

6.3 How Does Animal Classification Relate to the Classification of Neoplasms?

The taxonomy of neoplasms is finite and comprehensive. Every tumor that has ever occurred in man is unique. The uniqueness of the tumor is predetermined by the uniqueness of the person in whom the tumor develops (because, with the exception of identical twins, the gene sequence of an organism is different from the gene sequence of every other organism) and in the changes that occur in tumor cells during and following initiation. Every tumor can be named as a member of a particular species of tumor, and every species of tumor can be classified. The same principles of classification that apply to animal species will also apply to tumors because both animals and tumors develop through an ancestral lineage and have a defining set of inherited properties common to every member of the species; moreover, each species (animal and tumor) is composed of unique members.

6.4 What Is the Most Common Impediment to Classifying a Species?

Sometimes we know that two species are different, but we do not know why. Imagine two groups of frogs, one selected from the upper Amazon region and the other selected from the lower Amazon region. Both species look alike to amphibian experts, but males and females from the two groups cannot produce offspring. Scientists can infer that the two groups are different species, but they cannot specify any particular difference between the groups. This is an example in which there is a lack of informative markers to separate two inferred species.

The same situation arises daily in the cancer field. Two patients each have cancer. The tumors in the two patients have the same diagnosis (i.e., are the same tumor type) and have identical morphologies. One of the tumors grows aggressively, metastasizes, and kills the patient. The other tumor grows slowly and never metastasizes. We can infer that the tumors are biologically different, but we do not know why.

Samples of the tumors are sent for **molecular signature** analysis. Over 10,000 expressed genes are compared, and the two tumors are found to have similar profiles, confirming their original diagnosis; however, the expression level of thousands of different genes are slightly different in either tumor. This information we have collected is uninformative. The gene expression differences between the two tumors have no value unless specific differences can be directly related to the biological differences that consistently distinguish subclasses of a tumor. Much of the field of cancer research is devoted to finding informative markers that separate biologically distinct classes of morphologically uniform tumors.

6.5 The Sui Generis Taxon

When we think about the qualities of an ideal classification, the idea of a balanced distribution of classes is likely to enter our deliberations. We like to think that each class will be about the same size (i.e., have a nearly equal number of members). If one class of organisms contains a million members and another class of organisms has only one member, it may be difficult for anyone to think of these two classes as equivalent objects within the classification.

As strange as it may seem, the classes of animals have wildly different sizes. As vertebrates, we like to think that we occupy a popular class of animals, but 95% of the animal species of on earth are invertebrate. About 80% of the invertebrate species are Arthropods.

Among the mammals, the least populated class is certainly Prototheria. Class Mammalia branches into the Therians (marsupials or Metatheria, and placental mammals, or Eutheria) and the Prototherians (also called Monotremes). The Prototherians consist of two species, the duck-billed platypus and the spiny anteater.

Class Prototheria is not the smallest class of animals. In 1995, a new species was reported, and a new class was created to hold a **sui generis** (one of a kind) species. Symbion pandora was placed into the new class Cycliophora. Cycliophora may be closely related to class Bryozoa (microscopic animals that grow on algae and shore rocks) (110).

Does it really make sense to have classes of organisms that contain only a few members and other classes that contain hundreds of thousands of members? Yes, because the differences that distinguish one class from other classes can help us understand the basic biological characteristics of both the rare class and of the more populous neighbor class (from which it must be distinguished).

In Chapter 15, we discuss **rhabdoid tumors** as a sui generis class that does not readily fall into any of the major subclasses of neoplasms. Like the platypus, this class breaks almost all of the rules of classification. Currently, our understanding of this very rare tumor is meager. Ultimately, we may learn more about neoplasia by studying rhabdoid tumors than by studying any of the commonly occurring cancers.

6.6 The Classification of Organisms Is Not Strictly Cladistic

A **clade** consists of a class of organisms that share the same direct ancestor and that includes every direct descendant of that ancestor. **Cladistics** is the assignment of classes and subclasses within a classification by strict order of lineage (i.e., through clades) and is the idealized basis for biological classifications.

Because we speak of the classification of organisms as the **validation** of the evolutionary development of species, we tend to assume that classification is purely cladistic, showing the ancestral lineage of each extant species. This is not strictly true. The commonly used classification of living organisms assigns organisms to classes based on lineage and similarity. There are instances when an organism is assigned a particular class even when that organism does not share the same direct ancestor with the other members of the class. For example, class Reptilia excludes class Aves (birds), although we now know that birds are descended from reptiles. The reason for the noncladistic assignment is that birds share many characteristic with animals outside class Reptilia (such as warm bloodedness). If modern animal classification is phylogenetic (determined by ancestry), why would be break the rules for class Aves?

Animal classification has never been a straightforward exercise in cladistics because certain essential traits of an organism are sometimes not contributed by a direct ancestor. Here is an example. Imagine that Napoleon had three brothers. One of Napoleon's brothers marries his cousin and rears many children, all of whom marry other cousins. This practice continues for three generations. The other one of Napoleon's brothers freezes his sperm for 60 years, and the stored sperm is used to impregnate a Duchess of the realm. Napoleon himself has children with a series of women of various ethnic and geographic origins who bear Napoleon many children. His children, in turn, marry an international selection of spouses and bear a large set of grandchildren.

After 60 years (three generations), the descendants of Napoleon's brothers share many phenotypic features of Napoleon himself, although none of these offspring are direct descendants of Napoleon. The third generation of the direct descendants of Napoleon have scarcely any of Napoleon's genes and no phenotypic properties that can be specifically credited to Napoleon. If, from among all of the descendants, you needed to choose a successor to the Napoleonic monarchy, from which brother's descendants would you choose? You might choose one of Napoleon's direct descendants (even though they have none of the characteristic features that seemed to animate Napoleon himself). You might choose the offspring of the brother who froze his sperm (the genotypically closest descendant), or you might choose the descendant of Napoleon's inbreeding brother who looked and acted the most like Napoleon (the phenotypically closest descendant).

The moral of this story is that the best way to classify is not always based on direct descendancy. If your goal is to capture a set of genotypic and phenotypic properties that express the essential biological qualities of a biological class, you may sometimes need to abandon a strictly cladistic assignment.

In Chapter 16 and in Chapter 19, we revisit noncladistic classifications as they apply to cancer.

Chapter 6 Summary

Over the millennia, the most important sustained activity of biology has been the development of a classification for all organisms on earth. The classification of all living things permits us to create and test new hypotheses of life. Although many people today think that the classification of life is based on Darwin's theory of evolution, the opposite is true. Darwin based his theory on a hypothesis drawn from the existing classification.

A biological classification is a hierarchical grouping, with each group defined by the greatest number of **taxa** (informative features) that can apply to every instance of the group. In a classification, every instance must fit into the classification, and every instance must have exactly one slot in the classification.

The classification of organisms is not based exclusively on ancestral lineage and is not cladistic. The classification is based on both ancestry and genetic similarity among the classes of animals.

The same principles that apply to animal classification apply to the classification of human neoplasms adopted in this book. In Chapter 14, we discuss the ways we might use those principles to construct a neoplasm classification.

CHAPTER

7 Classification of Neoplasms

All this has been said before, but since nobody listened, it must be said again.

—Andre Gide

7.1 Background

Classifications are important. Simply knowing the class of a bacterium can provide a microbiologist with deep insights relating to the expected growth conditions for the organism and the types of antibiotics that may be effective against the organism. A classification encapsulates all knowledge related to a knowledge domain.

A modern tumor classification serves as the computational "key" that links neoplasms to tumor-related data held in biomedical databases. With a neoplasm classification we can determine whether new therapeutic agents target pathways that are present in classes of tumors, not just the tumor that the agent was designed to treat. We can predict the properties of individual neoplasms and classes of neoplasms if our tumor classification is built on sound scientific principles.

Most of the current ways of classifying tumors do not serve biologists well. Pathologists use tumor classifications that are nothing more than anatomic groupings. The molecular biologists prefer to group tumors based on shared gene expression patterns. Neither type of classification is satisfactory. The purpose of this chapter is to provide the principles of neoplasm classification.

7.2 The Challenge of Classifying Neoplasms

A classification is an organization of everything in a domain by hierarchical groups, according to features generalizable to the members of the groups. Four terms with distinctly different meanings have been used interchangeably with "classification," leading

List 7.2.1 Four easily confused terms.

Identification (also known as diagnosing or naming) is the act of placing something into its correct slot within an existing classification.

Discrimination is finding features that separate members of a group according to expected variations in group behavior. Examples of discrimination are **grading** and **staging**. Grading and staging involve reporting additional morphologic features (grading) or clinical behavior (staging) that help predict a particular tumor's clinical course or response to therapy.

Taxonomy is a complete listing of all the members of a classification. In the case of neoplasia, a taxonomy would be the complete listing of all the different named tumors. The included items in a tumor taxonomy would vary depending on the taxonomist's approach to splitting terminology. Should distinctive variants of a tumor be included as taxonomic species? For example, should "goblet cell **carcinoid** of appendix" be a taxonomic species distinct from "carcinoid of appendix"? Actually, goblet cell carcinoid of appendix is an aggressive tumor that receives different therapy than carcinoid of appendix, despite the similarity of their names (111). Most taxonomists would distinguish the two tumors.

Ontology is a rule-based grouping of some portion of a taxonomy. Ontologies support queries and logical inferences pertaining to the (ontologic) group members. This definition is elaborated in Chapter 22.

to considerable confusion among biomedical curators. These terms are identification, discrimination, taxonomy, and **ontology** (List 7.2.1).

7.3 Molecular Analyses of Neoplasms

In January 1999, the U.S. National Cancer Institute challenged the cancer research community to develop a molecular classification of tumors. The federal request for applications elaborated a plan "to harness the power of comprehensive molecular analysis technologies to make the classification of tumors vastly more informative. This challenge is intended to lay the groundwork for changing the basis of tumor classification from morphological to molecular characteristics" (112). Not surprisingly, this challenge resulted in lively debate over the relative value of morphologic and molecular classifications (113).

In 2008, we can look back and see that the responsive projects used molecular profiling techniques, particularly gene expression arrays, to discover subtypes of tumors that could not be distinguished by morphologic examination.

Gene expression arrays, also known as gene chips, consist of thousands of small samples of DNA arrayed onto a support material (usually, a glass slide). The samples of DNA selected for the gene chip are preselected for the intended use of the chip. For example, a gene chip may consist of human mitochondrial genes, all known mutations of the p53 gene, a large selection of genes involved in *apoptosis,* or a random collection of genes from the rat and so on. The gene chip serves as a substrate to which a DNA sample from a tissue of interest (usually a cancer) is hybridized. Hybridization involves the annealment of complementary single-stranded DNA. When the gene chip is incubated with a cell sample, hybridization will occur between molecules anchored on the array and single-stranded complementary (i.e., identically sequenced) molecules present in the cell sample.

Messenger RNA in the cell sample is transcribed into DNA molecules and tagged with a fluorescent molecule. The gene chip is incubated with the cell sample and hybridization reactions proceed. An instrument prepares an image of the array and quantifies the fluorescence in each array spot. High fluorescence in an array spot indicates a high concentration of the tagged DNA molecules in the cell sample. The data set comprised by the fluorescent intensity of each posthybridization array spot produces a gene expression profile characteristic of the cell sample.

For a given tumor, the gene chip produces a cancer profile, or signature. Researchers found that the gene chip signature of a cancer had diagnostic relevance (114). The signature of a tumor can be used to find biological variants of the tumor or to predict the clinical behavior of a tumor (115, 116). Methods for analyzing gene expression array data, to find the most informative genes in the array (117), or to enhance the selection of important subtypes of tumors (118) are active areas of cancer research.

Research articles often employ the word "classification" to describe the use of gene signatures to subtype cancers. The word classification can be applied to any activity that groups objects, but the biologist usually reserves "classification" to refer to the much more demanding activity of creating strict hierarchical divisions that hold every object in a biological domain. Classification provides a conceptual framework for understanding the relationships among the classes of objects. Gene analyses of tumors have enlarged the taxonomy of neoplasms (the number of species of lesions and the criteria by which we can discriminate the new species from the ancestor species) without producing a coherent molecular classification for neoplasms.

7.4 Formal Principles of Biological Classification

There are many definitions for "classification." Some people consider any grouping of objects to be a classification. Others use the term classification to refer to the act of assigning an object to its correct group (e.g., "I agree with your classification of this tumor as a squamous cell carcinoma"). In this book, the definition of classification is taken from the work of Ernst Mayr, one of the most influential evolutionary biologists of the 20th century (119). A classification is a hierarchical and comprehensive grouping of all of the members of a well-defined knowledge domain. In a classification, a member can only be assigned one class. Each class can be the subclass of exactly one immediate ancestral class. The top class in a classification is the class with no ancestor. This creates

List 7.4.1 Properties of a classification.

A classification is a grouped taxonomy (listing of objects in a knowledge domain) with the following properties:

Inheritance: Hierarchical structure, with each class inheriting properties of its ancestral lineage. Classifications have single-class inheritance in which a member of a class has exactly one direct ancestor class.

Uniqueness: Each member of the classification occurs in only one place in the classification.

Comprehensiveness: All members of the biological domain are included.

Intransitivity: A member of one class cannot change into a member of another class (e.g., an adenocarcinoma cannot transform into a lymphoma).

a simple lineage for each member that extends to the top class in the **hierarchy** (List 7.4.1).

To "classify" (verb form) is to find traits that reliably distinguish the members of a group from the members of all other groups. These differences apply to the group's hierarchical descendants.

In the prior section, we discussed gene array signatures. **Cluster analysis** of the gene expression array values can separate tumors into groups with common expression patterns. Some of these groupings will prove to have a specific biologic feature (e.g., increased tendency to metastasize, higher response to a chemotherapeutic agent, lengthened survival). Clustered groups do not qualify as new classes if they merely represent variations in the expected biology of a type of tumor. Variant groups are disqualified as classes if a tumor of a certain type may progress from one variant group to another variant group over time (e.g., slow-growing variant at one stage in development and fast-growing variant at another stage). This is because a class cannot change over time (e.g., a mammal never becomes a reptile). Clustering by variation only identifies differences among tumors. The data that come from the molecular analysis of tumors can be considered a tentative first step in the process of tumor classification.

After a complete taxonomy is created, the process of tumor diagnosis reduces to choosing a name from the taxonomy that fits the features of the tumor that is being diagnosed.

7.5 Current Status of Tumor Classification

Currently, there is simply no comprehensive modern tumor classification. Some pathologists believe that past attempts to build a comprehensive tumor classification have not been useful (120). This dour perspective may actually represent the modern scientist's perspective. Current tumor classifications suffer from the following limitations (List 7.5.1).

List 7.5.1 Limitations of current neoplasm classifications.

Classifications are created piecemeal for specific sites or organ systems. There is no standard comprehensive neoplasm classification, although near-comprehensive taxonomies are available.
Classifications sometimes group tumors according to medical specialties, rather than on any biologic principles (e.g., classification of dermatologic tumors or urologic tumors).
A given tumor will appear redundantly when subclassifications are merged (i.e., piecemeal classifications, such as tumors of the head and neck, contain some of the same tumors included in another piecemeal classification, such as tumors of the lymphatic system).
No tumor classification has been prepared in a standard format designed to exchange, merge, or analyze heterogeneous biological data.

The most widely used authoritative resources are the World Health Organization classifications, which list the tumors that occur at different body sites. The problem with an organ system approach to classification is that every organ contains organ-specific and organ-nonspecific cell types. The brain, for instance, contains connective tissue and lymphoid tissue and therefore is prone to tumors of connective tissue and lymphoid tissue. A listing of tumors that occur in the brain must include osteocartilaginous tumors, lipoma, fibrous histiocytoma, hemangiopericytoma, rhabdomyosarcoma, melanoma, lymphoma, and myeloma, among others. These same tumors will be included again and again in every site-specific classification (e.g., lymphoma of brain, lymphoma of stomach, lymphoma of skin, lymphoma of testis, and so on). Although each term may occur only once in each site-specific classification, the same lesion may occur a virtually limitless number of times when the site classifications are combined into a comprehensive classification of tumors. Cancer taxonomies that list most of the recognized types of tumors are available, but all current taxonomies fail to group tumors into a biological classification.

When designing a classification, there is a tendency to choose an organizing principle that has immediate utility. This seems imminently practical, but it is actually counterproductive. Here is an example. Oncologists generally treat cancers based on the aggressiveness and stage of the tumor. Often, tumors that are localized and slow growing will receive a different treatment than tumors that are widespread and fast growing. Tumors could be organized by their biological stage and rate of growth. After a tumor's stage and growth rate are known, the tumor could be subdivided into grade (the level of nuclear atypia), size, and so on.

The problem with this approach is that a tumor that is small and low grade in March may be large and high grade in June. When the members of a grouped domain freely transition among the different groups, you cannot make a sensible classification. The identity of the members of the classification will be constantly changing, and no class rules will apply at all times to all members of the class.

List 7.5.2 Ways of classifying tumors

By age of patient
By gender
By ethnicity
By cause
By genotype
By behavior (function and common pathways)
By morphology
By treatment
By anatomic site of occurrence
By cell of origin (**histogenesis**)
By developmental lineage

Animal classifiers faced the same dilemma. The classification of animals does not group organisms into Class Seed, Class Larva, Class Embryo, Class Fetus, etc. These are transitional phases in the life of an organism. The classifiers of animals needed to find taxonomic strategies that capture constitutive properties of each type of animal.

Over the past century, we have found that many of the properties of individual cancers are influenced by specific conditions, and these conditions have served as the basis for tumor classifications (List 7.5.2).

A condition that has a striking influence on tumor biology is seldom an adequate classifier. For example, gender greatly influences carcinogenesis. Aside from the gynecologic tumors and genitourinary tumors, some of which are gender specific, males are much more prone to cancer than females. With the exception of thyroid cancer, meningioma, and gall bladder cancer (all of which occur more often in women than men), virtually all other common tumors of humans occur more frequently in males.

The same applies to rodents. Almost every protocol for the production of tumors in mice and rats will produce more tumors in the male population than the female population.

Can we use gender to create a new classification of neoplasms? Of course not. Both men and women develop many of the same kinds of tumors, and the morphology and biology of the tumors that occur in men are equivalent to the morphology and biology of the tumors that occur in women (e.g., lung cancers in men look and behave like lung cancers in women).

To classify neoplasms, we need to find a biological principle that accounts for every stage of development of every tumor of humans, that distinguishes each tumor from every other tumor, and that unites similar tumors into hierarchical classes.

Chapter 7 Summary

There are about 6,000 different known tumors of humans. One approach to curing all of the different kinds of neoplasms in humans is to tackle all 6,000 tumors, one at a time, until the job is done, starting with the most common tumors. Remarkably, this approach is more or less the way we have been doing things. What have we gotten from this approach? The most common cancer killers are still killing us at nearly the same frequency as they had when the war on cancer began in 1971.

We really need a better approach, and this involves classifying tumors into groups that share common biological features that can be exploited to develop new methods of prevention and treatment that might apply to every member of the group.

The currently most popular method of tumor classification has no sound biological basis. When tumors are grouped by anatomic site (e.g., tumors of the head and neck and tumors of the abdomen) or system (endocrine tumors and tumors of the skin), tumors within a single class will include biologically unrelated neoplasms and will exclude biologically related neoplasms that arise from different locations.

In Chapters 8 through 13, we discuss how the members of a tumor taxonomy can be organized into a classification that allows us to make general assertions that apply to every tumor species in a class and its subclasses.

CHAPTER

8 Classification by Morphology (What the Tumor Looks Like)

It's not always about finding a simple solution to a complex problem; occasionally it's about simplifying the problem.

—Quotation has been attributed to Adam Bosworth, but has probably been stated by countless others through the years.

8.1 Background

As an anatomic pathologist and a cytopathologist, I was taught that you can diagnose any tumor just by staring at it through a microscope. For reasons that we do not fully understand, the molecular and biological properties that distinguish one cancer from another are distilled in the morphology of neoplastic cells.

There are some instances in which molecular biology provides diagnostic information unavailable to the morphologist. Die-hard morphologists believe that in these instances, the pathologist has simply failed to perceive some important clue that is present in the morphology. The morphologist believes that, in time, after careful correlation with all available data, the essence of every molecular aberration can be visualized as a histologic property of the tumor.

Skeptics would argue that the value of morphologic examination has reached its limits. The morphologic diagnosis of a lesion can be subjective. Different pathologists can look at the same tissue and reach very different conclusions. Almost every aspect of histologic diagnosis is rife for debate (List 8.1.1). Oncologists and other clinicians charged with treating patients become exasperated when pathologists cannot agree among themselves.

After the pathologist renders a diagnosis, will the pathologic evaluation provide the clinician with sufficient information to choose the correct treatment for the patient? The clinician wants to know, based on objective observations of the primary

List 8.1.1 The kinds of things over which pathologists sometimes disagree.

Whether the lesion is a neoplasm or a reactive process (e.g., atypical lymphoid hyperplasia or lymphoma)

If the lesion is a neoplasm, the particular type of neoplasm that applies (e.g., poorly differentiated carcinoma of lung or small cell carcinoma of lung)

If the lesion is a neoplasm, whether it is benign or malignant (e.g., epithelioid **nevus** or malignant melanoma)

If the lesion is a neoplasm of an agreed type, the grading of the tumor (e.g., Gleason score 5 adenocarcinoma of prostate or Gleason score 7 adenocarcinoma of the prostate)

If the lesion is of an agreed type, the extent of the lesion (e.g., basal cell carcinoma with positive margin versus basal cell carcinoma with negative margin)

tumor, whether the tumor has metastasized, how the tumor might respond to the available treatment regimens, and how the available treatments will be tolerated by the patient.

When the pathologist renders a morphologic diagnosis, does the rendered diagnosis provide information that relates to a biological class of tumors, or does the diagnosis only provide information related to the tumor that is being examined? If every tumor is unique, with no common biologic features shared among the members of a common class of tumors, how can we hope to develop treatments for every specific type of tumor?

One of the purposes of a neoplasm classification is to reduce the complexity of the thousands of different kinds of tumors known to occur in humans. Has the morphologic examination of tumors provided a sensible biologic classification in which all tumors of a common class are cured by the same treatment? No. Why have pathologists not followed the well-marked path trod by thousands of biologists over two millennia to create a classification that groups tumors under classes that have morphologic and biologic properties common to all the members of a class and absent from members outside the class? The answer is simple. Pathologists have opted for an identification system, not a classification. They have done this because the job of a pathologist is to identify a tumor so that it can be treated by the clinicians. The job of a pathologist has never been to create a biologic classification of tumors.

The purpose of this chapter is to show that the current morphological groupings of tumors are an identification system (i.e., a diagnosis method) and not a classification system. A valid classification for neoplasms will require the use of molecular techniques, coupled with morphologic expertise.

8.2 Epithelial and Nonepithelial Tumors

Epithelial cells are usually described as cells that have the shape of a polygon. Nonepithelial cells are usually described as spindle-shaped cells. A spindle is a long, thin, and smooth piece of wood, slightly rounded at each end. Spindles were used in spinning wheels (hence the name). Actually, both terms are slightly misleading.

Epithelial cells are round, not polygonal in shape. When you disperse so-called polygonal epithelial cells as a suspension of single cells, you see perfectly spherical cells, not floating collections of polyhedral shapes. Groups of round epithelial cells appear to be polygonal in tissues because spheres of equal size produce flat surfaces at their points of intersection (Figures 8.1 and 8.2).

Epithelial cells are spherical for the same reason that nuclei are spherical and for the same reason that soap bubbles are spherical. A sphere is the shape that develops when an outward force, that pushes equally at all points, is enclosed by a membrane. When epithelial cells are bunched together in a gland, the adjacent sides will flatten, and junctions (e.g., desmosomes and tight junctions) will form to "zip up" and stabilize the flat, intersecting surfaces. When you look at a cross-section of normal epithelial cells, you see a regular network of straight lines that look something like a beehive.

FIGURE 8.1 A single spherical object. Surface tension, exerted on a single cell, produce a spherical shape.

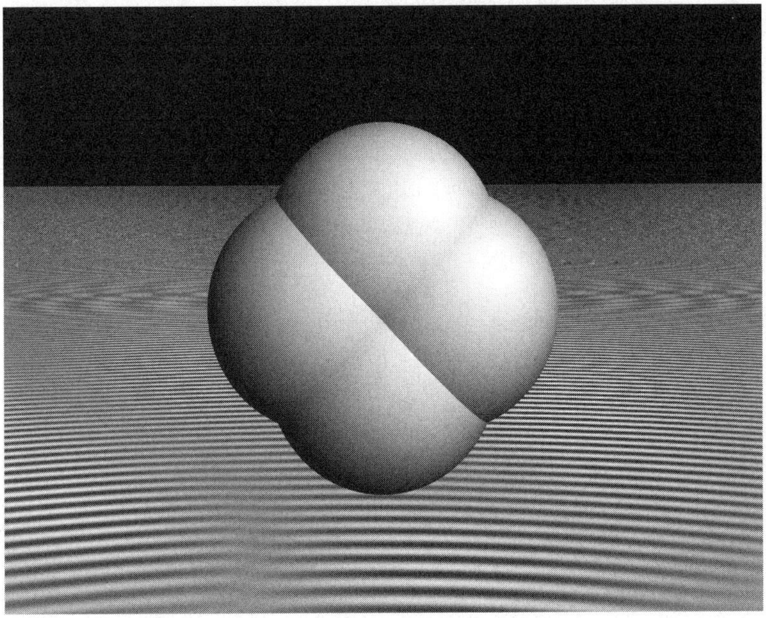

FIGURE 8.2 Four spheres of the same size that are touching resemble an epithelial gland. Surface tension, exerted on a group of spherical cells, produces flat surfaces where the cells intersect.

When epithelial cells divide within a polygonal glandular structure, they loosen their junctions, and each daughter cell molds itself into the available spaces, creating new flat surfaces (Figure 8.3). Following this, new junctions are formed, and the network of flattened surfaces that characterizes epithelial glands and ducts and pavemented surfaces is re-established.

When round cells of unequal size are pushed together, flat surfaces are not formed. Instead, the smaller sphere will bulge into the larger sphere. This phenomenon is called Ostwald ripening and is predicted by the Laplace equation. Neoplasia is often characterized by inequality of cell size. When an epithelial tumor is composed of cells of different sizes, the resulting clumps of cells are characterized by bulges, not polygonal networks (Figure 8.4).

Spindle cells are elongated because they contain a stiff protein that runs along the length (long diameter) of the cell. The protein forms the cytoskeleton of the cell and might be composed of actin–myosin filaments or desmin or any filamentous material that can stretch the cellular membrane. Even spindle cells have a spherical natural state. If there is a defect in the production of cytoskeletal protein, if the cytoskeletal protein is weakened, or if the cytoskeletal protein is disassembled, the spindle cell resumes a spherical shape. Many tumors of connective tissue are spindle shaped.

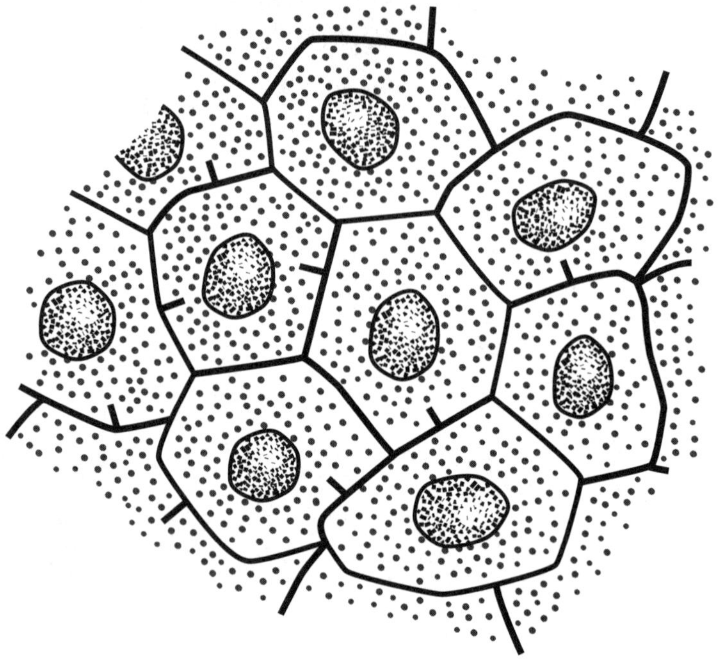

FIGURE 8.3 A simple epithelium.

The simple morphologic distinction between epithelial cells and spindle cells has served as the basis for the most important classifier of neoplasms. With few (but notable) exceptions, epithelial tumors derive from glands, ducts, mucosal surfaces, and epidermal surfaces (the carcinomas). Spindle cell tumors arise from connective tissue (the sarcomas). Every other tumor is an exception to the epithelial/spindle dichotomy and can be dealt with individually. This summarizes the morphologic classification of neoplasms!

The separation of tumors into two large groups, epithelial and nonepithelial, is a popular and practical conceit. If a patient has a brain mass that is stereotactically biopsied, the anatomic pathologist can make a quick, intraoperative consultation based on a small sampling of the specimen, reporting that the tumor is epithelial. An epithelial tumor is unlikely to be a primary brain tumor. It is much more likely to be a metastatic lesion from a primary carcinoma that arose somewhere else in the body. The treatment of a primary brain tumor is different from the treatment of a metastatic tumor to the brain. Hence, the intraoperative assessment of epithelial morphology can be very helpful.

Despite its clinical utility, the separation of tumors into epithelial and nonepithelial subtypes does not provide a consistent dichotomy for tumor classification.

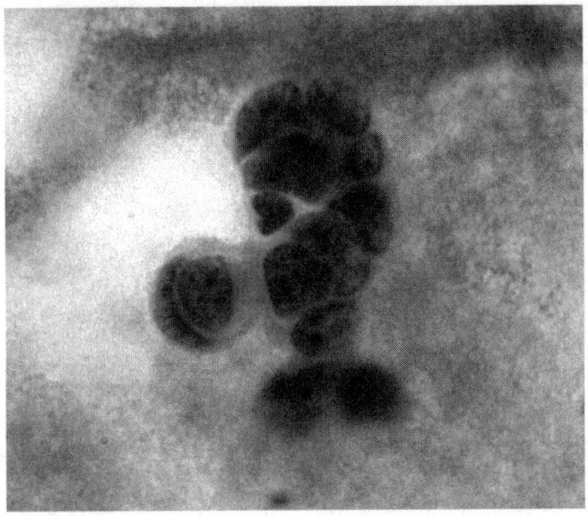

FIGURE 8.4 Cytologic preparation of small cell carcinoma. The intersections of cells are not characterized by straight lines, as you might expect in an aggregation of equal-sized cells. Instead, cells bulge into each other in curved intersecting lines. This is the cytologic embodiment of Ostwald ripening. It indicates the presence of a heterogeneous population of round cells.

Experts in the field of classification consider it poor form to create a class based on criteria of exclusion (such as nonepithelial tumors). Classifications are intended to identify properties common to classes (119). There is seldom much value in assigning class membership based on the absence of a property found in another class.

In the specific case of a morphologic separation of tumors into epithelial and nonepithelial classes, a problem arises when experts do not agree on the qualifications of epithelial tumors. Some might argue that epithelial tumors are tumors that arise from an epithelium (e.g., squamous cell carcinoma) or an epithelial glandular lining (e.g., adenocarcinoma). The qualifying features of an epithelium are controversial. Are mesothelial cells (which cover the serosal surfaces of the pleura and peritoneum) epithelial cells? Mesotheliomas arising from coelomic epithelium are classed as (nonepithelial) sarcomas, because coelom arises from mesoderm, the same embryonic layer that produces connective tissues. Are specialized mesothelial cells that cover the serosal surface of the ovaries epithelial? Ovarian tumors arising from specialized coelomic mesothelium, including papillary cystadenocarcinoma, are classed as epithelial tumors (carcinomas). Where is the consistency?

Some anatomic pathologists divide tumors into epithelial and nonepithelial classes based purely on morphology without regard to their tissues of origin. An epithelial cell is a round or polyhedral cell that adheres closely to other tumor cells, using specialized

List 8.2.1 Limitations of an epithelial/nonepithelial dichotomy.

Some tumors have mixed epithelial and nonepithelial morphologies (mixed tumor of salivary gland, fibroadenoma of breast, synovial sarcoma, carcinosarcoma of uterus).

Some tumors that are typically epithelial can have nonepithelial variants (e.g., sarcomatoid squamous cell carcinoma, spindle cell nevus, sarcomatoid melanoma).

Some tumors that are typically nonepithelial can have epithelial morphology (epithelioid leiomyoma) or morphologic features in common with epithelial tumors (clear cell sarcoma of soft parts).

Biochemical markers intended to distinguish epithelial from nonepithelial cells sometimes co-express in epithelial and nonepithelial tumors.

A single oncogene or molecular marker may characterize both epithelial and nonepithelial tumors (121). This means that the epithelial/nonepithelial tumor dichotomy conflicts with a molecular classification of tumors.

A genetically homogeneous class of tumors of a single cell type may span epithelial and nonepithelial morphologies. For example, **PEComa** (perivascular **epithelioid cell** tumors) is a rare, usually benign tumor that comprises a morphologically diverse family of neoplasms, including angiomyolipoma, lymphangioleiomyomatosis, clear cell "sugar" tumors, and clear cell myomelanocytic tumor.

cell junctions called desmosomes. Anatomic pathologists subscribing to a purely morphologic separation of tumors face an assortment of problems when they try to maintain a dichotomous classification (List 8.2.1).

The lesson learned here is that the essence of a lesion (i.e., features that characterize a lesion and distinguish the lesion from all other pathological entities) cannot be faithfully categorized by a single morphological criterion.

8.3 Tumor Variants

Sometimes it is difficult to categorize a lesion using the totality of morphologic features revealed by microscopic examination (List 8.3.1).

This list is just a small sample of the morphologic challenges that pathologists face every day. In most cases, a pathologist can provide a confident diagnosis when she correlates **histology** (the appearance of the lesion viewed through the microscope), with anatomic references (the location of the tumor), gross observations (the appearance of the tumor to the unaided eye), and clinical presentation (medical symptoms and growth history).

Sometimes molecular analysis can provide information about lesions that cannot be determined by morphologic examination.

List 8.3.1 Lesions with mimics.

Cloacogenic carcinoma of anus can look like a basal cell carcinoma of skin.
Cerebellar hemangioblastoma can look like **renal cell carcinoma** metastatic to cerebellum.
Spitz nevus can look like melanoma.
Keratoacanthoma can look like squamous carcinoma.
Lymphoid hyperplasia can look like lymphoma.
Chloroma (a leukemic infiltrate of tissue) can look like lymphoma.
Mycosis fungoides can look like chronic dermatitis.
Pseudoepitheliomatous hyperplasia can mimic invasive squamous carcinoma.
Gliosis (reactive hyperplasia of glial cells) can look like glioma.
Chronic sclerosing sialadenitis (also known under the highly misleading name, **Kuttner tumor**) can mimic carcinoma arising from salivary glands.

As an example, a morphologist would assume that a low grade fibromyxoid sarcoma would be a subtype (variant) of a fibromyxoid sarcoma. That does not seem to be the case. Molecular studies indicate that the chimeric FUS/CREB3L2 gene is specifically associated with low-grade fibromyxoid sarcoma and is not found in high-grade fibromyxoid sarcoma (122). This molecular information would suggest that the two tumors, despite their morphologic similarities, are distinct and that the high-grade tumor does not develop from the low-grade tumor.

In Chapter 20, we discuss partial and complete gestational moles. For now, suffice it to say that **complete mole** is a condition from which choriocarcinoma commonly arises. **Partial mole** is a condition from which choriocarcinoma rarely arises. It is important to distinguish between these two lesions when either occurs. Pathologists, using purely morphologic methods to diagnose gestational moles, will sometimes confuse complete and partial moles (123). Genetically, a complete mole is diploid without maternal contribution, whereas a partial mole is triploid with a maternal chromosome complement. A gene-based diagnosis will distinguish complete and partial moles with high accuracy. In this instance, genetic diagnosis transcends limitations in morphologic diagnosis (123).

When lung cancer occurs in a young adult, it may be a new case of midline carcinoma of children and young adults. This tumor has a characteristic gene arrangement involving the NUT gene (124). One day, this tumor may have a specific treatment in which an altered pathway is targeted, and this treatment would likely to be different from the treatment of smoking-related **bronchogenic carcinoma**. The only way of establishing the diagnosis with certainty is through genetic analysis.

After a cancer is diagnosed, subtyping the tumor into predictive categories may be a job for genetic analysis, not morphology. Currently, morphology-based tumor grading is the best overall **predictor** of treatment response, but a new generation of molecular profiling tests is now under development (125).

8.4 Determining Histogenesis

In 1926, Abrikossoff described a peculiar benign tumor of tongue, a tissue that consists largely of muscle (Figure 8.5). The cells in the tumor were large, eosinophilic, and granular (126). Abrikossoff thought these tumor cells sometimes contained muscle fibers and were probably the neoplastic analog of myoblasts (muscle precursor cells). He called the tumor "myoblastoma" and began a controversy that persisted until the 1980s.

The lack of resemblance of the so-called myoblastoma cells to true myoblasts and the occurrence of similar tumors at extralingual sites devoid of muscle caused the myogenous origin of granular cell tumor to be challenged as early as 1937 (127).

When electron microscopy became popular in the 1960s, pathologists studied the ultrastructure of granular cell "myoblastomas." The granules of the granular cells were found to be autophagosomes, lysosomes that filled much of the cytoplasm, engulfing cellular material produced within the cell. A Schwann cell origin for the granular cell myoblastoma was proposed based on ultrastructural findings (128, 129). Amid this controversy, the name of the neoplasm adopted by many pathologists became a noncommittal "granular cell tumor."

In 1978, while reviewing a series of **schwannomas** occurring in F344 rats, I found a tumor that was a mixture of two types of cells: typical schwannoma cells and typical granular cell tumor cells. Schwann cells are a specialized cell of the peripheral nervous system that wrap around the axons of neurons. By all analytic criteria available at the time, the granular cells were identical to the granular cells found in granular cell tumors. If granular cell tumor cells arise within schwannomas, they must have a related lineage. From this single observation, my co-workers and I published an article supporting a neurogenic origin of granular cell tumors (130).

Although human schwannomas never seem to produced mixed tumors of Schwann cells and granular cell tumor cells, the F344 rat provided a valuable new clue. Subse-

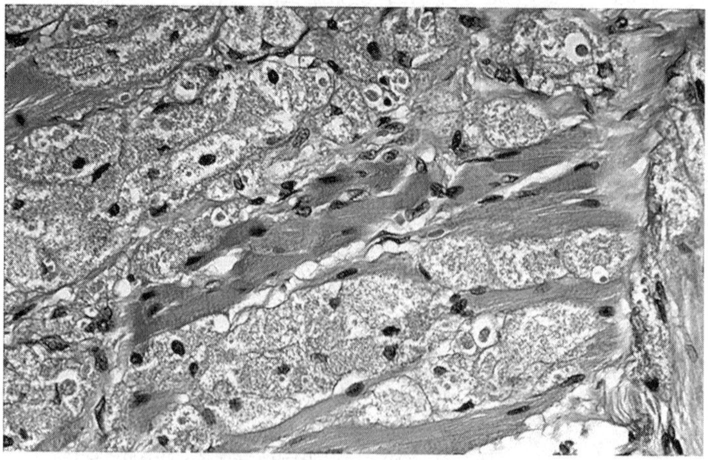

FIGURE 8.5 Granular cell tumor. The image is filled with large cells filled with granular cytoplasm.

quently, in the early 1980s, immunostains raised against specific cell proteins became available to pathologists. By examining cells for the presence of specific antigens reacting against purified antibody reagents, it became possible to identify cell types based on their characteristic products of differentiation.

Reports showing that granular cell tumors stained positively for S-100, an enriched protein fractionate obtained from nervous system tissue and thought to be specific for cells of nervous system origin, started to appear in the early 1980s (131).

This was the final nail in the coffin for the "myoblast" origin of granular cell tumors. Pathologists began to call these tumors "granular cell schwannomas."

Subsequently, it was found that S-100 stains for a wide variety of tumors and tissues of diverse developmental origins and has no real specificity for Schwann cells or any nervous system cell, for that matter.

After several decades of work with immunostains, we know that there are very few instances for which a single immunostain is, by itself, diagnostic of any tumor or cell lineage.

In the case of classifying granular cell tumors, careful observations made on rodent tumors were just as valuable as observations on human tumors.

8.5 Morphology and the Invention of Rare Tumors

Two tumors of skin are so common that their incidence is nearly equal to the incidence of all other cancers combined: squamous cell carcinoma of skin and basal cell carcinoma. Discounting these two tumors, just 15 kinds of tumors account for 85% of the cancer occurrences in humans (List 8.5.1).

If we had chosen to throw in squamous cell carcinoma and basal cell carcinoma of skin, then 17 tumors would account for approximately 92% of all cancer occurrences of humans.

Although 17 tumors account for over 90% of the tumors of man, pathologists have described the morphology and the biological features of thousands of different tumors. Consequently, the vast majority of the different kinds of tumors known to occur in man are rare tumors. Many tumors occur in just a few dozen patients each year in the United States.

Regardless, rare tumors must be taken seriously for two reasons: (1) If you are a patient with a rare tumor, you need to be treated with the best possible care, supported by the best possible research, and (2) much of what we know about the biology of common tumors is learned from the study of rare tumors.

8.6 What Is a Rare Tumor?

The definition of a rare disease is written into the U.S. Orphan Drug Act of 1983 (133).

> (2) For purposes of paragraph (1), the term "rare disease or condition" means any disease or condition which (A) affects less than 200,000 persons in the United States, or (B) affects

List 8.5.1 The 15 tumors that account for 85% of the cancer occurrences of humans (excludes squamous cell carcinoma of skin and basal cell carcinoma of skin). Numbers provided are age-adjusted incidences expressed as cases per 100,000 population. The total number of cancer occurrences is 470.1 (132).

Prostate carcinoma—73.9
Breast carcinoma (invasive and in situ)—85.3
Lung and bronchial carcinoma—64.5
Colon and rectal carcinoma—51.6
Non-**Hodgkin lymphoma**—19.3
Bladder cancer—21.1
Melanoma of skin—18.5
Renal cancer—12.8
Uterine cancer—12.3
Pancreatic carcinoma—11.4
Stomach carcinoma—8.1
Ovarian cancer—7.4
Liver carcinoma—5.6
Esophageal carcinoma—4.6
Laryngeal carcinoma—3.7

more than 200,000 in the United States and for which there is no reasonable expectation that the cost of developing making available in the United States a drug for such disease or condition will recovered from sales in the United States of such drug.

With just a few exceptions, all cancers occur in fewer than 200,000 people in the United States each year. Thus, the official U.S. definition of a rare disease is not particularly helpful. The National Cancer Institute defined a rare cancer as a cancer with fewer than 40,000 new cases in the United States each year. This definition also includes all but a handful of cancer types and does not help distinguish a special set of tumors.

In my opinion, a rare cancer is one that occurs with an incidence under 1 per 100,000 persons per year (i.e., under 3,000 new cases in the United States each year). A rare tumor may be encountered once every few years in a busy hospital. For any given rare tumor, the average oncologist might encounter a patient with the tumor once or never within the span of his or her career.

8.7 Why Are Rare Tumors Important?

We spend our cancer research money on the same diseases that we spend our cancer care dollars: lung cancer, colon cancer, breast cancer, and prostate cancer kill the most Americans and receive the most research funding; however, if you stop and look at the

List 8.7.1 Questions answered by the rare tumors.

- Why is the tumor rare?
- What special conditions must prevail before the rare tumor will occur?
- How does the ordinary phenomenon of common tumor development compare with the extra-ordinary phenomenon that causes rare tumors to develop?

cancers whose understanding has most advanced the field of cancer research, and in which fundamental research has actually led to cancer cures, you find a very different list: gastrointestinal stromal tumor, chronic myelogenous leukemia, acute lymphocytic leukemia, Hodgkin disease, and seminoma. Every rare tumor (and there are many of them) reveals new secrets (List 8.7.1).

8.8 The Importance of Uncommon Variants of Common Tumors

Pathologists can be divided into two groups based on temperament: lumpers and splitters.

Lumpers tend to reduce tumors to a few essential properties and provide a diagnosis that contains a minimal set of morphologic and clinical information. For example, a lumper might provide the following diagnosis "basal cell carcinoma, margins involved."

A splitter might try to assign each basal carcinoma to one of several dozen subtypes (List 8.8.1).

In addition to rare tumors, pathologists describe neoplastic entities that are variants of tumors (List 8.8.2). Tumor variants can be thought of as potential new tumors. If studies indicate that a tumor variant represents a homogeneous and identifiable group, with morphologic, clinical, or genetic features that distinguish the variant from the classic form of the tumor, pathologists will accept the variant as a new tumor type.

Pathology journals are filled with articles that carefully describe variant forms of tumors. Nonpathologists seldom appreciate the scientific and medical value of these tedious pursuits. Nonetheless, variant tumors teach us a great deal about cancer.

8.9 Example Case: Gastrointestinal Stromal Tumor

Most rare tumors were "discovered" as uncommon variants of common tumors. The story of GISTs (gastrointestinal stromal tumors) demonstrates the value of pursuing the study of morphologic variants. There are two common spindle cell tumors that occur in the gastrointestinal tract: leiomyomas, which arise from smooth muscle tumors, and schwannomas, which arise from Schwann cells. These two tumors have similar morphology, but a sharp pathologist can almost always

List 8.8.1 Some of the morphologic variants of basal cell carcinoma.

Adamantinoid basal cell carcinoma
Adenoid basal cell carcinoma
Adenoid cystic basal cell carcinoma
Basal cell carcinoma with eccrine differentiation
Basal cell carcinoma with granular cell change
Basal cell carcinoma with signet ring change
Basal cell carcinoma with sweat gland differentiation
Circumscribed solid basal cell carcinoma
Desmoplastic basal cell carcinoma
Fibroepithelial basal cell carcinoma
Follicular basal cell carcinoma
Infiltrative basal cell carcinoma
Infundibulocystic basal cell carcinoma
Keratotic basal cell carcinoma
Metatypical basal cell carcinoma
Micronodular basal cell carcinoma
Mixed basal and squamous cell carcinoma
Morpheaform basal cell carcinoma
Multicentric basal cell carcinoma
Nodular basal cell carcinoma
Noduloulcerative basal cell carcinoma
Pigmented basal cell carcinoma
Solid basal cell skin carcinoma
Superficial basal cell carcinoma
Superficial multifocal basal cell carcinoma

List 8.8.2 Examples of uncommon variants of common tumors.

Atypical melanocytic nevi simulating melanoma occurring on the vulva of young women (134)
Lung cancer occurring in young people (124)
Secretory beast carcinoma in young women (135)
M3 subtype of acute myelocytic leukemia (the promyelocytic subtype) (27)
Renal cell carcinoma occurring in children (121)
Angiolipomatous variant of lipoma (58)

distinguish the two neoplasms. In the 1980s, pathologists started to notice an uncommon morphologic variant of spindle cell tumors of the GI tract that had some morphologic features of smooth muscle cells and other features of Schwann cells. Even under electron microscopic examination, these tumors seemed to have features of both smooth muscle and Schwann cells. To deal with the ambiguity, some resourceful pathologists invented the concept of a spindle cell tumor "spectrum" occurring in the gastrointestinal tract (136). At one end of the spectrum is the leiomyoma. At the other end of the spectrum is the schwannoma. In between are mixed tumors. The spectrum of all the tumors was called GIST. When pathologists encounter such a tumor, they could merely label it a GIST, provide an estimate of the relative contributions of smooth muscle and Schwann cell component, and be done with the it.

There were several very important drawbacks to this "tumor spectrum" approach. First, neoplasms are clonal. Smooth muscle cells derive from mesoderm. Schwann cells derive from the neural crest. There is no reason to imagine that Schwann cells and smooth muscle cells would arise in the same neoplasm. To do so would require a progenitor cell that differentiates in two unrelated cellular lineages. Second, tumors that seemed to be of mixed muscle and Schwann cell phenotype had their own distinctive behavior. Unlike leiomyomas and schwannomas, the middle-spectrum tumors often metastasized when the tumor size exceeded 2 cm.

In the next decade, evidence mounted that the middle-spectrum tumors were biologically distinct from either leiomyomas or schwannomas. GISTs, it turned out, are derived from the interstitial cell of Cajal (ICC), a cell that is not a type of smooth muscle cell or a type of Schwann cell. GISTs can be distinguished from leiomyomas (of smooth muscle origin) and from schwannomas by immunostaining with CD-117, the c-KIT receptor kinase protein.

In 2001, Dr. Tim O'Leary and I convened a consensus conference on the pathology of GIST (137, 138). From this conference, recommendations for the diagnosis of GIST were suggested (139). These recommendations have been used to evaluate patients with new cases of GIST who could be enlisted in new therapeutic protocols testing the drug Gleevec, which inhibits the c-KIT tyrosine receptor kinase.

Gleevec was the first nontoxic, molecular targeted chemotherapeutic agent found to be effective against a **solid tumor** (it had already been found to be useful in the treatment of chronic myelocytic leukemia) (140). Before Gleevec, there was no effective treatment for metastatic GISTs. Today, thousands of patients with GIST have benefited from this novel and modern treatment. This advancement could not have been achieved if pathologists had not recognized that the GIST variant of GI spindle cell tumors is a specific neoplasm, growing from a specific type of cell, and having a narrow range of molecular alterations that drive its growth. We encounter GISTs again in Chapter 14, when we describe the developmental origin of the cell of Cajal.

8.10 What Do Rare Tumors Have in Common with Uncommon Variants of Common Tumors?

When an uncommon variant of a common tumor is shown to have a unique etiology, a characteristic genetic marker, or a distinctive biology, it can be elevated to the status of a rare tumor.

Currently, there are many tumors that are in a sort of limbo. Nobody is quite certain whether these tumors are distinct entities or just outliers of commonly occurring tumors. One example is medullary carcinoma of breast, formerly known as circumscribed tumor of breast. This rare tumor has its peak occurrence in a slightly younger age group than infiltrating ductal carcinoma of breast. As its former name implied, it is circumscribed. That means the pathologist is always certain of the point where the tumor stops and the normal breast tissue begins. This property does not hold for infiltrating ductal carcinoma of breast, for which it is difficult to assess a well-defined tumor perimeter. The cells of a medullary carcinoma of breast are big and round and interspersed with lymphocytic cells, morphologic features that do not characterize infiltrating ductal carcinoma of breast. Finally, patients with medullary carcinoma of breast have a better prognosis than patients with infiltrating ductal carcinoma.

Are these differences between medullary carcinoma of breast and infiltrating ductal carcinoma of breast sufficient to distinguish the former as a rare tumor? Maybe not. The problem is that medullary carcinoma of breast is composed of malignant ductal cells and may contain focal areas that morphologically resemble infiltrating ductal carcinoma of breast. Conversely, infiltrating ductal carcinoma of breast may have foci that resemble medullary carcinoma of breast. Until we have evidence that medullary carcinoma of breast has a genetic marker that distinguishes it from common breast cancer, its status as a distinct neoplasm is questionable.

8.11 What Can We Learn from Uncommon Variants of Common Tumors?

Lymphomas, for the most part, arise in lymph nodes or in mucosa-associated lymphoid nodules; however, in immunodeficient patients, for reasons that are not at all obvious, many lymphomas arise from extranodal tissues, such as brain, pleural cavity, or abdominal cavity. The extranodal development of lymphomas in immunodeficient patients is an example of an uncommon variant of a common neoplasm. A variant neoplasm makes us ask why the rules of neoplasia seem to be suspended for this particular instance.

In the case of lymphomas that arise in body cavity fluids (the so-called **primary effusion lymphomas** [PELs]), there is a tentative answer. Virtually all PELs are associated with, and presumably caused by, HHV-8 (human herpes virus 8), the same virus associated with Kaposi's sarcoma and with multicentric Castlemen's disease (141). If no one had noticed that effusion lymphomas are a rare variant of lymphomas and if they were simply reported as "lymphoma" without specifying the subtype, is it likely that anyone would have looked for (and found) the cause of these clinically and anatomically distinct variants of lymphomas?

Chordoid glioma is a recently described tumor. It occurs exclusively in women, particularly middle-aged women, and has a distinctive histological appearance (tumor cells in chords). It arises in the region of the third ventricle and the hypothalamus. Although technically a variant of glioma, most pathologists would categorize chordoid glioma as a unique tumor entity, based on its characteristic morphology, anatomic location, and gender specificity.

An example wherein an uncommon variant became a rare tumor is the previously discussed midline carcinoma of children and young adults. Lung cancer rarely occurs in nonsmokers and rarely occurs in children. This rare lung cancer of children was considered a variant of adenocarcinoma of lung until a specific rearrangement was discovered in the NUT (NUclear protein in Testis) gene (124). Most epithelial cancers and virtually every lung carcinoma studied to date are cytogenetically complex. Lung cancers are aneuploid, and they contain multiple genetic alterations. The midline carcinoma of children and young adults is one of only a few highly malignant epithelial tumors that can be diagnosed by a single rearranged gene. After the NUT gene rearrangement was found, this tumor ascended to the realm of a rare tumor.

8.12 Can We Understand Common, Sporadic Tumors by Studying Rare, Inherited Tumors?

Many of the greatest advances in our understanding of cancer have come through the study of familial cancer syndromes. Although familial cancers are almost always rare, they comprise a large group of syndromes (List 8.12.1).

Each inherited neoplasm syndrome from the list is accompanied by an OMIM (Online Mendelian Inheritance in Man) number. The OMIM number can be used to retrieve the updated OMIM record for the syndrome (81, 82). You can use this list as the starting point for your own investigations of these important cancer syndromes.

It is important to appreciate that there are many more kinds of rare cancers than there are common cancers. Although the list is far too long to permit description of each included syndrome, we can infer several important lessons from a cursory examination of its contents (List 8.12.2).

In most cases, the same genetic lesion that occurs in the germline of rare tumor syndromes is also found in a portion of the cases of the sporadically occurring (nonfamilial) common cancers (List 8.12.3).

List 8.12.1 Familial syndromes associated with neoplastic development.

101000 Acoustic schwannomas bilateral
102660 Adamantinoma of long bones
102700 Severe combined immunodeficiency due to a deficiency
103580 Albright hereditary osteodystrophy
105600 Anemia with multinucleated erythroblasts erythroreticulosis hereditary benign included
106070 Angioma hereditary neurocutaneous spinal arterial venous malformations with cutaneous hemangiomas included
106200 Aniridia
106400 Ankylosing vertebral hyperostosis with tylosis diffuse idiopathic skeletal hyperostosis included
109350 Gastroesophageal reflux pediatric adenocarcinoma of esophagus included
109390 Basal cell carcinomas with milia and coarse sparse hair
109400 Basal cell nevus syndrome
112200 Blue rubber bleb nevus
112250 Diaphyseal medullary stenosis with malignant fibrous histiocytoma
113705 BRCA1 breast cancer
113970 Burkitt lymphoma
114030 Cafe-au-lait spots multiple
114400 Lynch cancer family syndrome II
115310 Carotid body tumors and multiple extraadrenal pheochromocytomas
116860 Cerebral capillary malformations cerebral cavernous malformations 1 included
117550 Sotos syndrome
118350 Chemodectoma intraabdominal with cutaneous angiolipomas
118865 Choroidal osteoma bilateral
120435 Colorectal cancer hereditary nonpolyposis type 1
124030 Debrisoquine 4-hydroxylase codeine ultrarapid metabolism of included
126337 DNA damage-inducible transcript 3 (gadd153 myxoid liposarcoma and chop/fus fusion gene included)
130650 Wiedemann-Beckwith syndrome
131100 Multiple endocrine neoplasia type I
131440 Myeloproliferative disorder chronic with **eosinophilia**
131445 Ependymoma familial

(continued)

List 8.12.1 (continued)

132600 **Pilomatrixoma**
132700 Cylindromatosis familial
132810 Diphenylhydantoin defect lymphoproliferative disorders susceptibility to included
133100 Polycythemia familial
133180 Erythroleukemia familial
133450 Ewes gene neuroepithelioma peripheral included
133510 Xeroderma pigmentosum complementation group B
133700 Exostoses multiple type I
135150 Birt-Hogg-Dubé syndrome fibrofolliculomas with trichodiscomas and acrochordons
135300 Fibromatosis gingival hereditary
137357 Genitourinary dysplasia component of wagr syndrome
137360 Genochondromatosis
137550 Giant pigmented hairy nevus
137575 Gigantiform cementoma familial
137800 Glioma of brain familial **glioblastoma multiforme** included
138800 Goiter nontoxic with intrathyroidal calcification
141000 Hemangioma-thrombocytopenia syndrome Kasabach-Merritt syndrome
142981 Homeobox d4 hoxd4 acute lymphoblastic susceptibility to included
144700 Renal carcinoma familial associated 1 included
146510 Pallister-hall syndrome hypothalamic hamartoblastoma hypopituitarism imperforate anus and postaxial polydactyly
148000 Kaposi sarcoma
148500 Tylosis with esophageal cancer
149000 Klippel-Trenaunay-weber syndrome angioosteohypertrophy syndrome
150699 Leiomyoma uterine
150800 Leiomyoma hereditary multiple of skin
151623 Li-Fraumeni syndrome 1
153480 Bannayan-zonana syndrome macrocephaly pseudopapilledema and multiple hemangiomata Bannayan-Riley-Ruvalcaba syndrome included
153550 5q syndrome macrocytic anemia refractory due to 5q deletion
155240 Thyroid carcinoma familial medullary
155600 Dysplastic nevus syndrome hereditary b-k mole syndrome

(continued)

List 8.12.1 (continued)

158320 Muir-Torre syndrome cutaneous sebaceous neoplasms and **keratoacanthomas** multiple with gastrointestinal and other carcinomas
158350 Cowden syndrome multiple hamartoma syndrome Lhermitte-Duclos disease included
159550 Myelocerebellar disorder
159555 Myeloid/lymphoid or mixed lineage leukemia trithorax Drosophila homolog of mll/fbp17 fusion gene included
160980 Carney myxoma-endocrine complex carney syndrome
161550 Nasopharyngeal carcinoma
162091 Neurilemmomatosis congenital cutaneous
162200 **Neurofibromatosis** type I
162220 Nf3b
162240 Neurofibromatosis-pheochromocytoma-duodenal carcinoid syndrome
162300 Neuromata mucosal with endocrine tumors
163000 Nevi flammei familial multiple
163200 Linear sebaceous nevus syndrome
164330 Odontoma-dysphagia syndrome
166000 Osteochondromatosis
166950 Ovarian teratoma
168000 Paragangliomas familial 1
169545 Pelvic lipomatosis with crossed renal ectopia
171400 Multiple endocrine neoplasia type ii pheochromocytoma and amyloid-producing **medullary thyroid carcinoma**
171420 Pheochromocytoma—islet cell tumor syndrome
174800 Albright syndrome polyostotic fibrous dysplasia
175050 Juvenile polyposis/hereditary hemorrhagic telangiectasia syndrome
175100 Adenomatous polyposis of the colon apc
175200 Peutz-Jeghers syndrome
176450 Sacral agenesis hereditary with presacral mass anterior meningocele and/or teratoma and anorectal malformation included
180200 Retinoblastoma
180295 Rhabdomyosarcoma embryonal 2
184500 Steatocystoma multiplex
186600 Syringomas multiple
187300 Telangiectasia hereditary hemorrhagic of Rendu Osler and weber

(continued)

List 8.12.1 (continued)

188470 Thyroid carcinoma follicular
188550 Familial nonmedullary thyroid cancer
190345 Trichoepitheliomas multiple desmoplastic
190685 Leukemia megakaryoblastic of down syndrome
191100 Tuberous sclerosis
193300 Von Hippel-Lindau syndrome
194070 Wilms tumor 1
194080 Denys-Drash syndrome nephropathy Wilms tumor and genital anomalies
194400 Xeroderma pigmentosum autosomal dominant mild
201910 Adrenal hyperplasia congenital due to 21-hydroxylase deficiency
206550 Angiolipomatosis familial
208900 Ataxia-telangiectasia
210900 Bloom syndrome
211980 Lung cancer alveolar cell carcinoma included
227650 Fanconi anemia
228550 Fibromatosis congenital generalized
231100 Hemochromatosis neonatal
235200 Hemochromatosis
246400 Histiocytosis x acute disseminated
246470 Leukemia acute myelocytic with polyposis coli and colon cancer
254450 **Myelofibrosis** familial
254700 Myeloproliferative disease autosomal recessive
256700 Neuroblastoma
259500 Osteogenic sarcoma
259550 Osteoid osteoma
260500 Choroid plexus papilloma
263300 Polycythemia vera
267000 Nephroblastomatosis fetal ascites macrosomia and Wilms tumor
267700 Hemophagocytic reticulosis familial
273120 Teratoma pineal
273300 Testicular tumors teratoma testicular included
276300 Turcot syndrome malignant tumors of the central nervous system associated with familial polyposis of the colon
278700 Xeroderma pigmentosum I

(continued)

List 8.12.1 (continued)

300147 Prostate cancer hereditary x-linked
301845 Follicular atrophoderma and basal cell carcinomas
305350 Epidermodysplasia verruciformis x-linked
308240 Lymphoproliferative disease x-linked
308300 Incontinentia pigmenti familial male-lethal type
308940 Leiomyomatosis esophageal and vulval with nephropathy
312820 Sarcoma synovial x-chromosome–related 1
312870 Dysplasia gigantism syndrome x-linked
600048 Breast cancer 11-22 translocation associated
600080 Myelocytic leukemia-like syndrome
600139 Subependymoma
600185 BRCA2 breast cancer type 2 included
600195 Venous malformations multiple cutaneous and mucosal
600258 Mismatch repair gene pmsl1 colorectal cancer hereditary nonpolyposis type 3 included
600376 Osler-Rendu-weber syndrome 2
600542 Chondrosarcoma myxoid extraskeletal fused to ews
600678 Gtbp colorectal cancer hereditary nonpolyposis type 5 included
601223 Neuronal intestinal dysplasia type B
601228 Polyposis syndrome mixed hereditary
601321 Noonan-neurofibromatosis syndrome
601347 Myelodysplasia immunodeficiency facial dysmorphism short stature and psychomotor delay
601359 Sebaceous nevus syndrome and hemimegalencephaly
601363 Familial Wilms tumor 1
601399 Platelet disorder familial with associated myeloid malignancy
601518 Prostate cancer hereditary 1
601583 Wilms tumor 5
601606 Trichoepithelioma multiple familial
601650 Glomus tumors familial 2
601859 Canale-Smith syndrome (autoimmune lymphoproliferative syndrome)
602089 Hemangioma capillary infantile
602596 Pancreatic lymphoma familial
603372 Thyroid carcinoma with thyrotoxicosis included

(continued)

List 8.12.1 (continued)

603554 Reticuloendotheliosis familial with eosinophilia
603641 Neuroendocrine carcinoma of salivary glands sensorineural hearing loss and enamel hypoplasia
603656 Exostosis Dupuytren subungual
603669 Eccrine syringofibroadenomatosis with eyelid abnormalities
603670 Blue nevi familial multiple
603688 Prostate cancer/brain cancer susceptibility
603737 Ovarian germ cell cancer
603744 Papillary thyroid microcarcinoma
604287 Carney triad
604287 Gastric leiomyosarcoma pulmonary chondroma and extraadrenal paraganglioma
604370 Epithelial ovarian cancer
604373 Checkpoint kinase 2 s. pombe homolog of breast and colorectal cancer susceptibility to included
604442 Nonpapillary renal carcinoma 1
604451 Basal cell carcinoma with follicular differentiation
604856 Langerhans cell histiocytosis
604919 Becker nevus syndrome
605027 Lymphoma non-Hodgkin familial
605041 Spiegler-Brooke syndrome
605074 Renal cell carcinoma papillary
605075 Renal cell carcinoma papillary 3
605233 Autoimmune lymphoproliferative disease without fas mutations
605244 Carney myxoma-endocrine complex type II
605288 Primordial germ cell tumor susceptibility locus 1
605365 Breast cancer 3 brca3
605373 Paragangliomas 3
605373 Glomus tumors familial 3
605462 Basal cell carcinoma multiple
605642 Thyroid carcinoma papillary with papillary renal neoplasia
605827 Basaloid follicular hamartoma syndrome generalized autosomal dominant
605839 Leiomyomatosis and renal cell cancer hereditary
605982 Familial Wilms tumor 2
606179 **Aneurysmal bone cysts**

(continued)

List 8.12.1 (continued)

606240 Thyroid carcinoma nonmedullary 1
606445 Persistent polyclonal B-cell lymphocytosis
606519 Phace (posterior fossa brain malformations, hemangiomas of the face, arterial anomalies, cardiac anomalies, and eye abnormalities) association
606660 Melanoma uveal susceptibility to 1
606690 Lymphangioleiomyomatosis
606719 Familial atypical multiple mole melanoma-pancreatic carcinoma syndrome
606764 Gastrointestinal stromal tumor gist
606864 Paraganglioma and gastric stromal sarcoma
606893 Hemangioma intraosseous
607174 Meningioma familial meningioma caused by *somatic mutation* included
607248 Glioma familial 1
607278 Osteofibrous dysplasia
607464 Thyroid carcinoma Hürthle cell
607785 Juvenile myelomonocytic leukemia
607859 Tufted angioma
607893 Ovarian cancer epithelial susceptibility to
607907 Dermatofibrosarcoma protuberans
608035 Melanoma cutaneous malignant 4
608266 Parathyroid carcinoma
608354 Capillary malformation-arteriovenous malformation
608456 Colorectal adenomatous polyposis autosomal recessive
608615 Oligodontia-colorectal cancer syndrome
608623 Neuroectodermal tumors supratentorial primitive with cafe-au-lait spots
608656 Prostate cancer hereditary 20
608658 Prostate cancer hereditary 7
608812 Colorectal cancer susceptibility to on chromosome 9
608837 Carney complex variant
608935 Lung cancer 1
609048 Melanoma cutaneous malignant 3
609299 Prostate cancer hereditary 3
609310 Colorectal cancer hereditary nonpolyposis type 2
609322 Rhabdoid tumor
609363 Colloid cysts of third ventricle

List 8.12.2 General observations on the long list of inherited neoplastic syndromes.

1. In many cases, neoplastic changes occur in only a small minority of the patients who have inherited a cancer gene.
2. Inherited cancer genes are compatible with successful natal development. The cancer gene is developmentally silent, with the exception of specific abnormalities and predispositions characteristic of the syndrome. This means that cells can sometimes contain cancer genes without becoming cancerous.
3. Although the persons with inherited cancer syndromes develop normally, with the exception of specific syndrome-associated abnormalities, those abnormalities indicate that cancer genes play narrow roles in development.
4. Many of the inherited cancer syndromes are associated with commonly occurring tumors. The Li-Fraumeni syndrome is associated with breast cancer, lung cancer, colon cancer, pancreatic cancer, and prostate cancer, among a variety of rare tumors. Because not all people with common cancers have an inherited condition that caused the cancer, we can infer that common cancers must have multiple different causations.
4. Many cancer genes cause neoplasms that are specific for one germ layer (e.g., endoderm, mesoderm, or neural crest).
5. A single inherited cancer syndrome can cause all types of neoplasms: cancers, benign tumors, hamartomas, and precancers. This implies that the cause of all these different types of neoplasms must share at least some step or steps in common.

List 8.12.3 A few examples of rare germline mutations that also occur in sporadic tumors.

Germline mutations of the p53 tumor suppressor gene are present in the rare Li-Fraumeni syndrome. A somatic p53 mutation is found in about half of all human cancers (142).

Families with germline mutations of the KIT gene develop GISTs. Somatic mutations of KIT occur in the majority of sporadic GIST tumors.

Germline RET gene mutations occur in familial medullary carcinoma of thyroid and in sporadic cases of medullary carcinoma of thyroid (143, 144).

Germline RB1 gene mutations occur in familial retinoblastoma syndrome and in sporadic cases of retinoblastoma (145).

Germline patched (ptc) gene mutations occur in basal cell nevus syndrome and in sporadically occurring basal cell carcinomas (146).

In tumor after tumor, the genetic lesion found in the sporadically occurring cancers would not have been found without knowledge of the syndromic gene.

8.13 Will We Learn How to Prevent Common Tumors by Studying Rare Tumors?

Carcinoma of the scrotum is a very rare tumor in 2008, but it was a common tumor in 1775 (147). In 1775, scrotal cancer occurred frequently in chimney sweeps. Percivall Pott saw the association between exposure to soot and scrotal cancer. His observations led to the near-eradication of scrotal cancer, and to the beginning of the field of cancer prevention. Since 1775, many of the greatest advances in the field of cancer prevention have followed observations made on rare tumors or on uncommon variants of common tumors.

In the 1930s through the 1950s, it was common to use thorotrast, a colloidal suspension of radioactive thorium dioxide, as a radiological contrast agent (148). This agent is absorbed into the Kupffer cells of the liver. Kupffer cells are specialized phagocytic endothelial cells that line the hepatic sinusoids. Once absorbed in Kupffer cells, the radioactive compound emits alpha particles. Emission continues at a time-exponentially decreasing rate, for the lifetime of the patient. Alpha particles have low penetrance, and the tissue cells most likely to be damaged are the cells carrying the thorotrast and the cells immediately adjacent to these cells. Tumors caused by exposure to radiological thorotrast are hepatocellular carcinoma, cholangiocarcinoma, and hepatic angiosarcoma. Hepatocellular carcinoma is a common tumor, but cholangiocarcinoma is rare, and angiosarcoma of the liver is extremely rare. Hepatic angiosarcoma was a sentinel tumor, warning pathologists that thorotrast caused cancer. Today, barium is used as a contrast agent, and the incidence of liver angiosarcomas dropped until 1974.

In 1974, representatives of the rubber and tire manufacturing industries reported that several angiosarcomas had occurred in three workers who worked at the site of vinyl chloride polymerization reactor vessels (149, 150). This news arrived several years after studies on rats had indicated that aerosolized polyvinyl chloride was carcinogenic. Once again, the occurrence of just a few rare tumors, in the setting of a specific type of occupational exposure, was sufficient to alert the public that another carcinogen had entered the workplace environment (149).

In a landmark article published in 1971 by Herbst and coworkers, the authors reported an increase in the number of young women who developed an extremely rare cancer: clear cell adenocarcinoma of the cervix or of the vagina (151). The mothers of most of these young women had ingested a nonsteroidal synthetic estrogen (diethylstilbestrol [DES]) during their pregnancies. Somehow, in utero exposure to a drug caused a specific rare tumor to occur in the offspring. The oldest patient was 27 years old at the time of diagnosis. The youngest was 7 years old. Ninety-one percent were 14 years old or older (152). Men exposed in utero to DES seemed to suffer no increased cancer risk (153). The findings of Herbst and coworkers provided a major advancement to the field of **perinatal carcinogenesis** and arose from the chance observation of a very rare tumor.

It is worth remembering that the rare tumor encountered today may be a common tumor in the near future. Before smoking became a national pastime, lung cancer was rare. If the link between cigarettes and cancer had been established when the number of new tumors (and new smokers) was still small, a vigorous effort to curtail smoking might have staved off the current worldwide epidemic of lung cancer.

8.14 Why Are We Likely to Prevent and Cure Rare Tumors Before We Cure Common Tumors?

Most of the cancers that oncologists can cure are rare tumors. They include several tumors of childhood, plus gestational cancer, acute promyelocytic leukemia, GIST, Hodgkin's lymphoma, seminoma, and hairy cell leukemia.

Although we have learned how to prevent some cases of common cancers (invasive cervical squamous cell carcinoma, smoking-associated lung cancer), we have failed to come up with a bona fide cure for any common, advanced cancer.

Why have we done so much better for rare tumors than for common tumors? First, there are thousands of different tumors, and all but a few handfuls of them are rare. By chance alone, if you reach into the taxonomy of all tumors and pull out an entity that wins the "cure" status, you will likely pull a rare tumor.

Aside from this numbers bias, rare tumors are much more likely to have a single cause or a single carcinogenic pathway or a single inherited gene or a single acquired marker than are any of the common tumors.

When we look at past triumphs in the field of cancer causation, most are examples for which a small population of humans was exposed to a specific carcinogen that produced a rare type of tumor (List 8.14.1).

List 8.14.1 Epidemiologic triumphs in cancer research.

Angiosarcoma of liver (thorotrast exposure in the 1940s, polyvinyl chloride exposure in the 1970s)
Mesothelioma (in World War II asbestos-exposed navy shipbuilders, 20+ years after exposure)
Scrotal cancer (in chimney sweeps in 1700s)
Esophageal adenocarcinoma (late 20th century, particularly among obese, middle-aged Caucasians)
Oral cancer in teenagers (tobacco chewers late 20th century)
AIDS-related cancers (beginning about 1980)
Lung cancer (yes, this was a rare tumor in the 1800s)
Leukemia (from benzene exposure)
Thyroid cancer in Chernobyl (pulse exposure to a carcinogen)

Rare tumors tend to result from a single identifiable genetic cause or by exposure to a single identifiable carcinogen. In many instances, rare tumors are characterized by a small number of genetic mutations (sometimes only one) deviating from normal cells by changes in a small number of cellular pathways.

Common tumors develop within a large, diverse population, not a population with a unique genotype. When the prevalence of a tumor is high, it is difficult or impossible to find a single carcinogen to which all of the cancer patients were exposed. Whenever we study the common cancers, we find multiple classes of genetic markers that characterize different examples of the same type of tumor. For example, colon cancers may contain any of several different classes of tumor suppressor genes resulting in loss of genetic stability (25). This plurality of factors makes it difficult to find a single cause for common tumors. Indeed, there is little reason to assume that common tumors have a common cause.

The common tumors (skin, lung, colon, pancreas, and esophagus) arise from tissues that are directly exposed to the environment, either through air (the skin and lungs) or through diet (the gastrointestinal tract). These tissues are exposed to a wide variety of chemical, biological and physical agents. The resulting tumors have phenotypic and genotypic complexity. It is not surprising that common cancers are difficult to understand and cure.

8.15 Are Sporadic Tumors Always Characterized by the Same Mutation that Is Present in the Germline of Patients Who Develop These Tumors as Part of an Inherited Condition?

In general, a sporadically occurring tumor of a commonly occurring cancer will not have the mutation that characterizes a familial syndrome that predisposes persons to that cancer. The opposite is true for a sporadically occurring rare tumor.

The reason for this is simple. There are many causes of common tumors (that is why they are common). There are few causes of rare tumors (that is why they are rare). This means that any one cause of a very common tumor is unlikely to be present in any given occurrence of the tumor. It also means that any known cause of a rare tumor is likely to be the cause of any example of the rare tumor.

A variety of mutations in the ***BRCA*** gene are associated with breast cancer. If a woman inherits a mutated *BRCA* gene, she may have an increased risk of developing breast cancer; however, most women who develop breast or ovarian cancer will not have mutated *BRCA* genes in their tumor cells. Breast cancer can occur in any of several genetically distinct inherited syndromes, such as **Lynch syndrome** (MSH), Li-Fraumeni syndrome (p53), Cowden's syndrome (PTEN), Peutz-Jeghers syndrome (STK11 gene), and ataxia-telangiectasia (ATM). Sporadic breast cancers have thousands of different alterations. It is unlikely that any common cancer has a single genetic cause.

A contrasting example would be the RB gene in retinoblastoma, a very rare tumor. Virtually all cases of retinoblastoma have biallelic inactivation of the RB gene. This is true whether the tumor is of the hereditary form (where one allele is inactivated through the germline) or of the sporadic form (where both alleles are inactivated in somatic cells).

The restricted pathway that leads to any inherited rare tumor is likely to be the same pathway that leads to the sporadic tumor of the same type.

Chapter 8 Summary

Morphologic examination has been the most effective and least expensive method for identifying (diagnosing) and classifying (categorizing) neoplasms. The gross and microscopic morphology of a tumor often captures the functional (e.g., rapid growth, invasiveness) and molecular (e.g., aneuploidy and nuclear atypia) changes that accompany the neoplastic phenotype. Using standard methods for morphologic examination, a pathologist can determine whether a sampled lesion is neoplastic, the diagnosis of a neoplastic lesions, its grade, and in many cases, its clinical stage. In addition, morphologic approaches have yielded many new neoplastic entities, distinguishable from histologically similar lesions by subtle, morphologic clues.

The kinds of general class divisions that come from morphological examination, particularly the dichotomy between epithelial and nonepithelial tumors, may have outlived their value. Nonmorphologic methods, particularly molecular analyses coupled with epidemiologic studies of families in which rare tumor variants occur, have uncovered the characterizing biological processes that correlate with morphologic observations. Increasingly, the role of careful morphologic examination is to provide a starting point for other analytic studies.

Morphologic examination will always play an important role in diagnosing, describing, and discovering new neoplastic entities. Today, morphology is a taxonomic tool used in neoplasm classification, but it is no longer the best organizing principle for neoplasm classification.

CHAPTER

9 Classification by Cause

Nobody goes there anymore. It's too crowded.

—Yogi Berra

9.1 Background

For the past few decades, most deaths in the United States have been due to vascular disease and cancer, with the former edging out the latter by a small margin.

In the United States, in 1996, among the million Americans 65 years or older, there were 1,717,218 deaths. Deaths from cancer were 386,092.

Also in 1996, among the 33.9 million American children 1–4 years of age, there were just 440 deaths caused by malignant neoplasms. There were 71 deaths from benign neoplasms or neoplasms of unspecified nature (132).

There is no doubt that cancer is a disease of the aged. Cancer deaths among children represent a small fraction of American cancer deaths.

Why does cancer target the older population? Actually, cancer targets all ages. Indeed, there is overwhelming reason to believe that early carcinogenic events favor the young over the old. Virtually every animal model for carcinogenesis demonstrates that the younger an animal, the easier it is to induce tumors.

The reason that cancer occurs much more frequently in older persons than in younger persons relates to two phenomena:

1. Carcinogenesis is a long-term process. If a cancer requires 20 years to develop, from the time of first exposure to a carcinogen to the time that the tumor is clinically detected, you would expect that no tumors would occur in humans under the age of 20. This is one of the reasons that tumors occurring in children seem to have a different biology than tumors occurring in adults. Pediatric tumors all share one property—short latency period.

2. Cancer begins with a rare initiating event. As we live longer and longer, we extend the period in which the rare event may occur, thus increasing our changes of developing cancer. This is a fundamental property of all stochastic processes. A simpler way of explaining the situation is that as we live longer and longer, our luck eventually runs out.

9.2 Determining the Cause of a Cancer

Is there such as thing as a spontaneously occurring tumor? Throughout my career, I have been approached by people with cancer who ask, "What caused my tumor? What could I have done differently to avoid all this?" Sometimes the answer is obvious. Cigarette smoke is the major identifiable cause of cancer. For most cancers, a specific cause is unknown. We seldom know whether anyone's cancer arose from their genes, their environment, or just bad luck.

Consider this thought experiment, loosely adapted from an essay written by Geoffrey Rose, in 1985 (154). Smokovia is a land where the most important crop is tobacco, and everyone in the country smokes cigarettes. Families, children included, are smokers. Children are permitted to smoke cigarettes in school. Cigarettes are smoked in every hospital corridor.

In Smokovia, there is a very high rate of lung cancer. One of five Smokovians dies of the disease. Nobody has a clue why this might be. Because there is no subpopulation of Smokovians who do not smoke, there is no indication that the risk of lung cancer might be reduced in nonsmokers.

The Smokovian scientists decide that there must be some defect in the gene pool that predisposes some Smokovians to cancer. They have compared the genomes of Smokovians who develop lung cancer from those Smokovians who have lived a full life without developing lung cancer. Every patient with lung cancer had a characteristic gene variation that was lacking in the cancer-free population. The scientists developed a method for detecting the cancer-prone gene variant from a simple blood test. They suggested that, as a stop-gap solution, Smokovians carrying the lung cancer gene might choose not to procreate. In the meantime, Smokovian scientists would work on a technique to replace the cancer-prone gene with the cancer-resistant gene in early embryos.

While this work was proceeding, Smokovia was hit by a tobacco blight. In a matter of weeks, the entire crop of tobacco was destroyed. It will take scientists years and years to develop a blight-resistant tobacco. Because tobacco was the most important crop in Smokovia, all scientists were inducted into the service of the tobacco industry.

Years passed in Smokovia. The nation endured the symptoms of cigarette withdrawal. Economic times were bad, but the farmers eventually switched to alternate crops. Smokovia slowly recovered, and people noticed that the rate of lung cancer was falling. After about 50 years, lung cancer had become a rare disease. At first, people

List 9.2.1 What we have learned from the Smokovians.

1. The concept of "cause" in the field of cancer is fundamentally confusing.
2. Sometimes the answer is not as important as a precise statement of the question.
3. Science cannot always help us choose how to live.

thought that their genetic selection project had paid off, but a genetic census of Smokovians showed that lung cancer had vanished even in the small subpopulation of Smokovians who had the cancer-prone gene. Nobody could understand what had happened.

Fifty years after the blight, Smokovian scientists found a blight-resistant tobacco plant. Once more, tobacco was planted, cigarettes were produced, and Smokovians resumed their national pastime. High lung cancer rates returned. Again, tumors occurred in the subpopulation of Smokovians that carried the cancer-prone gene.

This story has several morals (List 9.2.1).

The Smokovian scientists were looking for a single factor that correlated with the occurrence of disease, when they should have been thinking in terms of the process of carcinogenesis. Many events precede the appearance of an invasive cancer, and there may have been multiple points at which the process could have been stopped. Breeding out a cancer gene from the population was not humane, nor fair, nor necessary. The Smokovian scientists lacked imagination. They did not anticipate that a cancer that occurs in a vulnerable, genetically-defined sub-population may be initiated by an environmental agent.

9.3 Carcinogen Exposure and Type of Cancer

Ultraviolet (UV) light has limited tissue penetration. The longer UV wavelengths (about 400 nanometers) penetrate deeper than the shorter UV wavelengths (about 200 nanometers). Most UV light does penetrate more than about 2 mm into the skin. The epidermis is, on average, about half a millimeter in thickness. Consequently, UV light can damage the entire epidermis and some of the superficial dermis. The most frequently occurring tumors caused by UV radiation are squamous carcinoma of skin and basal cells carcinoma of skin. Neither of these tumors is likely to result in death. The most lethal tumors produced by UV light are melanomas.

Does UV light cause cutaneous lymphomas? The most common skin lymphoma is mycosis fungoides. Unlike the epithelial cancers of skin, which have a very low incidence in black patients, mycosis fungoides occurs more commonly in blacks than whites (155). This clinical observation casts doubt on an association between UV light and lymphoma of skin. The simplest explanation for the cause of a cancer (UV light in the case of skin cancer) is not always the correct explanation.

9.4 How Does Sunlight Produce Melanomas in Locations that Are Not Exposed to Sunlight?

If melanomas are caused by UV light, why do some melanomas occur in locations that are not exposed to sunlight? I have heard some very inventive answers for the question. The most popular is that melanocytes, as neural crest derivatives, have the ability to migrate throughout the body. Initiated melanocytes in sun-exposed skin occasionally migrate a few centimeters and settle in a nonexposed site (such as oral mucosa, vulva, or anus), where they eventually develop into cancers.

I find this explanation neither plausible nor necessary. Melanocytes and primary melanomas are found, although rarely, great distances from skin (including gallbladder and prostate). The simplest explanation is that not all melanomas are caused by sun exposure. The melanomas that occur in the gallbladder or the prostate are almost certainly produced by carcinogens that we have not identified. There is no reason to assume that UV light is responsible for all cases of melanoma in humans.

9.5 Are the Steps in Chemical Carcinogenesis Cell Type Dependent?

Metabolism is the process whereby chemicals in the body are transformed to other chemicals. Sometimes the process of metabolism builds large, complex molecules from simple molecules. At other times, metabolism breaks down complex molecules into simple constituents (such as carbon dioxide and water). Sometimes molecules are conjugated with a moiety that confers water solubility, permitting the modified molecule to be excreted through the kidneys. In the act of transforming, decomposing, or conjugating chemicals, many biologically active chemicals are deactivated. Some biologically inactive chemicals are activated.

The most potent metabolizing organ in the body is the liver. Any molecule that can be metabolized by human cells can be metabolized by the liver. Other organs that are potent metabolizers of exogenous chemicals are the kidney and the lungs. The kidney metabolizes chemicals encountered as it filters perfused blood, as do the lungs. The organs that are exposed to the greatest burden of exogenous chemicals are the organs that are the best metabolizers. As a general rule, cancers emerge from the same cell populations that metabolize the carcinogen to its mutagenic form (lung, gastrointestinal tract, kidneys, pancreas, and liver).

9.6 Why Do Liver Tumors Occur Less Frequently than Tumors of the Lung and Colon?

If carcinogens need to be metabolically activated, would you not expect that the organ that is the strongest carcinogen activator would be the organ most susceptible to carcinogenesis? Somehow, it does not work that way.

Among all the different types of cells in the body, the liver is by far the most powerful activator of carcinogens. Metabolizing chemicals is the most important job of the liver, and it is specially designed to receive exogenous chemicals. The liver has two blood supplies: the portal (portal vein) and the systemic (the hepatic artery and vein). The portal vein receives nutrients absorbed from the gut and sends them to the liver for metabolism. Along with nutrients are a wide variety of chemical compounds (hundreds of thousands of chemical species) that are contained in the food we eat, but that have no nutrient value. So active are the oxidative pathways in liver that up to 6% of the body's oxygen can be used for metabolic functions of the hepatic endoplasmic reticulum (156).

Under normal circumstances, the liver is a nondividing tissue (157). Nondividing cells are poor targets for carcinogens, which require mitotic activity to replicate mutations. In the 1970s, my coworkers and I showed that dividing liver epithelial cells in culture are much more sensitive to mutagenicity than are nondividing cell populations. When rat or mouse hepatocytes are stimulated to divide, the liver becomes exquisitely sensitive to carcinogenesis (158). Similarly, fetal mouse liver, a rapidly dividing organ, is much more susceptible to experimentally induced hepatocellular carcinoma (HCC) than is the adult organ (159). The heightened sensitivity of dividing liver cells to carcinogens is one of the most dramatic phenomena in the field of experimental chemical carcinogenesis and is reproducible with a wide variety of chemical agents and experimental conditions in mice and rats.

In humans, circumstances that produce a sustained hepatocellular proliferation are associated with an increased risk of HCC. These include inborn errors of metabolism (e.g., hemochromatosis, Wilson's disease, and hereditary tyrosinemia), alcoholism, and chronic hepatitis, all of which may lead to the development of cirrhosis.

Cirrhosis is a liver condition in which fibrous tissue proliferates in liver acini (the glandular portion of the liver) and around ducts and vessels. This results in the death of liver cells and a marked reactive overgrowth of hepatocytes with distorted and enlarged liver acini. The reactive proliferation of hepatocytes in cirrhosis is a persistent process that can extend over the life of the cirrhotic patient.

In the United States, about 80% to 90% of HCC cases arise in cirrhotic liver. Johnson and Williams have argued that cirrhosis, regardless of etiology, strongly predisposes the liver to the development of HCC (160).

HCC develops in 20% to 30% of hemochromatosis patients (161). Hemochromatosis is a genetic disease that results in increased iron storage within hepatocytes. If left untreated, hemochromatosis leads to cirrhosis. Hepatocellular iron, by itself, is not a cause of HCC, as patients diagnosed in the precirrhotic stage and treated with phlebotomies (routine blood drawings to reduce the levels of circulating iron) have a normal life expectancy (162); however, after cirrhosis develops, phlebotomy does not reduce the risk of HCC (162).

Ethanol abuse is associated with an increased risk of HCC. Estimates vary greatly, but the frequency of HCC in livers of alcoholics that have developed cirrhosis is about

10% to 20%. It is important to consider that the statistics of the occurrence of HCC in cirrhotic livers are skewed to underestimate the true biologic risk, as cirrhotic patients have a life expectancy (following diagnosis) of only 5 years, thus drastically limiting the time in which HCC is permitted to develop.

In a transgenic mouse model, there is overproduction of the hepatitis B virus envelope protein within hepatocytes. The accumulating intracellular protein produces chronic hepatitis, regenerative liver nodules, and hepatocellular carcinomas in every mouse (163). These findings suggest that prolonged reactive hyperplasia (caused by a viral protein) is a potent predisposing factor in hepatic carcinogenesis.

All of these findings suggest that liver cells, the primary activator of carcinogens, are normally protected from carcinogenesis because, under normal conditions, they do not divide. After they begin to divide, their protection from carcinogenesis is lost, and they become exquisitely sensitive to carcinogens.

9.7 Variations in Tumor Progression

Tumor progression is the acquisition, over time, of additional genetic mutations and epigenetic alterations in a developed cancer, leading to increased heterogeneity and aneuploidy in the tumor population, often accompanied by a more aggressive phenotype.

The reverse phenomenon, tumor regression, would involve the reduction of size of tumors or a reduction in the rate of growth of tumors over time, the reversal of cytogenetic abnormalities, a reduction in the number of mutations found in tumor cells, a reduction in cellular heterogeneity, and so on. As discussed in Chapter 2, regression of fully developed and clinically detectable cancers is extremely rare.

The observations that may best explain tumor progression are (1) tumor cells constantly undergo mutations, (2) tumor cells that have a high proliferative rate or a low rate of mutation repair or poor mutation exclusion (i.e., impaired apoptosis) will preserve mutations in their subclonal populations, and (3) if a subclone has a mutation that confers a growth advantage, the subclone will increase in size. This means that during tumor progression, there will be many subclones that disappear because they have acquired a mutation that confers a growth disadvantage, whereas other subclones with mutations that confer a growth advantage will enlarge. Enlarging subclones may be replaced by new subclones emerging from their own populations, and these new subclones will have greater growth advantages than the parental subclone.

Tumor progression seems to be a snowballing phenomenon. Tumors that are growing slowly and that have diploid or near-diploid **karyotypes** are likely to progress slowly over a period of years. Tumors that are growing quickly, that are aneuploid, and that have genetic instability are likely to progress quickly.

Progression can occur suddenly from an otherwise indolent lesion. For example, a patient with chronic myelocytic leukemia, may enjoy a period of remission, during which the neutrophil count is relatively low and karyotype abnormalities are restricted to the **Philadelphia chromosome**. The sudden appearance of circulating blasts heralds the appearance of a rapidly expanding population of immature cells with a complex karyotype. This phenomenon, called **blast crisis**, carries a poor prognosis.

Likewise, patients with chronic lymphocytic lymphoma may enjoy relatively good health for years without treatment. Occasionally, a large lymphoid (lymphoblastic) population will emerge that replaces the small lymphocytes of chronic lymphocytic leukemia. The sudden progression from chronic lymphocytic leukemia to an aggressive diffuse large cell lymphoma is called Richter syndrome.

In both blast crisis chronic myelogenous leukemia and in Richter syndrome, progression seems to follow a prescribed biological pattern. These tumors are exceptional. In most cases of tumor progression, the biologic changes occurring in cells seem to develop randomly. The cells in a tumor become heterogeneous, as many different subclones emerge, all with different cytogenetic markers.

Currently, we know very little about how tumor progression can be predicted, controlled or even mitigated.

9.8 Clastogens and Leukemias

Clastogens are a class of mutagenic agents (chemical or physical) that break chromosomes. Examples of clastogens are benzene and ionizing radiation. Clastogens are different from most other mutagens. Clastogens produce chromosomal lesions that are visible under a light microscope (chromosome aberrations, sister chromatid exchanges, and micronuclei). All other types of mutagens produce small changes in the sequence, structure, or physical environment of individual nucleotides.

The biologically intriguing property of clastogens is that they all seem to induce leukemias. Some produce leukemias as well as other types of cancer (e.g., breast and ovarian cancers), but all clastogens produce leukemias, a type of cancer that is actually quite uncommon. In the United States, the incidence of leukemia is about 13 cases per 100,000 population, and that incidence has barely changed since 1975 (the earliest provided SEER data) (164).

9.9 Why Do Chromosome Breakage Syndromes Often Cause Leukemias and Seldom Cause Carcinomas?

If you simply ignore the clastogen-associated leukemias and only examine clastogen-associated tumors of other organs, you would probably conclude that clastogens, unlike other mutagens, are either very weak carcinogens, or possibly noncarcinogens. What is the special association between clastogens and leukemias?

The answer may become apparent if we study the different conditions under which clastogenic leukemias occur (List 9.9.1).

List 9.9.1 Clastogenic conditions associated with leukemogenesis.

Fanconi anemia (165, 166)
Bloom's syndrome (167)
Ataxia telangiectasia
Benzene toxicity (168)

List 9.9.2 Common features of clastogens.

- They all cause leukemias, a rare type of cancer that is seldom caused by nonclastogenic carcinogens.
- They are all weak carcinogens (i.e., they do not cause other types of cancer, or if they do, they produce other types of cancer in a lower incidence than leukemias).
- They cause bone marrow failure and/or aplastic anemia.
- They are sometimes associated with nonneoplastic clonal hematopoietic disorders (such as paroxysmal nocturnal hemoglobinuria).

Fanconi's anemia is an inherited condition. The incidence of acute myelogenous leukemia in children with Fanconi's anemia is 15,000 times the risk of children in the general population (165). Leukemia occurs in about 15% of people who have Fanconi's anemia. Fanconi's anemia, along with Bloom's syndrome and **ataxia telangiectasia** are inherited (germline) diseases characterized by a low tolerance for clastogenic DNA damage. In these diseases, the bone marrow seems to be particularly sensitive. Aplastic anemia (bone marrow failure) occurs and leukemia follows with incidences that far exceed those observed in the general population.

Benzene is a ring-shaped aromatic hydrocarbon (C6H6) derived from petroleum and used in a wide variety of synthesized chemical products. It is produced in great quantities and is now ubiquitous in our environment. It is metabolized almost exclusively by the liver but does not produce much hepatotoxicity. Instead, the biological effects of benzene are targeted toward the bone marrow, where it causes aplastic leukemia, myelodysplasias (precancers of blood cell origin or preleukemias), and leukemias. Nobody fully understands why a chemical that is metabolized to its active form in the liver produces its biological effect in the bone marrow (168). We do know that benzene is fat soluble. It is likely that benzene or its active metabolite is absorbed into marrow adipocytes (the fat-storing cells of the bone marrow), where it leeches into adjacent hematopoietic stem cells. Presumably, the levels of metabolized benzene, leeching from marrow adipocytes, are sufficiently high to wreak havoc on proliferating blood elements. The tumors that arise within the bone marrow are leukemias (not liposarcomas of adipocyte origin).

The clastogenic conditions seem to have several common features (List 9.9.2).

Organs can die. Massive necrosis of an organ is often a fatal event for the organism, unless the organ can recover. Occasionally, the liver may suffer a toxic chemical or viral event that leads to massive necrosis followed by a gradual recovery from a small number of regenerating hepatocytes. Toxic epidermal necrolysis is a rare condition in which large areas of epidermis are lost through a generalized blistering process. Generalized and bilateral acute tubular renal necrosis may occur, and this condition is usually due to a toxic event or to ischemia (loss of oxygenated blood flow) and is also often fatal.

These rare organ failures are not associated with the emergence of cancer. In the bone marrow, however, organ failure is commonly followed by leukemia. This is true regardless of the cause of the aplastic anemia.

Because clastogens cause aplastic anemia and are not associated with oncogene mutations (166), it is tempting to infer that clastogens cause leukemias through their ability to destroy the population of normal hematopoietic marrow stem cells. When patients recover from aplastic anemia, the bone marrow is repopulated from a small number of surviving stem cells. If the small surviving stem cell population contains abnormally proliferative cells, a leukemia may follow. As noted earlier, normal blood cells have several properties of the malignant phenotype: the ability to invade tissues and to seed (metastasize) through vessels. Whenever a hematopoietic stem cell produces a sustained, expanding clonal population of proliferating cells, it will manifest as a leukemia. Is there any reason to think that abnormal clones of cells may populate the bone marrow? In Chapter 1, we discussed paroxysmal nocturnal hemoglobinuria. In this instance, an acquired abnormal clone of cell expands to pollute the normal population of red blood cells with abnormal cells that are prone to lysis. It is not difficult to imagine that individual stem cells with abnormal growth properties may inhabit an otherwise normal marrow.

9.10 Myeloproliferative Disorders, Myelodysplasias, and Preleukemic Conditions

In addition to the leukemias, there are two general conditions characterized by the growth of abnormal populations of blood cells: myelodysplasias and **myeloproliferative disorders**.

The myelodysplasias (formerly known as preleukemias) consist of several closely related diseases characterized by anemia and pancytopenia, disorders of myeloid growth and maturation, the appearance of circulating **blast cells**, chromosomal abnormalities, and the frequent progression to leukemia. The myelodysplastic entities are (listed as increasing grades of dysplasia): refractory anemia, refractory anemia with ringed sideroblasts, refractory anemia with excess blasts, and refractory anemia with excess blasts in transformation. Chronic myelomonocytic leukemia is sometimes included with the myelodysplasias, but currently this disease is included among the myeloproliferative disorders. Myelodysplasias occur most often in older people and sometimes follow the occurrence of aplastic anemia.

The myeloproliferative disorders are neoplastic conditions characterized by the increased growth of a stem cell subpopulation of myeloid origin. The myeloproliferative disorders include chronic eosinophilic leukemia, chronic myelocytic leukemia, chronic myelomonocytic leukemia, chronic neutrophilic leukemia, essential thrombocythemia, hypereosinophilic syndrome, juvenile myelomonocytic leukemia, polycythemia vera, primary myelofibrosis, and systemic mastocytosis. The list includes disorders of diverse biological properties and diverse etiologies. With the exception of myelofibrosis, all of the myeloproliferative disorders are characterized by an increase in nonproliferating differentiated myeloid cells (e.g., platelets and red blood cells, both of which lack nuclei and cannot divide). Myelofibrosis is included because it is characterized by a sustained hyperplasia of extramedullary hematopoietic stem cells (myeloid growth outside of the bone marrow) secondary to primary bone marrow fibrosis. Most of these disorders have characteristic mutations: JAK2 mutations in polycythemia vera, essential thrombocythemia, and primary myelofibrosis; c-KIT mutation in systemic mastocytosis; rearrangements of PDGFRB in chronic eosinophilic leukemia and chronic myelomonocytic leukemia; and RAS/PTPN11/NF1 mutations in juvenile myelomonocytic leukemia (169).

If we include animal models, many more myeloproliferative disorders would need to be listed. One particularly informative disease is mouse erythroleukemia induced by a polycythemic strain of Friend Murine Leukemia Virus (F-MuLV). F-MuLV does not encode an oncogene or a tumor suppressor gene. A viral protein mimics erythropoietin and produces a profound erythrocytosis (hyperplastic increase in red blood cells) and splenomegaly (enlarged spleen) (170, 171). Over time, a monoclonal population may emerge. When this happens, the condition has become a true erythroleukemia.

The myeloproliferative disorders remind us that any myeloid stem cell disorder that yields increased numbers of circulating differentiated cells sustained over time will yield a hematologic disease that is indistinguishable from leukemia.

9.11 What Are the Chief Differences Between Myelodysplasias and Myeloproliferative Disorders?

Two features distinguish the myelodysplasias from the myeloproliferative disorders: **dyserythropoiesis** (abnormal cellular functions and morphology) and pancytopathic (reducing the numbers of several different blood cell types) effects are characteristic of the myelodysplastic diseases and absent from the myeloproliferative disorders. Dyserythropoiesis indicates that the progeny of the myelodysplasia stem cell are malfunctioning and that the **myelodysplastic syndrome** is not simply an overproduction of otherwise normal cells. The presence of pancytopenia indicates that the myelodysplastic clone is primitive, resulting in changes in more than one type of blood cell, and growth-impaired. In contrast, the myeloproliferative disorders can usually be characterized by one observed defect involving one blood cell component.

A reasonable, but unproven, hypothesis to explain the myelodysplasias and leukemias that occur after clastogen exposure is that clastogens act through toxicity, not mutagenicity. When the bone marrow suffers a toxic event, the population of blood cells is restored by the clonal expansion of primitive bone marrow stem cells. Any pre-existing oncogene mutations in primitive bone marrow stem cells will be expressed in dividing cells during bone marrow recovery. This simple process is unique to bone marrow cells, which have the ability to invade and metastasize as part of their normal cellular phenotype (i.e., they do not need to acquire metastasis and invasiveness through cumulative mutagenic events).

9.12 What Are the Features Shared Among Tumors Produced by a Particular Carcinogen?

The mechanism of action of a carcinogen provides some information about the behavior of the resulting tumors (List 9.12.1).

9.13 Do Some Carcinogens Cause Only Benign Tumors and Others Cause Only Malignant Tumors?

Carcinogens can cause either benign or malignant neoplasms. There is no current evidence that the mechanism of initiating a benign tumor is any different from the mechanism of initiating a malignant tumor.

Everyone knows that cigarette smoking causes lung cancer. Fewer people know that cigarette smoking is associated with at least one benign tumor: Warthin tumor of salivary glands.

List 9.12.1 Generalizations about the tumors produced by classes of carcinogens.

Carcinogens that act by mutating DNA tend to produce malignant tumors.

Carcinogens that act by stimulating cell proliferation tend to produce benign tumors (although not exclusively). Examples are hormones and growth factors.

Clastogens tend to produce leukemias.

Oncogenic viruses can produce either benign tumors (e.g., warts) or malignant tumors (e.g., cervical carcinoma). Tumors that appear in immunodeficient patients, especially if they occur in groups of synchronous tumors, or quickly following the time when immunodeficiency begins, are almost always caused by endogenous viruses.

Radiation usually causes malignant tumors, though benign tumors may also occur (e.g., benign thyroid tumors and benign meningiomas).

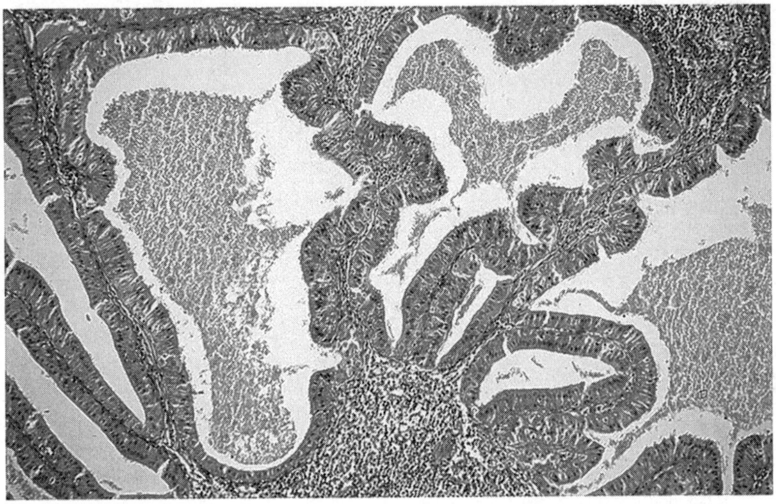

FIGURE 9.1 Warthin tumor. This benign tumor of salivary gland has a histological appearance that never seems to deviate. Glands lined by a two-cell layer of so-called oncocytic epithelial cells and filled with the debris that maintains the shape of necrotic oncocytic remnants (so-called ghost cells). The epithelial glands are embedded in a stroma containing unremarkable but densely packed lymphocytes.

Warthin tumor is a benign salivary gland tumor that occurs most often in the parotid gland (Figure 9.1). Men who smoke have nearly eight times the risk of developing Warthin tumor compared with nonsmoking men. Women who smoke have about 17 times the likelihood of developing Warthin tumors than nonsmoking women (172).

9.14 Tumors Caused by Viruses

Tumors of viral origin, unlike tumors induced by chemicals or tumors resulting from inherited genetic alterations, seem to share one telltale clinical feature: rapid occurrence following **immunosuppression**.

People who have AIDS, and people who have received immunosuppressive regimens to reduce the risk of transplant rejections are prone to a variety diseases caused by endogenous infectious agents. Endogenous infectious agents are organisms that live in our bodies throughout most of our lives without causing clinical disease, or they are organisms that create a mild form of disease that does not normally lead to our deaths. These same organisms may, when the immune system is suppressed, break out as fulminant, life-threatening infections. The long list of opportunistic infections includes protoctists (*Toxoplasma, Plasmodium,* and various amoeba), fungi (*Candida, Aspergillus, Mucor, Pneumocystis*), bacteria (including several *Mycobacteria* species), viruses (including several Herpes species), and even parasitic animals.

List 9.14.1 Tumors that may occur rapidly after immunosuppression.

Kaposi's sarcoma (a type of angiosarcoma)
Primary central nervous system lymphoma
Cervical squamous cell carcinoma
Squamous cell carcinoma of skin (173)
Keratoacanthoma (of skin) (174, 175)
Dermatofibroma (of skin) (176, 177)
Verrucae (common warts) (178)
Primary effusion lymphoma (141)

Among the viruses that lurk in man are a variety of tumor viruses. Shortly after a patient is immunosuppressed, sometimes in just a few months, any of a variety of neoplasms may occur (List 9.14.1).

9.15 Immune Surveillance and the Role of Immunosuppression in Carcinogenesis

For many decades, a popular theory held that in normal individuals the growth of cancers is kept in check by the immune system. In this theory, the job of the immune system is to find and destroy any new cancer cells that might grow within the body. The idea is that immune surveillance for cancer cells works just like immune surveillance for pathogenic microorganisms. The theory is based in part on the following observations (List 9.15.1).

List 9.15.1 Observations that support an important role for immune surveillance in carcinogenesis.

Cancer cells contain antigens that are not present in normal cells. Every cancer is immunologically unique.
Certain tumors are known to regress spontaneously due to an inflammatory reaction that seems to have an immunologic origin (e.g., halo nevi and some cases of primary melanomas).
Certain immunocytes (lymphocytes) can recognize and destroy cancer cells.
Immunosuppressed patients are prone to develop a wide range of rapidly growing tumors.
The tumors that develop in immunosuppressed patients will often regress after normal immune function is restored in the patient.

List 9.15.2 Observations that refute the role of immune surveillance in carcinogenesis.

Despite decades of research, attempts to cure cancer by stimulating the immune system have had limited success. With very few exceptions, such as the use of interleukin-2 for renal cell carcinoma and melanoma (179) and the use of BCG for the treatment of transitional cell tumors of bladder, immune-related therapies have miniscule importance in modern cancer therapy.

The observation that certain immunocytes can kill cancer cells is not a compelling argument for the importance of immune surveillance in carcinogenesis. There are many diseases (e.g., tuberculosis, leishmaniasis, and leprosy) for which the body launches a toxic immune response but for which the immune response is inadequate to stop the disease. There is a huge difference between killing cancer cells and eradicating a cancer.

The cancers that typically arise in immunosuppressed patients all seem to be virally induced cancers. The normal immune system seems to limit the growth of endogenous viruses. When the body is immunosuppressed, carcinogenic viruses can rapidly proliferate. This is not evidence that the normal immune system kills cancer cells produced by oncogenic viruses.

Mice with inborn immune deficiencies (so-called nude mice) are not particularly prone to cancers induced by chemical carcinogens.

Neoplasms occur in all multicellular animals, regardless of whether the animal has an immune surveillance system.

If you believe that immune surveillance is nature's way of suppressing the growth of neoplasms, you might also think that the best way to conquer cancer is by enhancing the host immune system. Although the immune surveillance theory of cancer has been around for many decades, its popularity has steadily declined (List 9.15.2).

There is a large literature, including thousands of research papers, on the role of the immune system in cancer. Many lines of research must be followed, and many questions must be answered, before we gain even a modest understanding of this important subject; however, the conclusion drawn back in 1974, by the cancer immunologist Robert Prehn, seems to apply today, "Immunologic surveillance, as originally conceived, probably does not exist" (180).

9.16 Why Do the Tumors that Occur in Some Tumor Inheritance Syndromes Cross Multiple Cell Lineages?

The Li-Fraumeni syndrome is an example of an inherited genetic syndrome that is accociated with many different types of cancer. The Li-Fraumeni syndrome is caused by mutations in the p53 tumor suppressor gene. As we discussed, inactivation of a

List 9.16.1 General properties of inherited tumors.

Tend to occur in younger age group than **sporadic tumors**.
Occur in families.
Tend to be multiple or bilateral but may be discovered either synchronously or asynchronously.
Tend to include nonendoderm/nonectoderm tumors.
Tend to occur in pleiotropic syndromes and may occur with other kinds of tumors or malformations, often of the same developmental class.
Tend not to be associated with exposure to any known carcinogen.
Tend to be rare tumors.

tumor suppressor gene increases the likelihood that additional mutations will occur in other genes.

Unlike tumor syndromes caused by inherited oncogenes, which produce a specific range of tumors, usually confined to a single embryonic lineage, tumors associated with tumor suppressor genes produce tumors from many different tissues.

Here are the diagnostic criteria for the LFS, from OMIM:

> Classic LFS is defined as a proband with a sarcoma before the age of 45 years, a first-degree relative with any cancer before the age of 45 years, and 1 additional first- or second-degree relative in the same lineage with any cancer before the age of 45 years or a sarcoma at any age. Li-Fraumeni-like syndrome (LFL) is defined as a proband with any childhood cancer, or a sarcoma, brain tumor, or adrenocortical tumor before the age of 45 years, plus a first- or second-degree relative in the same lineage with a typical LFS tumor at any age, and an additional first- or second-degree relative in the same lineage with any cancer before the age of 60 years.

The tumors associated with the Li-Fraumeni syndrome may involve almost any cell type from any embryonic lineage, and this is precisely what we would expect from an inactivating mutation in a gene that regulates the recovery from DNA damage.

The tumors that occur in inherited syndromes often have features that distinguish them from nonhereditary tumors (List 9.16.1).

Chapter 9 Summary

To a limited extent, a few generalizations relate types of tumors to specific etiologies (List 9.16.2).

Beyond these generalizations, for which there are many exceptions, there are no unifying principles that would justify a neoplasm classification based on etiology.

List 9.16.2 Generalizations for a neoplasm classifications based on etiology.

1. Tumors caused by chemicals arise from the dividing cell populations that activate the carcinogen. Cells that are poor activators of chemical carcinogens seldom yield cancers. In general, the epithelial tissues (liver, kidney, and lung) are better activators of carcinogens than the connective tissues.
   Cells that are high activators of carcinogens, but which do not divide, do not yield tumors unless they are somehow compelled to divide. This explains why liver, the foremost metabolizing organ in the body, is an infrequent site of cancer except in those conditions that cause a sustained hyperplasia of hepatocytes (such as cirrhosis).
2. Cells that cannot repair mutations caused by chemical and ionizing radiation are prone to develop cancer. This explains the high incidence of skin cancer in repair-deficient xeroderma pigmentosum cells.
3. Tumor caused by chemicals have a long latency period, and are seen primarily in adults. Even when exposure to chemicals is high and continuous, the time between the first exposure and the clinical manifestation of cancer is usually a matter of years. One of the shortest and most disturbing intervals between carcinogen exposure and cancer occurrence is seen in adolescents who chew tobacco. Chewers habitually push the tobacco against preferred sites in the mouth, and these are the sites where the precancers are most likely to arise (181). The occurrence of cancers in teenagers just a few years after they start to chew tobacco indicates that latency does not necessarily extend over decades.
4. Most of the well-studied precancers are the so-called intraepithelial neoplasms (small epithelial dysplasias growing from a surface epithelium) and precede the development of epithelial tumors that have a long tumor latency.
5. Tumors caused by inherited genes are often seen in young people, are often multiple, are often bilateral, are usually familial, and are often seen in conjunction with malformations and tumors of other types (syndromic).
6. Tumors caused by viruses are often seen in immunosuppressed individuals and can have very short latencies and high rates of regression.
7. Tumors caused by clastogens are primarily leukemias and lymphomas.
8. Tumors caused by radiation can be divided into external and internal (absorbed radiochemical) radiation. The skin is the target for superficially penetrating external radiation. Neoplasms caused by absorbed internal emitters occur in or near the cells that contain the emitter (e.g., thorotrast-induced hepatic angiosarcomas and radioactive iodine-induced thyroid malignancies).

CHAPTER

10 Classification by Topography (Where the Tumor Is Located)

When you get to a fork in the road, take it.

—Yogi Berra

10.1 Background

The most common approach to organizing tumors is by anatomic location. The World Health Organization (WHO) prefers this method and publishes books where tumors are classified by anatomic site. When a tissue is distributed over multiple anatomic locations, the classification is mapped to an anatomic system (e.g., the nervous system or the digestive system) (List 10.1.1).

List 10.1.1 Some classifications published by the WHO.

Tumors of the Nervous System
Tumors of the Digestive System
Tumors of Hematopoietic and Lymphoid Tissues
Tumors of the Breast and Female Genital Organs
Tumors of Soft Tissues and Bone
Skin Tumors
Tumors of the Urinary System and Male Genital Organs
Tumors of Endocrine Organs
Head and Neck Tumors
Tumors of the Lung, Pleura, Thymus, and Heart

List 10.1.2 Problems with classifying neoplasms by their anatomic site.

1. It adds to the complexity of neoplasms. It requires us to think of a single tumor as a member of many different tumor groups.
2. It violates the rules of classification. In a true classification, each classified member must belong in one and only one class. For example, a classification cannot place a dermatofibroma into class "Skin Tumor" and class "Soft Tissue Tumor."
3. It ignores biological properties shared among tumors that arise from different anatomic sites.

The advantage of a topographic classification (based on the tumor's location in the body) is that when you know the anatomic site or the tissue of origin of a tumor, you can choose the proper WHO book that is likely to contain the pathologic description of the tumor.

One problem with this kind of classification is that it groups together tumors that have no common biology, that happen to be located in the same tissue (e.g., salivary gland tumors and lymphomas arising from tonsils and oral squamous cell carcinomas and granular cell tumors of the tongue). Classification by site fails to group tumors that arise from different anatomic locations but that share common properties (e.g., melanomas of skin and pheochromocytomas of adrenal gland, both of which derive from neural crest).

Listing tumors by topographic site adds enormous redundancy to our understanding of human tumors. A gastrointestinal lymphoma would be redundantly included among "Tumors of the Digestive System" and "Tumors of the Haemopoietic and Lymphoid Tissues". Soft tissue tumors arising in the gastrointestinal tract would need to be included in "Tumors of the Digestive System" and "Tumors of Soft Tissues and Bone." In fact, virtually every tumor included in any WHO classification will be included over and over again in many other WHO classifications.

This approach is very unsatisfying for several reasons (List 10.1.2). Each day, we learn more and more about the pathways that control the growth of tumors. This knowledge will lead to targeted therapies directed toward tumors that share specific sets of molecular pathways. Classifications based on anatomic site will fail to group tumors by shared biological properties.

The purpose of this chapter is to examine the strengths and limitations of classifying tumors by their anatomic locations.

10.2 The Range of Tumors Occurring in a Single Organ

Within any organ is a diverse set of tissues from which tumors may arise. The thyroid gland is an example. Teratomas, squamous cell carcinomas, medullary carcinomas, and lymphomas can all occur in the thyroid gland, and all of these tumors are biologically unrelated to one another.

10.3 Subsets of Anatomic Classifications

When anatomic location is used as the basis for a tumor classification, there is a tendency to subclassify tumors through increased attention to anatomic detail. An example would be tumors of the lip. The lip is a sun-exposed tissue. The most frequently occurring cancer of the lip is squamous cell carcinoma. In most people, the lower lip juts out to catch light and cast a shadow. This results in the lower lip having a bit more sun-exposed surface area than the upper lip, and this translates into a higher incidence of squamous cell carcinoma in the lower lip (182). This is not the only difference between upper and lower lip. Tumors of the minor salivary glands occur more often in the upper lip than the lower lip. Minor salivary gland tumors of the lower lip are more likely to be malignant than the more frequently occurring minor salivary gland tumors of the upper lip (183).

Does this signify the importance of subclassifying tumors by their subanatomic location within an organ? Let us hope not.

Categorizing tumors can impose a set of logical dilemmas on **biocurators.** Imagine that you have a dermatofibroma biopsied from a patient's thigh. A dermatofibroma is a benign, fibrous tumor that grows in the dermis of the skin, just between the epidermis and the subcutis (the fatty layer). This lesion is traditionally classified as a soft-tissue tumor because it grows from soft connective tissue and has morphologic similarities to other soft-tissue tumors.

Some people classify dermatofibroma as a tumor of skin because it occurs only in the skin. Still others would classify this particular dermatofibroma a tumor of the thigh. The thigh is an anatomic compartment that includes skin and muscle and nervous tissue and vessels and so forth. When we choose an anatomic classification for dermatofibroma (skin, soft tissue, or thigh), how can we expect every pathologist to make the same choice?

Does it really make any difference whether we classify dermatofibroma as a soft-tissue tumor, or as a skin tumor, or as a tumor of the thigh? Maybe we would all be better off if we chose not to categorize this lesion and moved onto a more productive activity! The problem is that tissue databanks, tumor registries, certifying agencies, research protocols, code systems for billing, and legal regulations may require anatomic pathologists to group tumors so that the members of the groups can be counted, compared, retrieved, and organized for indeterminate future uses. Hospitals cannot perform these tasks if the method for organizing tumors puts unique tumor entities into any one of several different anatomic groups.

Chapter 10 Summary

The anatomic classification of tumors has enormous clinical utility. If you know that a neoplasm has arisen from a particular anatomic location, you can narrow your differential diagnosis to a short list of neoplasms that are known to occur at the site. You can even prioritize your differential diagnosis based on the relative incidence for each tumor type at the location. Nonetheless, the process of identifying a neoplasm is fundamentally different from the process of classifying a tumor into a biologic group.

Anatomy-based classifications ignore shared properties among tumors that may occur in different organs; therefore, anatomy-based classification will not help cancer researchers find class-specific properties that extend to tumors of many different anatomic origins, and it will not help trialists find class-specific cancer cures.

CHAPTER

11 Classification by the Age at Which the Cancer Occurs

Cancer is a disease that primarily affects the elderly, but carcinogenesis is a disease process that primarily affects the young.

—Anonymous

11.1 Background

Tumors discriminate among people based on age (List 11.1.1).

Age seems to be a major determinant of tumor biology. Should we classify tumors by the age at which they occur? No. Every kind of tumor has its own age-related biology, including the age when the patient was exposed to an initiating carcinogen, the length of tumor latency, and the age at which the resulting tumor became clinically apparent. The observation that certain tumors occur at certain ages follows from the uninformative observation that tumors are diverse.

The purpose of this chapter is to examine the biological properties that characterize tumors that have a particular age of onset. We discover whether those properties are age determined. If so, then age is a proper basis for tumor classification. If not, then age is just one of many thousands of features that help distinguish one tumor from another.

1.2 What Tumors Occur in Infants and Children?

The U.S. National Cancer Institute's Surveillance, Epidemiology and End Results (SEER) program has published an excellent monograph on cancer occurring in children. It covers data collected from U.S. Cancer Registries between 1975 and 1995 (184). The only drawback of this monograph is that it is restricted to cancer diagnoses and omits benign tumors, hamartomas, and precancers. An excerpted summary of the age-stratified incidence of cancer in children is shown (Figure 11.1).

List 11.1.1 Age-based cancer observations.

Cancer occurs commonly in adults but only rarely in infants and children.

The most common tumors of adults are almost never seen in children.

The most common tumors of childhood are almost never seen in adults.

All congenital tumors take nine months or shorter to develop.

The common tumors occurring in adults usually require many years to develop.

The common tumors of adults arise from surface epithelium, mucosal surfaces, or from glands that directly connect to mucosal surfaces.

The common tumors of children originate from almost every tissue except those that give rise to the common tumors of adults.

Tumors that occur in infants sometimes regress spontaneously.

Tumors that occur in the elderly almost never regress spontaneously.

Some metastatic cancers occurring in children have a high cure rate.

No metastatic cancers occurring in adults have a high cure rate.

There is no evidence that childhood cancers are caused by the same chemical carcinogens that are responsible for the common cancers occurring in adults.

The only human carcinogens shown to act in utero are diethylstilbestrol and ionizing radiation.

Age (in years) at diagnosis / Tumor category	<5 Rate	5-9 Rate	10-14 Rate	15-19 Rate	% of Total for 15-19 Group
All Sites	199.9	110.2	117.3	202.2	100.0%
Acute lymphoblastic leukemia (ALL)	58.2	30.3	17.8	12.9	6.4%
Acute myeloid leukemia (AML) (Ib)	10.1	4.5	5.7	8.5	4.2%
Hodgkin's disease (IIa)	0.8	3.9	11.7	32.5	16.1%
Non-Hodgkins lymphoma (NHL) (IIb,c,e)	5.9	8.9	10.3	15.3	7.6%
CNS tumors (III(total))	36.0	31.9	24.6	20.2	10.0%
Ependymoma (IIIa)	5.6	1.6	1.3	1.1	0.5%
Astrocytoma (IIIb)	15.0	15.9	15.1	12.3	6.1%
Medulloblastoma/PNET (IIIc)	9.6	7.3	4.0	2.5	1.2%
Neuroblastoma & ganglioneuroblastoma (IVa)	27.4	2.6	0.8	0.5	0.2%
Retinoblastoma (V(total))	12.5	0.5	0.0	0.1	0.0%
Wilms', rhabdoid, clear cell sarcoma (VIa)	18.0	5.8	0.6	0.4	0.2%
Hepatic tumors (VII(total))	4.8	0.4	0.4	1.0	0.5%
Hepatoblastoma (VIIa)	4.6	0.2	0.1	0.0	0.0%
Osteosarcoma (VIIIa)	0.3	2.8	8.3	9.4	4.6%
Ewing's sarcoma (VIIIc)	0.3	1.9	4.1	4.6	2.3%
Soft tissue sarcoma (IX(total))	10.9	8.3	10.9	15.9	7.9%
Rhabdomyosarcoma and embryonal sarcoma (IXa)	6.5	4.4	3.5	3.9	1.9%
Non-rhabdo soft tissue sarcoma (IXb-e)	4.4	4.0	7.4	11.9	5.9%
Germ Cell, trophoblastic, & other gonadal tumors (X (total))	6.9	2.4	6.7	30.8	15.2%
Thyroid carcinoma (XIb)	0.1	1.0	4.1	14.6	7.2%
Malignant melanoma (XId)	0.8	0.6	2.8	14.1	7.0%
Other and unspecified carcinomas (XIf)	0.4	0.8	2.8	10.5	5.2%

FIGURE 11.1 Age-specific cancer incidence rates per million by age group (all races and both sexes using SEER data covering 1986–1995), public domain chart from (184). Source: Ries LAG, Smith MA, Gurney JG, Linet M, Tamra T, Young JL, Bunin GR (eds). *Cancer Incidence and Survival among Children and Adolescents: United States SEER Program 1975-1995,* National Cancer Institute, SEER Program. NIH Pub. No. 99-4649. Bethesda, MD 1999.

The most common tumors in adults (squamous cell carcinoma of skin, basal cell carcinoma of skin, bronchogenic carcinoma, and adenocarcinomas of colon, pancreas, breast, and prostate) are virtually nonexistent in the pediatric population. Tumors of the brain and the hematolymphoid/lymphoid tissues, both uncommon in adults, account for about half of the malignant tumors occurring in children (184).

Ominously, the incidence of almost every type of childhood cancer has increased during the years from 1976–1994 (Figure 11.2).

Cancer type	1976-84	1986-94	% Change
All cancer	197.9	269.3	36
Neuroblastoma	55.2	74.4	35
All leukemia	35.9	45.4	26
All CNS	23.3	36.5	57
Retinoblastoma	22.1	31.5	43
Wilms'	21.4	23.6	10
Germ cell	9.5	21.3	124
Soft tissue	13.6	16.6	22
Hepatic	7.6	11.4	50
Lymphoma	4.5	4.2	- 7
Epithelial	2.6	3.0	15
Other/Unspecified	1.4	1.0	-29
Bone	0.5	0.5	0

FIGURE 11.2 Average annual incidence rates per million by type, age less than 1 years (all races both sexes, SEER, 1976–1984 and 1986–1994), a public domain chart from (184). Source: Ries LAG, Smith MA, Gurney JG, Linet M, Tamra T, Young JL, Bunin GR (eds). *Cancer Incidence and Survival among Children and Adolescents: United States SEER Program 1975-1995,* National Cancer Institute, SEER Program. NIH Pub. No. 99-4649. Bethesda, MD 1999.

11.3 The PNET Tumors

The **PNET** tumors are rare tumors that occur in infants and children. They provide a challenging example of the complex and unique issues of tumor development and classification that arise in the tumors of childhood. PNET stands for primitive neuroectodermal tumor and constitutes about a half dozen different tumor entities. PNETs all fall into an even broader group of rare tumors collected under the morphological banner of "small round-cell tumors of childhood."

The PNET term applies to a group of childhood tumors of primitive neural cells occurring in the brain and in other parts of the body. The PNET terminology is somewhat archaic, referring to a time when the neural crest and the central nervous system (CNS) were both thought to be equivalent derivatives of **neuroectoderm**. As we describe in Chapter 19, the neuroectoderm produces the neural tube. The neural crest develops from cells lying adjacent to the neural tube and subjacent to ectoderm. In other words, the neural crest is different from neuroectoderm. The derivatives of the neural crest are anatomically distinct from the derivatives of neural tube (i.e., the brain and spinal cord). Neural tube derivatives are only found in the brain and the spinal cord. Neural crest derivatives are found everywhere except the brain and the spinal cord. The neural tube-derived cells include neurons of the CNS. The neural crest-derived cells include the neurons of the peripheral nervous system. Because both neural tube and neural crest produce a type of neuron, the two anlagen have been confused by generations of biologists and pathologists. Beyond these functionally and cytologically similar cells (i.e., neural tube neurons and neural crest neurons), the cellular derivatives of the neural tube and the neural crest are distinctly different.

This brings us to the classification of PNETs. The tumors that are called PNETs include neoplasms that are now known to be of neural tube origin (true descendants of neuroectoderm), along with neoplasms that are now known to be of neural crest origin. Not surprisingly, the two groups of PNETs contain very different kinds of tumors.

11.4 What Is the EWS Family of Tumors?

The peripheral PNETS are a subset of the Ewing tumor family (List 11.4.1).

All of the peripheral PNETs occur outside of the CNS, and all of the Ewing family of tumors derive from the neural crest. A translocation, t(11;22)(q24;q12), is frequently seen in the Ewing family of tumors. This translocation is sometimes seen in neuroblastoma, another primitive tumor of neural crest origin (185). In the Ewing family of tumors, a translocation produces a fusion gene involving EWS and a member of the ETS family of transcription factors (FLI1, ERG, ETV1, E1AF, or FEV).

The EWS gene is also involved in several non-PNET, nonneural crest cancers of childhood. EWS fused with ATF1 is found in clear cell sarcoma, a tumor of young adults that often arises in the lower extremities. EWS fused with CREB1 is found in a variant of clear cell sarcoma found in the gastrointestinal tract and lacking

List 11.4.1 The Ewing family of tumors.

Ewing's sarcoma
Peripheral PNET
Askin tumor

melanocytic differentiation. EWS fused with CHOP is found in a small subset of myxoid liposarcomas (186). EWS fused with Wilms tumor 1 gene (WT1) has been found in intra-abdominal desmoplastic small round cell tumors (185). These finding indicate that tumors with EWS fusion genes are not limited to tumors of the neural crest. With the exception of the involvement of an EWS fusion gene in myxoid liposarcoma (a tumor of adults), the EWS fusion gene tumors occur in childhood tumors.

The other PNETs are the so-called CNS PNETs (or the central PNETs). These tumors arise from the neural tube. Unlike the peripheral PNETs, they do not include the Ewing family tumors and are not characterized by a unifying cytogenetic marker. The CNS PNETs include medulloblastoma (usually occurring in the cerebellum) and pineoblastoma (usually occurring in the pineal gland).

Medulloblastoma is the most common CNS tumor of childhood and accounts for about 20% of childhood brain tumors. No known cytogenetic lesion characterizes all medulloblastoma, but many of these tumors have a loss of 17p. A recent study comparing the gene expression profiles of medulloblastomas with those of mouse cerebellum at different stages of development shows similarity between human medulloblastoma and mouse cerebellum at postnatal days 1 to 10 (187). This indicates that it is feasible to map a tumor of primitive cells against a normal tissue containing the putative progenitor cells (even when the normal tissue comes from a different organism).

In Chapter 5, we discussed trilateral retinoblastomas. Sometimes in children, the occurrence of bilateral retinoblastomas is followed by the occurrence of a pineoblastoma, and this is referred to as trilateral retinoblastoma. Pineoblastomas have a developmental anlage identical to that of retinoblastomas of the eye. Both derive from neuroectoderm and have no developmental relationship with PNETs of neural crest origin.

The lesson learned from the PNETs is that there are tumors that occur in childhood that do not occur in adults and that their biology closely mimics primitive embryonic development. Advances in our knowledge of normal developmental biology greatly inform our ability to understand and classify the childhood tumors. Conversely, attempts to classify childhood tumors based purely on morphologic features (e.g., small round-cell tumors of childhood) produce an amalgam of unrelated tumors with unpredictable biologies.

11.5 Leukemias

Leukemia is not a tumor restricted to childhood. Although 33% of childhood cancers (in the 0–14 age group) are leukemias or lymphomas, most leukemias occur in adults. Leukemia strikes 10 times as many adults as children (about 40,440 adults compared with 3,800 children, ages 0–19) (188).

Because a tumor that has a high incidence in children has an even higher incidence in adults, it would be difficult to construct a neoplasm classification based strictly on age.

11.6 Transplacental Carcinogenesis

The term **transplacental carcinogenesis** describes the initiation of cancer during gestation. It is a somewhat mammal-centric term. Many animals have no placenta. The term "transplacental carcinogen" sometimes includes agents that act directly on the fetus without actually crossing the placenta (such as ionizing radiation), and it seems to ignore the preimplantation period and the immediate postimplantation period during which there is no formed placenta. With these quibbles aside, how do the youngest of the young (the embryos and fetuses) deal with environmental carcinogens?

One might think that organisms in the embryonic and fetal stages would be highly susceptible to carcinogens. Lucy Anderson and coworkers have compiled a list of factors that might influence transplacental carcinogenesis (List 11.6.1) (189).

Almost every observable property of fetal development seems designed to enhance the carcinogenic process. Fetuses have a rapidly dividing stem cell population, and their cells have diminished capacity to repair DNA. Every dividing cell that is mutated can pass its mutation into all their descendant cells and lineages; most of the chemicals that enter the mother's blood stream will be distributed by the placenta into the fetal system. One would predict that the embryo and the fetus are highly susceptible to carcinogens.

Actually, we know very little about transplacental carcinogenesis, and the few things that we know are enigmatic (List 11.6.2).

The listed items all seem to be counterintuitive. You need to really stretch your preconceived notions of cancer epidemiology and of carcinogenesis as you ponder transplacental carcinogenesis.

List 11.6.1 Factors that might influence carcinogenicity during fetal development (189).

Ability to repair DNA damage
Development of differentiated characteristics, including the ability to carry out metabolic activation of chemicals
Effects of rates of cell division on the passage of a mutation to descendant cells
The selective expansion of clones of mutated cells as part of normal **ontogeny** (i.e., organ development selecting one set of mutated cells over another)
Immaturity of the endocrine and immunological systems
Metabolic detoxification by placenta and/or maternal tissues
Metabolic detoxification by the perinate herself
Numbers of target cells at risk
Presence of undifferentiated stem cells
Sensitivity to cell killing

List 11.6.2 Things we seem to know about transplacental carcinogenesis.

The only known human carcinogens that are known to target the embryo or fetus are DES (diethylstilbestrol) and ionizing radiation (151, 190, 191).

Congenital tumors and tumors of childhood are extremely rare, accounting for less than 1% of the human cancer burden. This would seem to indicate that human embryos and fetuses are highly resistant to carcinogenesis.

In most animal systems (marsupials are an exception), the tumors produced by exposing embryos and fetuses to carcinogens are tumors of differentiated tissues, not tumors of primitive stem cells. Most transplacentally induced tumors of rodents are tumors of differentiated tissues and are the types of cancers that occur commonly in adult animals.

There seems to be no observable system, in man or animal, whereby exposure of the preimplantation or immediate postimplantation embryo will cause tumors. For all we know, the early embryo is completely resistant to carcinogenesis.

11.7 There Are Only a Few Known Human Transplacental Carcinogens

Maternal exposure to x-rays during gestation has been associated with an increased incidence of leukemia (190, 191). As described in Chapter 8, exposure of mothers to DES during pregnancy produces a rare tumor, clear cell carcinoma of the cervix and of the vagina, in the adolescent or young adult female children. In both cases, the tumors that occur are rare tumors. Other than these two examples, we have almost no knowledge of the types of carcinogens that produce childhood tumors in humans.

11.8 Why Are Congenital and Childhood Tumors Extremely Rare?

As discussed, fetuses have many factors that enhance their sensitivity to carcinogens (exuberant cell division of stem cells, deficient DNA repair, and an ability to pass mutations to cell and lineage descendants). Should we not expect many tumors to occur in neonates? What we observe is that childhood cancers account for less than 1% of human cancers. It is possible that the fetus is much more resistant to carcinogens than we might imagine. It is also possible that the fetus is exposed to very few carcinogens. A third possibility is that the fetus is highly sensitive to carcinogens, but that the resulting tumors do not occur until the fetus has grown into an adult.

In rodents, transplacental carcinogenesis results in fully differentiated tumors that occur in adults. The same may apply to humans. The known human transplacental carcinogens produce rare tumors. If a human fetus were exposed to a carcinogen that produced a common tumor, such as colon cancer, arising years later, in the adult, we would have no way of finding the connection between transplacental exposure and the occurrence of cancer. Because most animal models of transplacental carcinogenesis produce common tumors occurring in adult animals, we can hypothesize that there are human transplacental carcinogens that go undetected. Furthermore, we can guess that among the cancers that occur in adult humans, there are tumors that were initiated during gestation. If true, it would be impossible to build a comprehensive tumor classification based on the age of occurrence of tumors.

11.9 Why Does Transplacental Carcinogenesis Seldom Produce Tumors of Primitive Cells?

When animals are transplacentally exposed to carcinogens, the tumors that develop are almost always tumors of differentiated cells. Because the fetal cells exposed to carcinogens include primitive stem cells, would you not predict that the resulting tumors would be tumors of primitive stem cells (e.g., neuroblastoma, retinoblastoma, PNETs, and so on)?

Carcinogenesis is a multistep process that occurs while initiated cells continue to replicate. When initiation occurs in an embryo, normal development does not stop. During development (which may extend many years after birth), the population of primitive cells declines while the population of differentiated cells increases (List 11.9.1). The descendants of primitive initiated cells produce a differentiated population, and the resulting tumors are differentiated (i.e., not primitive tumors).

In cancer, very few generalizations are always true. As discussed in Chapter 5, rainbow trout eggs are exquisitely sensitive to carcinogens. The tumors that develop, following a single exposure to a low dose of aflatoxin, are hepatocellular carcinomas or mixed hepatocellular/cholangiocarcinomas. Bailey and coworkers report finding only two hepatoblastomas (tumors of primitive liver stem cells) in 30 years of trout carcinogenesis studies (95). The story for trout kidney carcinogenesis is quite different. Nephroblastoma, a primitive tumor of renal stem cell phenotype, is regularly produced in trout by a host of different carcinogens. Trout is the only animal in which nephroblastomas can be produced in high incidence.

Bailey and coworkers report that MNNG, a direct-acting carcinogen, induces nephroblastomas after trout fry are exposed to a solution of 50 ppm MNNG for 30 minutes. About

List 11.9.1 Why we infer that adults have very few primitive stem cells.

Adult humans cannot regenerate damaged limbs.
There is no pathogenic process that occurs in adult humans that leads to developmental anomalies.
Routine **histologic examination** of adult tissues does not reveal identifiable foci of stem cells.
Tumors of primitive stem cells are exceedingly rare in adults (although they do occur).

half of the fish will develop nephroblastoma as adults, 6 to 9 months after exposure (95). Why do nephroblastomas occur in trout, and not renal cell carcinomas, a tumor of differentiated renal epithelial cells? Nobody knows, but there is opportunity for speculation.

In humans and other animals, the nephrogenic **blastema** persists after birth, just as though it were a valued component of the postnatal kidney (192, 193). Persistent nephrogenic rests are found in about 40% of human nephroblastomas (also called Wilms tumors). It is possible that the occurrence of chemically induced nephroblastomas in greater than 50% of trouts may relate to the unique role of persistent nephrogenic blastema as a cancer target.

11.10 Why Are the Early-Stage Embryos Resistant to Carcinogenesis?

To the best of my knowledge, there are no known instances in which tumors developed following the exposure of a carcinogen to a preimplantation or recently implanted embryo. As far as anyone knows, these early embryos are fully resistant to carcinogens. It may be that the early embryo has so few cells that the likelihood of a cancer-causing mutational event occurring in any of the target cells is too small to detect through an experimental study. It may be that any mutational damage occurring in an early embryo will lead to embryonic death, thus precluding the development of cancer. Nobody really knows.

11.11 Congenital Tumors in Humans

Congenital tumors are tumors that are present at birth. As discussed, congenital tumors are very rare (184). Congenital tumors obviously have the shortest latencies of all tumors. Whereas tumors in adults may require decades to develop, every congenital tumor develops in under 9 months.

There is a fundamental difference between prenatal and postnatal development. At birth, the human body has its full complement of organs. The organs that we are born with are the same organs that we die with. The opposite is true during prenatal development. Many of the anatomically recognized embryonic tissues are absent from the postnatal organism. The allantois, the yolk sac, the notochord (except for vestigial cells), and the external granular cell layer of the cerebellum all disappear. The result of all this reconstruction is the absence of certain fetal organs, the persistence of small, vestigial organs, the development of homologous organs that have different functions and anatomic features in men and women, and the maintenance of primitive cells within the adult organism as stem cell populations capable of generating cell lineages appropriate for the progenitor cell. This somewhat complex and confusing array of biological opportunities contributes to the strange assortment of tumors that occur in the postnatal period.

Neuroblastoma accounts for 28% of cancers occurring in the first year of life. Leukemias comprise 17% of infant cancers. Central nervous system malignancies comprise 13% of infant cancers. Astrocytomas and other gliomas accounted for half of the CNS malignancies, followed by primitive neuroectodermal tumors and medulloblastomas. Retinoblastoma accounted for about 12% of infant cancer and Wilms tumor an additional 9%. The incidence rates for malignant germ cell tumors and malignant soft tissue tumors both accounted for about 6% of infant cancer. Fibrosarcomas and rhabdomyosarcoma together also accounted for another 6%. The rate for malignancies of the liver, almost exclusively hepatoblastoma, was about 4% of all tumors of infancy (184).

Congenital tumors do not represent a biologically homogeneous assemblage of tumors, and they seem to have no common **pathogenesis**. If anything, they are more complex, and more difficult to classify, than tumors that occur in adults.

11.12 Why Do Pediatric Tumors Respond So Well to Chemotherapeutic Agents and Adult Tumors Respond So Poorly?

The decline in mortality from childhood cancers has been one of the greatest achievements in the field of cancer research. Between 1975 and 1995, the overall mortality of childhood cancer has declined 40%. The declines ranged from 2.0% to 3.2% per year over this period (184). Although the death rates declined, the overall incidence of childhood cancers increased 0.8% per year. There were 2,275 cancer deaths among children in 1995, representing less than a half-percent of all cancer deaths.

Why have we succeeded to markedly reduce deaths from pediatric tumors, whereas progress against the common adult cancers have been disappointing (List 11.12.1)?

11.13 Why Can We Infer that Most Tumors Occurring in Adults Have a Long Latency?

Several observations suggest that cancers have a very long latency period following an initiating event (List 11.13.1).

Patients with the Li-Fraumeni syndrome have an inherited mutation in the tumor suppressor gene p53. Gene carriers can develop cancer throughout their lives, and the cancers can occur in many different organs; however, the tumors do not occur in children. Likewise, p53 knockout mice develop multiple tumors, but the tumors they develop occur in adult mice.

Why is the latency period so long? The acquisition of a tumor suppressor gene (such as p53) increases the likelihood that the process of carcinogenesis will advance. Single events seldom confer the full malignant phenotype.

List 11.12.1 Putative reasons for the breakthroughs in cancer treatment in pediatric tumors.

1. Protocol design of clinical trials now has a solid, scientific grounding that supports rapid advancement and successful new drug therapies.
2. Children are better subjects than adults, withstanding doses of chemotherapeutic agents and surviving bone marrow transplant procedures that would kill older patients.
3. Pediatric tumors are often derived from primordial germ cells or from somatic primitive (e.g., PNET tumors) cells or stem cells. The normal developmental fate of nongonadal primordial germ cells and of primitive cells is extinction. Because childhood tumors often derive from cells with a built-in tendency toward developmental extinction, they respond well to toxic agents that reduce proliferation.
4. Pediatric tumors are often characterized by a single gene rearrangement present in nearly all examples of the tumor. Chemotherapeutic agents that inhibit the key pathways activated by the tumor's single genetic aberration will be highly effective in every tumor of the same type.
5. Pediatric tumors, unlike tumors occurring in adults, are each composed of a homogeneous population of cells, often characterized by high genetic stability. A chemotherapeutic agent that kills one cell in a homogeneous tumor is likely to kill every cell in the tumor. Furthermore, homogeneous tumors are less likely to develop drug resistance than tumors with heterogeneous population.

List 11.13.1 Observations suggesting that tumor latency is long.

1. Most tumors occur in older adults. Cancers in infants (i.e., tumors of short latency) account for less than 1% of cancer deaths.
2. Cancers take a long time to develop after an exposure. Many people exposed for a brief period to asbestos developed cancers decades later.
3. Though patients with inherited cancer syndromes may develop childhood tumors, most tumors occurring in inherited cancer syndromes will arise during adulthood, indicating a process that extends over decades.
4. Rodents exposed to carcinogens in the perinatal period develop tumors as adults.

11.14 What Do We Learn from Age at Diagnosis and Synchronous Appearance of Tumors in Germline Mutation Syndromes?

Tumor synchrony (two tumors occurring at the same time) sometimes occurs by chance. More often, it occurs after a single carcinogenic event (List 11.14.1).

List 11.14.1 Some causes of tumor synchrony.

1. A germline mutation that confers an increased likelihood of cancer in may types of cells
2. Exposure of the organism to a strong carcinogen or virus that can target many different cells or types of cells
3. A physiologic event that triggers an end to tumor latency (e.g., immunosuppression and the rapid emergence of cancers induced by endogenous viruses)
4. Chance

Chapter 11 Summary

There are huge differences between the types of cancers that occur in children and the types of cancers that occur in adults. The reason for the differences in tumors among children and adults relates to developmental biology. Many of the tumors occurring in infants and very young children develop from cells and pathways that are not biologically available to adults.

The tumors occurring in adults derive primarily from ectodermal and endodermal lining cells and appear at very high incidence many years after more or less continuous direct exposure to environmental carcinogens. This mode of carcinogenesis is not available in infants, who have a compressed timeline for tumor development. Adult cancers of ectodermal and endodermal are usually characterized by multiple cytogenetic abnormalities, aneuploidy, tumor heterogeneity, and genomic instability. Tumors with these properties have low cure rates.

Most of the tumors occurring in infants are benign tumors of mesenchymal origin and are characterized by very simple genetic alterations, often a single translocation resulting in a characteristic fusion gene product. Many of the tumors of childhood result from inherited or sporadic germ line mutations that yield growing tumors soon after birth.

Although tumors have characteristic age specificity, the reason for most differences among tumors of children and adults relates to developmental biology (not to the passage of time). Some tumors, such as leukemias, occur in both adults and children. We can expect that some of the common tumors occurring in adults were initiated during gestation. Today, we know almost nothing about the non-inherited causes of cancer in children. For these reasons, it would be impossible to build a neoplasm classification based on the age of occurrence of tumors.

CHAPTER

12 Classification by Function and Common Pathways

O hell! What have we here?
A carrion Death, within whose empty eye
There is a written scroll! I'll read the writing.
All that glitters is not gold.

—William Shakespeare, *The Merchant of Venice*, 1596

12.1 Background

Anatomy books, physiology books, and pathology books all contain a chapter devoted to the endocrine system. All of the endocrine glands share a common function. They secrete chemicals (hormones) into the bloodstream, and those chemicals exert an array of physiological actions on distant organs. For the most part, all endocrine organs share a similar morphology. They are composed of epithelial cells often aggregated in small nests, or follicles or islands. Beyond these features, there is a wide array of diversity among the endocrine glands. Some are large, like the thyroid, and some are small, like the islets of the pancreas that fit snugly between the parenchymal acini of the exocrine pancreas. Some endocrine glands, like those in the gastrointestinal tract, are single cell structures sprinkled throughout the gut mucosa.

The hormones secreted by endocrine glands have no consistent chemical theme. Some are polypeptides. Some are steroids. Some are catecholamines, and some are iodinated conjugates. It is hard to imagine a common genetic or biochemical pathway that could account for this chemical diversity. Furthermore, the mode of action of the different hormones varies greatly, some acting on cell surface receptors, some acting on cytoplasmic receptors, some acting within the nucleus. With all of these different activities, it is highly unlikely that the endocrine glands share a common evolution or a common set of biochemical pathways.

It is an error to confuse shared behavior with shared class. A wide variety of animals from diverse evolutionary classes share the power of flight (Figure 12.1). Although all of these animals can fly, they are fundamentally different types of organisms. They do not share metabolic similarities or the same diseases or the same internal anatomy. They do not even use the same mechanisms to fly.

FIGURE 12.1 Unrelated organisms can fly.

12.2 Endocrine Neoplasia

Consider the paired adrenal glands. The adrenal is a composite gland, composed of cells with unrelated embryonic origins. The adrenal medulla derives from the neural crest. The adrenal medullary cells are a specialized type of paraganglia cell (cells that develop alongside the ending of a peripheral neuron). The medullary cells produce catecholamines. The cells of the adrenal cortex derive from mesoderm. They produce steroids, chemicals related to lipids, that have no chemical similarity with catecholamines. In function and in morphology, the cells of the adrenal cortex share similarities with Leydig cells of the testis. Leydig cells, like adrenal cortical cells, derive from mesoderm.

The tumors arising from medullary endocrine cells are pheochromocytoma, neuroblastoma, and ganglioneuroma. All of these tumors are rare. The tumors that arise from the adrenal cortex are adenoma of adrenal cortex, carcinoma of adrenal cortex, and myelolipoma of adrenal cortex. Adrenal cortical adenoma is a common tumor of humans that is often noticed as an incidental finding on abdominal radiologic images. Adrenal cortical carcinoma and myelolipoma are both rare tumors. Tumors of the adrenal cortex have virtually nothing in common, biologically or morphologically, with tumors that arise from the adrenal medulla.

The endocrine tumors exemplify the complexity and confusion that arise when you attempt to classify tumors by function. Examples of endocrine organs include the thyroid gland and the adrenal gland. There are eight major endocrine glands and several other tissues that secrete physiologically active hormones (List 12.2.1).

List 12.2.1 Endocrine glands and tissues.

- Pituitary gland
- Pineal gland
- Thyroid gland
- Parathyroid gland
- Adrenal gland
- Pancreatic islets of Langerhans
- Gonads (testes and ovaries)
- Other tissues
  - Thymus
  - Stomach
  - Small intestines
  - Heart
  - Placenta

List 12.2.2 Types of thyroid neoplasms (196).

Follicular adenoma
Papillary adenoma
Atypical adenoma
Papillary carcinoma
Follicular carcinoma
Undifferentiated anaplastic carcinoma
Medullary carcinoma
Squamous cell carcinoma
Metastatic carcinoma
Fibrosarcoma
Angiosarcoma
Lymphoma
Teratoma
Other sarcomas

Tumors of the endocrine glands do not comprise a class of tumors with shared neoplastic properties. In addition, a would-be class of endocrine tumors will exclude tumors of nonendocrine origin that happen to secrete hormones (i.e., have an endocrine function). Small cell carcinoma of lung and renal cell carcinoma are both known to secrete physiologically active hormones, yet neither of these tumors is ever classed with the endocrine tumors (194, 195).

There is no common morphologic appearance of the endocrine tumors. Some of the endocrine glands have epithelial morphology (e.g., thyroid); others have spindle morphology (e.g., ovarian stroma). Like the adrenal, other endocrine glands may contain diverse specialized cell types. The pituitary gland is divided into adenohypophysis (glandular part of the pituitary) and neurohypophysis (neural part of pituitary). The thyroid contains follicular cells and small nests of so-called C cells that produce calcitonin. Thyroid follicular cells have an endodermal origin, and thyroid C cells have a neural crest origin.

Within any endocrine organ are a variety of specialized cells that give rise to morphologically and clinically diverse neoplasms. The above list of tumors comprises a published "Classification of thyroid neoplasms" (196) (List 12.2.2).

This classification of thyroid neoplasia includes tumors from several different lineages, including endodermal (e.g., follicular carcinomas of thyroid), neural crest (e.g. medullary carcinoma), mesenchymal (e.g., fibrosarcoma), and germ cell (e.g., teratoma). The most common type of thyroid cancer is papillary carcinoma of thyroid, a

List 12.2.3 Excerpt from OMIM entry 162300 for multiple endocrine neoplasia, type 2B, MEN2B (neural crest origin) (81).

A single identical point mutation (exon 16) in the catalytic core of the tyrosine kinase domain of the ret gene has been found to be associated with both inherited and de novo MEN2B/
Ganglioneuroma (ganglioneuromatosis)
Pheochromocytoma
Calcitonin secreting medullary thyroid carcinoma
Parathyroid hyperplasia/adenoma (probably secondary to calcitonin secretion)

tumor that arises from the follicle lining cells. It seems unlikely that a single carcinogenic process could be responsible for the diverse tumors occurring in the thyroid.

In inherited cancer syndromes, the endocrine tumors co-occur with nonendocrine tumors that all share a common developmental lineage. Multiple endocrine neoplasia, type 2B (MEN2B), is an inherited condition in which affected persons develop several different types of endocrine tumors, but they also develop ganglioneuroma, a nonendocrine neoplasm. The neoplasms that occur in MEN2B have in common an origin from neural crest (List 12.2.3).

In multiple endocrine neoplasia type 1 (MEN1), another inherited tumor syndrome associated with the development of endocrine tumors, affected persons inherit endocrine tumors that derive from endoderm (List 12.2.4). The only exception to the endodermal origin of tumors in MEN1 is the occurrence, in some cases, of **adrenocortical adenoma.** Adrenocortical adenomas have a mesodermal origin. In this case, the tumors are most likely physiological hyperplastic adrenocortical nodules due to Cushing syndrome (caused by a pituitary adenoma) and should not be included with the neoplasms produced through an inherited germline mutation (197).

Mutations that predispose to cancer do not select for target tissues that share a common function. Selection seems to favor tissues that share a common developmental lineage.

Other than their common functionality (i.e., hormone secretion), tissues of the endocrine system may have nothing in common morphologically, genetically, or developmentally. A molecular classification of neoplasms tied to function does not serve as a consistent class scaffold.

12.3 Pathways

Pathways are functional subsystems. A single pathway does not define the functionality of cells, but a narrow combination of pathways may contribute to functions that are characteristic of the cell. Cancers can be classified by their functional pathways. Classi-

List 12.2.4 Excerpt from OMIM entry 131100 for MEN1 (endodermal origin) (81).

The MEN1 gene contains 10 exons and encodes a ubiquitously expressed 2.8-kb transcript. The predicted 610-amino acid protein product is termed menin, a putative tumor suppressor protein inactivated by MEN1 mutations.

Pancreatic **islet cell adenoma**

Parathyroid adenoma

Pituitary adenoma

Prolactinoma

Glucagonoma

Insulinoma

VIPOMA (vasointestinal peptide tumor)

Gastrinoma

Carcinoid tumors

Adrenocortical adenoma (secondary to Cushing syndrome) (197)

fying tumors based on the pathways that they employ to grow, invade, and metastasize is highly tempting. It offers the possibility of developing new drugs targeted against specific pathways that can treat all tumors of a common class (198).

The notion that pathways can be targeted for treatment is relatively new and replaces the earlier notion that the specific mutations that cause cancer can be corrected or targeted by modern gene-specific approaches (e.g., gene therapy, gene silencing, and gene-specific antibodies).

If the malignant properties of a tumor cell are driven by an activated oncogene, would it not make sense to develop drugs that specifically inhibit the activity of the abnormal gene? You would think so, but there are several problems with this approach (List 12.3.1).

As an example, consider GISTs, described earlier in Chapter 8. In GIST tumors, mutations may occur in either the c-KIT or the homologous kinase platelet-derived growth factor receptor-a gene (PDGFRA) oncogenes. A drug that is active against the altered protein product of one mutated gene may not be active against all of the variant mutant forms of the gene or of mutations in alternate genes associated with the same tumor. This means that even when an agent is targeted against a particular mutant oncogene, it may not be active against all of the naturally occurring variants of oncogenes known to produce the tumor (199).

It actually makes much more sense to target the normal pathway that is dysregulated by a cancer-causing mutation. The next generation of effective anticancer agents will target normal pathways, not mutated cancer genes (List 12.3.2).

List 12.3.1 Reasons that drugs will not be tailored to mutations.

A single type of cancer may result from one of many distinct molecular defects.

A single neoplasm may be associated with a dysregulation of a single gene, but this gene dysregulation may be achieved through different molecular mechanisms.

During tumor progression, cancers acquire new genetic alterations involving heterogeneous subclones. A treatment that targets a particular mutated gene in a subclone of cells may have no effect on other subclones in the same tumor.

It takes about a billion dollars to bring a new drug to market. Although cancer is a common disease, the occurrence of all but a few common tumors is rare. The economics of drug development will never allow companies to develop drugs that target the individual unique gene mutations that may characterize a fraction of the occurrences of rare tumors.

List 12.3.2 Reasons that drugs will be tailored to pathways.

Tumors are composed of cells that have a common developmental lineage. The pathways that are activated by oncogenes are pathways that are normal and preexistent, whose regulation is altered as the pleiotypic response to the activated oncogene.

We know that drugs can modulate the activity of pathways by targeting one or more of the normal enzymes of the pathways.

Tumors that are morphologically and biologically identical, even if associated with different altered genes, are likely to achieve their phenotype with the same pathways (199).

Even if there are multiple pathways that account for the malignant phenotype of a cell, it may be possible to modulate the activity of these pathways using drug combinations.

Drugs that are active against a normal pathway in a tumor are likely to be active against that pathway in all of the cells of the tumor.

Drugs that are active against a normal pathway in a tumor are likely to be active against that pathway in tumors of the same developmental lineage, because pathways are, to some extent, linage specific.This means that drugs that target pathways may have generalized utility, alone or in combinations with other pathway-specific drugs, in classes of tumors.

If pathways will be important targets for the next generation of chemotherapeutic drugs, should we not classify tumors by the pathways that they share? Two considerations make this approach currently untenable. The first problem relates to the definition of a pathway. The term "pathway" means something different today than it meant a few decades ago. According to traditional thinking, a pathway was a sequence of biochemical reactions involving a specific set of enzymes and substrates that produced a chemical product. The classic pathway was the Krebs cycle. It was common for students to be required to calculate the output of the cycle (in moles of ATP) based on stoichiometric equations employing known amounts of substrate. Today, in the field of cancer research, pathways operative in tumors usually involve observed linked actions in receptors, activators, enzymes, and sites in macromolecules. Their products are often a cellular action, rather than a chemical product. These pathways do not typically occur in an anatomically sequestered area of the cell. The pathways may interact with other pathways, and their direction and biological consequences may be complex and variable among different cell types or different physiological states of one cell type. Still, the term "pathway" is a convenient conceptual device to organize classes of molecules that interact with a generally defined set of partner molecules to produce a consistent range of biological actions. Because pathways in cancer cells are opportunistic creations, it may prove impossible to create a useful classification of tumors based on shared pathways.

The second problem relates to the complexity and heterogeneity of tumor cells. As tumors grow, they become more and more complex. It is likely that a heterogeneous population of genomically unstable cells will not have a single constitutive pathway for its cancer phenotype.

Chapter 12 Summary

Cell types that share a common functionality may be otherwise unrelated. The most striking example is the endocrine tumors, all of which secrete biologically active molecules into the blood stream. The endocrine tumors do not all secrete chemically related molecules, and they do not all have a common developmental lineage. Familial endocrine tumors are found to occur in syndromes that include nonendocrine tumors that share the same developmental lineage as the included endocrine tumors (e.g., MEN). These observations indicate that a neoplasm classification cannot be sensibly organized by grouping tumors by shared biological functions.

Pathways within cells may be shared among types of cells that have a common developmental lineage. Classifying tumors by their contained pathways may help us find new chemotherapeutic agents that will treat a variety of tumors that employ the same targeted pathways.

CHAPTER

13 Molecular Classification of Neoplasms

An ounce of gene expression is worth 200 pounds of pathologist.

—Anonymous

13.1 Background

If you speak with just about any molecular biologist and ask how tumors should be classified, he or she will answer that tumor classification should be based on tumor genetics. On the face of it, the arguments for a genetic classification of neoplasms are overwhelming (List 13.1.1).

List 13.1.1 Reasons for a genetic classification of cancer.

1. Genetic research has yielded many instances of cancers that have a characteristic genetic lesion that determines the phenotype of the tumor.
2. The genetic findings in cancers have led to a deepened understanding of tumor biology that was not accessible through the morphologic classification of tumors.
3. Gene expression studies have successfully subcategorized tumors of a given diagnostic type into clinically distinct groups that respond characteristically to different treatment protocols.
4. Gene expression studies have detected new pathologic variants of neoplasms for which no morphologic distinction could be made or for which morphologic distinctions were occasionally incorrect.
5. There is good reason to hope that future advances in the genetic characterization of tumors will clarify our understanding of every different type of cancer.
6. Traditional pathologic classifications have failed to predict observed clinical differences among tumors of identical morphologic appearance.

The purpose of this chapter is to show that, despite these considerations, a classification based on the genetic properties of tumors is highly problematic and is not currently feasible.

13.2 Fundamental Properties Are Not Always Good Classifiers

The physical properties of matter determine the qualities of all materials. If I have a collection of model cars, I could develop a sensible classification based on the materials that compose the model. I could put all of the iron cars in one class, the plastic cars in another, the wooden cars in a third, and so on; however, the class of iron cars and the class of plastic cars may each contain a replica of a 1967 Ford Mustang. The class of wooden cars may contain only model trucks, and the plastic class may contain one truck and several motorcycles. Although the strength and melting points and flammability of each car are determined by its composite materials, the purpose of each object might be better determined by the make and model of vehicle. This is why people who classify objects search for essential properties that relate to the purpose of the classification.

13.3 Limitations of a Molecular Classification of Neoplasms

Because neoplasms result from genetic aberrations, a classification based on molecular biology will encapsulate the fundamental properties of cancer but may fail to produce a sensible, nonredundant hierarchical structure. There are many reasons that tumors cannot be classified by their molecular alterations (Lists 13.3.1 and 13.3.2).

List 13.3.1 Arguments against a molecular classification of cancer.

1. The genetic changes in tumors are complex and may not be amenable to simple genetic classification. An advanced cancer may contain thousands of different genetic and epigenetic alterations.
2. The genetic features of any cancer change over time, during the steps of carcinogenesis and during tumor progression. Because the genetic features of a tumor may change over time, it may be impossible to determine a time-invariant set of genetic features for every tumor of humans, which would be necessary to build a tumor classification.
3. A single type of cancer may result from one of many distinct molecular defects. For instance, breast cancer may result from inherited defects in p53 (Li Fraumeni syndrome) or PTEN (Cowden syndrome) or STK11 (Peutz-Jeghers syndrome). MYC amplification may be seen in some neuroblastomas and may even correlate with prognosis, but it is not seen in all neuroblastomas

(continues)

List 13.3.1 (continued)

and is seen in tumors other than neuroblastoma. Colon cancer can be associated with any of several distinct molecular classes of tumor suppressor genes, all of which seem to confer increased genetic instability (25). Likewise, angiomyolipomas are characteristic of **von Hippel Lindau** syndrome and of tuberous sclerosis, each with a distinct genetic cause. This means that classes of cancers will not have a specific molecular alteration as a shared feature.

4. A single neoplasm may be associated with a dysregulation of a single gene, but this gene dysregulation my be achieved through different molecular mechanisms. For example, a Burkitt lymphoma is characterized by dysregulation of the c-MYC gene that controls lymphocytic proliferation and cell death. In Burkitt lymphoma, heightened c-MYC activity can be achieved by translocating the c-MYC gene to a highly expressed Ig locus, by mutating c-MYC regulatory regions, or by mutating a transactivation domain rendering the gene refractory to normal modulation. This means that even when a neoplasm is associated with a single gene, the association may have many different genetic mechanisms.
5. An altered gene may be characteristic of many different types of cancer. People with inherited defects of p53 (Li-Fraumeni syndrome) develop tumors of many different tissues (List 13.3.2). Mutations involving the ret gene are found in medullary carcinoma of the thyroid and in papillary carcinoma of thyroid, but the specific mutations of the ret gene are distinct and different for these two types of tumors. Medullary carcinoma is unrelated to papillary carcinoma of thyroid, except for an accident of anatomic proximity. FUS has been found in Ewing's sarcoma fused with the ERG gene, but a similar FUS-ERG fusion has been described in some acute myeloid leukemias. A FUS-CHOP fusion gene is found in myxoid liposarcoma. This means that classes of molecular alterations will not have a specific tumor as a shared feature.
6. Some characterizing molecular alterations are simple accidents of gene proximity. For instance, aniridia sometimes accompanies inherited Wilms tumors. In these cases, the chromosomal deletion that causes aniridia is the same chromosome deletion that causes Wilms tumor. Surprisingly, aniridia has no biological properties in common with Wilms tumor! The explanation is that co-occurrences of aniridia and Wilms tumor result from a large chromosomal deletion that happens to include both the aniridia gene and the Wilms tumor gene. If humans were designed with the aniridia gene and the Wilms tumor gene on different chromosomes, the association between aniridia and Wilms tumor would not occur. If two lesions share a genetic cause but lack common biological properties, a genetic classification of neoplasms will not make biological sense.

(continues)

List 13.3.1 (continued)

7. Some gene products are involved in complex interactions with other proteins, and their effects may involve many different tissues with no discernible pattern. An example is the von Hippel Lindau (VHL) protein. The VHL protein has regulatory functions for many pathways and interacts with many different protein factors and in the regulation of transcription of many different genes (200). It is not surprising that VHL gene familial syndromes produce a wide spectrum of neoplasms crossing different developmental lineages (e.g., hemangioblastomas, renal cell carcinomas, and pheochromocytomas). In general, oncogenes produce pleiotypic responses, and the response to a specific oncogene activation may differ from cell type to cell type. If mutations are associated with complex changes in other gene products, a classification based on mutations will not yield a simplified view of neoplasms.
8. Despite rapid advances in the past decade, we have only scratched the surface of knowledge pertaining to the genetic basis of neoplasia. It will take decades before we can examine all of these tumors using genetic techniques. Without detailed molecular data characterizing every tumor, it is impossible to assign a tumor to any particular molecular class. This means that a molecular classification will contain thousands of unclassified neoplasms.

List 13.3.2 Biologically distinct tumors may be associated with the same genetic abnormality. Example: the Li-Fraumeni syndrome (81).

Breast cancer
Soft-tissue sarcomas
Osteosarcomas
Brain tumors
Acute leukemias
Adrenocortical carcinomas
Lung adenocarcinoma
Colon cancer
Pancreatic cancer
Prostate cancer
Wilms tumor
Phyllodes tumor

List 13.4.1 Generalizations on the genetics of neoplasms.

1. The simplest dichotomy for all human tumors reduces to epithelial tumors (carcinomas) and nonepithelial tumors (sarcomas). Epithelial tumors are characterized by complex changes of the genome at the level of specific genes (multiple mutations in genes and gene products) and at the level of the chromosome (complex karyotype).
2. The sarcomas tend to be less genetically complex than carcinomas and can often be characterized by translocations that fuse a transcription factor activation domain to a transcription factor DNA-binding domain (201). This generalization extends to all of the tumors of nonepithelial origin, including pediatric sarcomas and pediatric primitive tumors. It also applies to benign tumors of mesenchymal origin, which seem to be characterized by the simplest of genetic aberrations.
3. Molecular studies of tumors may provide taxonomic features that can separate a tumor into subtypes that may have distinctive morphologic or clinical features. For example, angiolipomas have a normal karyotype, whereas spindle-cell lipomas, pleomorphic lipomas, hibernomas, and lipoblastomas have characteristic clonal chromosomal aberrations. As a general rule, different karyotypes among tumors that were once considered morphologic variants of the same tumor might suggest a different approach to diagnosis and therapy for each genetic variant (58).
4. Genetic studies can clarify the relationships among tumors. For example, the fusion gene ETV6-NTRK3 is found in infantile fibrosarcoma. The identical fusion gene is found in cellular variant of congenital mesoblastic nephroma (and absent from the classic variant of congenital mesoblastic nephroma). This finding suggests that these two tumors are closely related if not equivalent (202).

13.4 A Genetic Classification for the Major Categories of Human Tumors

A few general observations seem to apply to the genetics of tumors (List 13.4.1).

13.5 Complexities in the Genetic Approach to Cancer Classification

One of the simplest and most appealing theories of cancer is that the malignant phenotype is acquired by accumulating mutations in oncogenes and tumor suppressor genes. There is a tendency to think of oncogenes and tumor suppressor genes as func-

List 13.5.1 Generalizations on tumor oncogenes.

There is a tendency to have a short latency period following oncogene activation.
Oncogenes are more involved with tumor initiation than with tumor progression.
All oncogenes have homologs in animals.
The tumors associated with homologous oncogenes genes in animals bear a striking resemblance to the tumors produced by the human oncogene.
A mutation in one allele is sufficient to account for a neoplastic phenotype.
Oncogenes are captured in acutely transforming oncogenic viruses.
Oncogenes may produce extremely high incidences of tumors in persons who inherit the activated oncogene. For example, it is estimated that 90% of MEN2 (multiple endocrine neoplasia type 2) carriers will eventually show evidence for medullary thyroid cancer (203).
Several different mechanisms of oncogene activation have been described. For example, in the c-MYC oncogene in Burkitt lymphoma, heightened c-MYC activity can be achieved by translocating the c-MYC gene to a highly expressed Ig locus, by mutating c-MYC regulatory regions, or by mutating a transactivation domain rendering the gene refractory to normal modulation.
As discussed in Chapter 2, oncogenes tend to produce tumors in a very narrow range of cell types.

tionally equivalent, almost interchangeable terms, the only difference being that oncogenes cause cancer by being activated and tumor suppressor genes cause cancer by losing activity. It would be similar to the idea that being "on" is the same thing as being "not off."

Tumor suppressor genes have very different biological roles than oncogenes, and the expected consequences of tumor suppressor gene inactivation are different from the consequences of oncogene activation (Lists 13.5.1 and 13.5.2). Because each different oncogene and tumor suppressor gene can be altered by different types of genetic mutations, and because any gene in one individual may be subtly different from the same gene in another individual, and because the activities of genes vary from cell type to cell type, it is impossible to distill the behavior of oncogenes and tumor suppressor genes into a few biological rules; therefore, a classification based on the oncogenes and tumor suppressor genes that are found in a tumor may fail to yield a simple conceptualization of the tumor's biological properties.

Tumor suppressors, unlike oncogenes, are sometimes interchangeable within different specimens of the same tumor. A good example is **oligodendroglioma**, which is associated with several different genetic markers that characterize variant forms of the same tumor (List 13.5.3).

List 13.5.2 Generalizations on tumor suppressor genes.

Tumor suppressor genes have recessive-type expression (both copies of gene must be inactivated)

They tend to precede the appearance of activated oncogenes.

They tend to be involved in cell cycle arrest or control, apoptosis, DNA repair, genomic stability, or epigenomic stability.

They have less lineage specificity of the resulting tumors than is seen with oncogenes.

They are somewhat interchangeable, with one tumor suppressor gene substituting for another in variant forms of the same tumor (25, 204).

They tend not to be specific for a single type of tumor (205). For example, the Li-Fraumeni syndrome is caused by a germline mutation in the p53 gene. The tumors that occur in probands run the range of developmental lineages and include brain tumors, sarcomas, adenocarcinomas of breast, colon, prostate and pancreas, and Wilms tumors.

They are associated with increased sensitivity to chemical carcinogens.

A loss of tumor suppressor genes increases the likelihood (not the certainty) of developing cancer (205).

They have a long latency period (i.e., there are no acutely transforming tumor suppressor genes).

There seem to be many more tumor suppressor genes than tumor oncogenes.

Malignant tumors tend to increase their number of tumor suppressor inactivations during tumor progression.

Familial syndromes characterized by the inheritance of tumor suppressor gene mutations are usually associated with the occurrence of tumors in adults (RB tumor suppressor gene is an exception).

Familial syndromes characterized by the inheritance of oncogene mutations are usually associated with the occurrence of tumors in infants and children (e.g., the EWS fusion gene tumors).

List 13.5.3 Tumor suppressor losses associated with specific chromosome marker variants of oligodendroglioma.

1. 9p—p16 (206)
2. 10—PTEN (204)
3. 17p—p53 (207)

List 13.5.4 Germline tumor suppressor genes found in breast cancers (in adults).

Li-Fraumeni syndrome due to germline mutations in p53
Cowden syndrome due to mutations in the PTEN gene
Peutz-Jegher syndrome due to mutations in the STK11 gene
Ataxia-telangiectasia due to ataxia-telangiectasia mutated gene (ATM)
Bannayan-Riley-Ruvalcaba syndrome due to PTEN gene

It seems as though it does not matter which tumor suppressor has a role in oligodendrogliomas (p16, PTEN, p53). Interchangeable tumor suppressor genes among the variant types of oligodendrogliomas may contribute to the assortment of different cytogenetic markers found in these tumors.

Similarly, several different tumor suppressor genes are found in inherited conditions that are associated with high rates of breast cancer (List 13.5.4).

These findings would indicate that tumor suppressor genes have a very important role in carcinogenesis, but that a particular tumor suppressor gene is seldom necessary for the development of any particular neoplasm.

13.6 Proteomic Classification of Cancer

Despite the emphasis on cancer genetics, it is worth remembering that genes exert their effects through the creation of proteins. It is the collection of proteins within the cell that is responsible for every cellular process and every physical structure. It would seem self-evident that the malignant phenotype must be caused by a specific collection of proteins that is not representative of normal cells.

The profile of proteins within a cancer cell may include altered proteins that are not found in normal cells, or it may include a balance of proteins that favor growth, invasion, and metastasis that would not be found in normal cells. Wouldn't you think that it would be a simple matter to find a profile of proteins that is specific for cancers?

As usual, the problem has been that in cancer cells, everything changes. Every cellular pathway, every metabolic product, every organelle, and every cellular structure is changed in the cancer cell. Also, every type of cancer is different from every other type of cancer in its cellular phenotype. When you look for altered proteins in cancer cells, you find more changes than you can enumerate, and you cannot distinguish the changes that are important from the changes that are inconsequential. You encounter splice variants galore, and you encounter proteins that are observed in normal embryonic and fetal cells but not normally observed in any differentiated cells.

List 13.6.1 Requirements for a proteomic classification of neoplasms.

1. Every type of neoplasm would need to be examined.
2. The different stages of development of each neoplasm would need to be examined (e.g., an atypical nevus may have a very different profile than a metastatic melanoma).
3. The physiologic alterations that occur in a tumor would need to be examined to determine its effect on the tumor's protein profile (e.g., the protein profile of ischemic tumor cells might need to be compared with the protein profile of oxygen-rich tumor cells located adjacent to a vessel, and slow-growing tumor cells in the center of a mass might need to be compared with fast-growing tumor cells at the periphery).
4. The protein profiles of noncancerous but abnormal cells would need to be compared against the protein profiles of tumor cells to ensure that diseased cells do not have a profile that can be confused with cancer cells (e.g., fast-growing repair cells might be compared with tumor cells of related cell types).

Despite this overabundance of proteomic changes in cancer cells, a new subfield has emerged: cancer proteomics. The basic belief that fuels this new field is that the protein profiles of tumors differ considerably from the protein profiles of normal cells. Tumors secrete and leak proteins directly into the blood. When tumor cells die, some of the proteins in the tumor cells will also pass into the blood. If we can examine blood for all of the protein species, we will find that people who have tumors will also have proteins in their blood that are not present in people who do not harbor tumors. In some instances, the profile of proteins in the blood will identify the tumor that produced the proteins. Based on this belief, which is backed by some credible early observations in ovarian cancers, there is currently underway a new effort to classify tumors by their proteomic profiles.

A classification of cancer based on proteomic profiles would be a huge undertaking (List 13.6.1).

13.7 Epigenomic Classification of Cancer

In a fascinating article, mentioned briefly in the Introduction, Dr. Manel Esteller, a leading scientist in the field of epigenomics, called on the cancer research community to launch an International Human Epigenome Project (4). The reasons for supporting such an effort follow (List 13.7.1).

With all of these observations, should we support an international effort to build an epigenomic database of all human cancers? With such a a cancer epigenome database, every cancer could be identified by its epigenomic signature, and tumors with similar signatures could be treated with drugs that target specific epigenomic alterations.

Well, there are many problems with this approach. Before we decide, we should review some of the most important advances in epigenomics.

List 13.7.1 Reasons for supporting a Human Epigenome Project.

1. Cancers have an epigenome that is markedly different from the epigenome of normal cells.
2. Epigenomic alterations can modify the phenotype of cells, and these modifications can be inherited in much the same way that changes in the genome can be inherited.
3. Inherited epigenomic changes are known to play a role in cellular differentiation, and there are examples of epigenomic changes that cause loss of function in tumor suppressor genes.
4. Cancers seem to have epigenomic signatures that are unique for each cancer.
5. Epigenomic changes may be important in tumor progression.
6. Agents that modify the epigenome have importance as cancer chemotherapeutic agents.

List 13.7.2 The currently known mechanisms for epigenomic modification of phenotype (modified from [4]).

DNA Methylation
Histone modification
Chromatin remodeling

There are several mechanisms by which the expression of genes can be modified without actually altering the structure of DNA and without using genetic control loci, such as promoters, or molecular agents that modify transcription, such as transfer factors and micro-RNAs (List 13.7.2).

DNA methylation is the best understood mechanism for epigenomic control of DNA transcription. Methylation occurs at CpG (cytosine guanosine dinucleotides) throughout the genome. Approximately 3% to 4% of cytosines are methylated in normal DNA. **CpG islands** are concentrations of CpG dinucleotides that have a GC content over 50% and that range from 200 base pairs to several thousand base pairs in length. There are about 29,000 to 50,000 CpG islands, and most of these are associated with a promoter (208).

DNA transcription usually proceeds when you have a promoter region (for a gene) that is associated with an unmethylated CpG island; when there are appropriate transcription factors available near the **promoter site**; and when the histones attached to the DNA are in a permissive state.

If the CpG island is methylated, transcription of the gene is usually stopped. Methylation of CpG islands is a powerful mechanism for closing off transcription of selected genes (List 13.7.3).

Gene suppression occurs more often within the genome than does gene expression. More than a third of the human genome is filled with unwanted gene passengers

List 13.7.3 Gene control functions of CpG island methylation.

1. During cellular differentiation to constrict cells to a lineage-specific behavior.
2. During imprinting, to restrict the expression of particular alleles.
3. During x-chromosome inactivation to restrict the expression of a redundant sex chromosome.
4. As a mechanism of suppressing excess genetic baggage collected over the eons.

(including endogenous retroviruses and **transposons** and repetitive sequences), inexplicable amplified genes, and genes that have been duplicated for no apparent reason in some ancient ancestral cell. Many of the alternate variants of currently expressed gene products are now suppressed. It seems as though humans would benefit from a genetic yard sale in which we could strip down our genome to its bare essentials.

Other species that are just as complex as humans (based on their contained cell types, organ architecture, anatomy and physiologic pathways) have much less DNA. *Drosophila* cells have about 120 million base pairs. The fugu fish gets by with just 365 million base pairs. Humans carry about 3,000 million base pairs of DNA.

Our cells could not function if they did not have efficient methods of suppressing genes. In the cancers that have been studied, there is a massive genome-wide demethylation accompanied by hypermethylation of specific CpG islands. The pattern of hypermethylation in specific CpG islands may be characteristic for particular types of tumors (209, 211).

There is conflicting evidence supporting both causative and curative roles for hypomethylation in cancer. Mice with germline mutations of the APC tumor suppressor gene have a high incidence of colon tumors. If these mice are provided with a mutation in the methyltransferase gene, resulting in decreased genome-wide methylation, their incidence of colon tumors decreases, suggesting a protective effect of hypomethylation (212). Furthermore, DNA hypomethylating agents are useful treatments for myelodysplasia. What is the role of hypermethylation in cancer? We know that hypermethylation of an allele can suppress transcription and that this can suffice to cause the loss of function of a tumor suppressor gene where another allele is inactivated through mutation. This occurs in some cases of Von Hippel-Lindau disease (211). Hypermethylation of CpG islands is a suggested mechanism for colon carcinogenesis. A subset of human colon cancers has a high frequency of methylated CpG islands (the so-called CpG island methylator phenotype, or CIMP). Colon tumors with the CIMP epigenotype are associated with mutations of the BRAF oncogene (215). This suggests that the hypermethylation of specific CpG islands may enhance the carcinogenic process.

As noted previously, cancer cells are different from normal cells by almost every measurement of every cellular property. It is not surprising, or particularly informative, to find differences in genome methylation (increases and decreases) in cancers. To my way of thinking, it makes little sense to launch a massive Human Epigenome Project until the role of epigenomic modification in tumors is better understood.

13.8 Is a Human Tumor Genome Project Feasible?

> "NCI [the National Cancer Institute] and NHGRI [National Human Genome Research Institute] have agreed to examine the best approaches to achieve the goal of building a comprehensive collection and analysis of mutations found in human cancers. Through a multi-phase approach, NCI and NHGRI are examining the major issues and barriers associated with compiling a complete catalog of the genetic changes that characterize cancer (216)."

Early evidence indicates that the genome of cancer is incredibly complex. Loci for functional gene classes may have thousands of different mutations in different cancers (217). Likewise, the expression of genes differs enormously between tumor cells and their normal cellular counterparts (218). Furthermore, cancers splice exons differently than normal cells. A relatively small number of alternatively spliced exons can yield thousands of different transcripts (219).

Putting aside the many genomic and epigenomic alterations that occur in a tumor, when you consider that tumors change their genome throughout carcinogenesis and tumor progression, and that there are many thousands of different tumors and morphologic variants of tumors, "compiling a complete catalog of the genetic changes" seems an impossible pursuit.

If there were a classification of human cancer, with a few dozen major classes of cancer, the task might be feasible by narrowing attention on a few types of tumor from each major class of cancers.

Chapter 13 Summary

Molecular biology provides important information about tumors and has helped us to identify characteristic genetic alterations among tumors that lack a distinctive morphologic appearance. However, a method that identifies new taxonomic species will seldom suffice as the central principle upon which a classification is built. Currently, we have found diagnostic molecular markers for only a small fraction of the different tumor types. It is unlikely that molecular biology will form the exclusive basis for a classification of all neoplasms.

CHAPTER

14 Classification by Developmental Lineage Is Best

Omnes exeunt in mysterium (All Things Lead into Mystery)

—Anonymous

14.1 Background

In the mid 20th century, tumor biologists developed a very interesting theory that seemed to explain why cancers occur in humans. It was observed that Urodele organisms (newts and salamanders) have a remarkable ability to regenerate tissues (and limbs) after injury. It was further observed that when newts are injected with carcinogens, they develop accessory, newly regenerated limbs at the site of injection (88, 89). This observation led to the following hypothesis: lower animals can regenerate limbs. Higher animals (i.e., humans) cannot. When a lower animal is exposed to a carcinogen, it responds by forming a new organism from blastemal cells. When a human is exposed to a carcinogen, it tries unsuccessfully to form a new organism from blastemal cells. In a failed attempt at regenerating a limb, humans produce a poorly organized uncontrolled growth, which we call cancer. This theory is sometimes called the blastemal theory of cancer. Robert Bolande, from the 1970s through the 1990s, published a series of brilliant articles that explained and updated relationships between embryonic development, **teratogenesis**, and carcinogenesis (220–222).

According to the blastemal theory, all human cancer is just the **forme fruste** (wrong form) of urodele blastemal regeneration. Multiple observations of tumors seemed to confirm this hypothesis. As we have already seen, virtually every lineage of cell development in humans is represented by a tumor. Tumors of primitive cells may differentiate in multiple directions, and they conform to a pattern of differentiation that mimics normal lineage ancestry.

This blastemal theory of human cancer was popular for much of the 20th century. It ran into some problems in the past 30 or 40 years (List 14.1.1).

List 14.1.1 Observations that do not fit well with the blastemal origin of cancer.

We now know that all animals (including urodeles, other amphibians, and even insects) and trees can develop tumors.

Virtually any type of proliferating cell, when treated with carcinogens in tissue culture, can be transformed into aneuploid cells that produce malignant tumors when injected back into the host animal. A blastema is not required.

Some cells can be transformed by activation of a single oncogene (again, no blastema is required).

Today, most cancer researchers are focused on the molecular events that lead to cancer, and this almost always involves finding altered cancer genes in the different types of human cancers.

As discussed in Chapter 13, a classification of neoplasms cannot be based on molecular alterations found in tumors. The purpose of this chapter is to propose a classification for neoplasms based on developmental lineage.

The fundamental hypothesis underlying the developmental lineage classification is that tumor cells achieve the neoplastic phenotype using a set of pathways and physiologic processes that are determined by a cell's lineage and that are available to every type of cell within a cell lineage. The lineage of a cell is determined by the hierarchy of epigenomic modifications that occur through every stage of normal development after conception.

Accepting this fundamental hypothesis permits us to construct a classification with some very useful features (List 14.1.2).

List 14.1.2 General features of a developmental lineage classification of neoplasms.

The developmental lineage of every tumor is comprehensive and hierarchical, providing a true classification of entities that belong to a system of coherent classes and subclasses (i.e., a formal classification).

A classification based on developmental lineage does not change the currently used taxonomy of neoplasms. Pathologists will not need to change the names of tumors diagnosed in the past, present, or future.

A classification based on developmental lineage can be easily understood and used by developmental biologists.

A classification based on developmental lineage can be annotated with genetic and other biological and clinical data to yield taxonomically distinguishable subclasses of tumors.

(continues)

Lis 14.1.2 (continued)

Organs may have multiple embryologic derivatives (e.g., skin contains tissues of ectodermal, neuroectodermal, and mesodermal lineages), but any given cell has only one lineage. This means that a lineage-based classification can assign any tumor to a unique position within the classification.

The lineage of a cell never changes. A tumor that is assigned a developmental lineage will never change into a tumor of a different cell lineage. This feature satisfies the nontransitive property of class membership.

For the most part, tumors have a cell developmental stage. For instance, blastomas are thought to arise from a cell type that precedes organ differentiation. Squamous cell carcinomas of skin are tumors that have features of cell type that developed from an ectodermal cell capable of post-mitotic differentiation (into a non-dividing keratinized cell). An embryologic approach permits us to assign a cell type and a developmental stage to tumors.

A classification based on developmental lineage is relevant to the behavior of the tumor. For example, cancers of the classes endoderm and ectoderm metastasize via lymphatics. Cancers of class mesoderm tend to metastasize by **hematogenous dissemination**. Mode of metastasis is an example of a behavioral property that is also a class property.

A lineage classification is organism independent because the class structure is valid across animal species that share the same developmental lineages. All mammals share the same developmental lineage classes (endoderm/ectoderm, mesoderm, neuroectoderm, neural crest, germ cell, and **trophectoderm**), and all mammals develop the same classes of tumors.

14.2 Lineage Can Be Determined with Certainty

Determining the embryonic lineage of cell types has been one of the chief activities of developmental biologists, and authoritative works describing the histogenesis of virtually every cell type of the body are available (223, 224).

New molecular techniques have improved the process of assigning cell lineage (225). A recent article by Frumkin and coworkers has shown that a cell's lineage can be determined by tracing **microsatellite** variability inheritance arising through acquired mutations after conception (226).

In Chapter 8, we described the strange history of GISTs (gastrointestinal stromal tumors). These tumors have features that seem to cross the boundary between smooth muscle tumors and Schwann cell tumors. Recent studies have shown conclusively that GISTs arise from the interstitial cells of Cajal (ICCs). ICCs have autonomic activity and muscular activity, both contributing to gut peristalsis. Functionally and anatomically, ICCs have some features of neural crest origin and some features of mesenchymal origin. Recent work by Young indicates that ICC cells have a purely mesenchymal origin

(227). ICC cells grew in gut embryonic organ cultures established at a moment in embryological development that preceded the arrival of neural crest cells. The cultured ICC cells and smooth muscle cell precursors expressed KIT, a marker for GIST tumors. During organ culture, fully differentiated smooth muscle cells downregulate KIT, whereas the differentiating ICC cells maintain KIT expression (227). In the case of GIST, and in the case of every other carefully studied tumor, a single developmental lineage can be determined.

14.3 Developmental Lineage Pathways Are Often Preserved in Tumors

Well-differentiated neoplasms sometimes function much like normal cells. For example, in chronic myelogenous leukemia, the neoplastic neutrophils that circulate in the blood perform many of the physiologic activities that are observed in normal neutrophils.

In addition to behaving like normal differentiated tissues, neoplasms sometimes behave like their developmental ancestors. For example, hepatocellular carcinomas typically secrete alpha-fetoprotein, an albumin-like protein normally secreted by fetal hepatocytes. Colon cancers will often secrete carcinoembryonic antigen, a protein that is synthesized in the fetal colon but that is not synthesized to any degree by normal adult colon cells.

Occasionally, we see neoplasms that yield atavistic cell types consistent with their particular developmental lineage. Congenital hemangiomas are a common lesion found at birth and occur in about 10% of infants. These tumors can grow rapidly after birth, but they tend to regress over the next few months or years. In most cases, they have no clinical significance. Ritter and coworkers examined congenital hemangiomas and found that they contained myeloid precursor cells. Furthermore, the endothelial cells that composed the hemangiomas co-expressed immunohistochemical markers for endothelium and for myeloid cells (228).

Why would an infantile hemangioma have myeloid differentiation? Although adult myeloid cells (erythrocytes, neutrophils, monocytes, and lymphocytes) come from bone marrow elements, fetal myelopoiesis has a different anatomic source. The earliest myeloid cells in the embryo derive from primitive vessels called blood islands (List 14.3.1).

List 14.3.1 Embryology of vasculature, from *Nomina Embryologica Veterinaria* (224).

- Textus angioblasticus
  - Insulae sanguineae (Blood island)
    - Endothelioblasti
    - Hemocytoblasti (Haemocytoblasti)

It may seem out of the ordinary for a hemangioma to produce blood cells, but the behavior of a congenital hemangioma is consistent with its normal developmental lineage. A primitive vascular tumor behaves like an embryonic blood island, producing capillary endothelium and hematopoietic cells.

14.4 Are There Examples of Tumors of Unknown Lineage?

The developmental classification is designed on the assumption that every tumor of the body derives from a cell with a single developmental lineage that determines a set of common properties shared among all of the members of the same developmental class.

In general, neoplasms do not have a mixed developmental lineage. Even multipotent teratomatous tumors derive from a single cell type, the primordial germ cell.

Some **blastomas** are composed of more than one cell type (e.g., hepatoblastoma, nephroblastoma, and pancreatoblastoma). Here, the lineage of the tumor is capable of differentiating along several **phenotypic pathways**. The phenotype may be diverse, but the lineage is unmixed.

It is likely that every tumor begins as a clone of single lineage. In the process of tumor progression, some tumors undergo enormous genetic and epigenetic instability. It is theoretically possible that the genome can be altered sufficiently to convert a cell's phenotype to that of a different lineage. For example, there are very rare cases of gastric adenocarcinomas (endodermal origin) that seem to convert into choriocarcinomas (trophectodermal origin) (229). In these cases, it is possible that during tumor progression, a mutation occurred in a **homeotic gene** (controlling tissue type).

The occurrence of tumors with a true, mixed lineage is exceedingly rare and seems to occur only during the progression phase of established cancers.

14.5 Are There Tumors That Arise Through Transdifferentiation?

Transdifferentiation is a biological phenomenon whereby a cell changes its morphology and functionality from one type of cell to another. If a retinal photoreceptor was to become a ductal cell lining a tear duct, that would be an example of transdifferentiation. Most people restrict the use of the word transdifferentiation to changes that exceed normal modes of differentiation. During renal organogenesis, when a metanephric mesodermal cell condenses around a ureteric bud to produce a renal epithelial cell, that process should not be considered transdifferentiation because it is the normal and obligatory pathway by which the kidney is created. In this book, the word transdifferentiation refers to a process by which a cell becomes a type of cell that lies outside its normal developmental lineage.

The basic premise of the Developmental Classification is that tumors derive from a lineage-appropriate progenitor cell and use cellular pathways that are common to the members of the lineage. If transdifferentiation were a common phenomenon in normal and cancer cells, a classification based on developmental lineage would have no value because tumors could change from one lineage-based class to another. It should be of no surprise that I tend to discount most scientific accounts of transdifferentiation.

Many reports of transdifferentiation in tumor cells should be interpreted as lineage-faithful **metaplasias**. When an endometrial carcinoma develops features of carcinosarcoma, this is not transdifferentiation. The uterus is derived entirely from mesoderm, and malignant mesodermal cells can be differentiated into either epithelial cells (of endometrium) or mesenchymal cells (of myometrium) without violating their lineage (Chapter 17). When a lung cancer is shown to contain cells of glandular origin admixed with cells of squamous origin and cells of neuroendocrine origin, this is not transdifferentiation because all three cell types are now known to derive from foregut (endodermal) epithelium (Chapter 5).

Observations in two recent articles seem to suggest that normal stem cells have enormous plasticity and can transdifferentiate across lineages. Houghton and coworkers report that gastric epithelium in C57BL/6 mice derives from bone marrow stem cells. In conditions where the stomach is exposed to chronic inflammation, the authors have shown that bone marrow stem cells repopulate the gastric epithelium. These bone marrow-derived stem cells may subsequently undergo metaplasia, dysplasia, in situ cancer, and eventually invasive gastric adenocarcinoma (230). Because bone marrow stem cells are of mesenchymal origin, this would seem to imply that mesenchyme may give rise to endoderm, which can give rise to an adenocarcinoma. Other workers, including Theise and coworkers, have shown in humans that endodermal hepatocytes may also derive from bone marrow stem cells (231).

These observations, taken at face value, would indicate that leukemias and adenocarcinomas share the same lineage, a conclusion that undermines the Developmental Lineage classification of neoplasms. Other work in the same field has shown that the bone marrow is a home for stem cells of non-mesenchymal origin (232–234). An endodermal stem cell may live in the bone marrow and migrate to the stomach or the liver to provide gastric enterocytes and hepatocytes. When this occurs, lineage fidelity is preserved, and transdifferentiation does not occur.

In Chapter 15, we discuss the strange biology of malignant **rhabdoid tumors**, which seem to regress to a pluripotent state after loss of the tumor suppressor gene INI1. This tumor may represent a rare exception to the rule forbidding transdifferentiation in tumors. Only a few dozen rhabdoid tumors occur in the United States each year, and their presence does not have much influence on the utility of the Developmental Lineage Classification of Neoplasms; however, it serves as a reminder that biological systems are complex and that tumors are unpredictable.

Chapter 14 Summary

Neoplasms do not "invent" their own phenotype. Neoplasms use the existing pathways that are permitted within their developmental lineage. This feature permits us to use developmental lineage as the primary classifier for neoplasms.

CHAPTER 15

The Six Major Classes of Neoplasms

I learned very early the difference between knowing the name of something and knowing something.

—Richard Feynman

15.1 Background

Class Craniata includes fish, rodents, and all mammals. Every **neoplasm** that occurs in any animal descending from class Craniata can be assigned to one of the following six developmental classes (List 15.1.1).

List 15.1.1 The six major classes of neoplasms.

1. Endoderm/Ectoderm
2. Mesoderm
3. Neuroectoderm
4. Neural Crest
5. Germ Cell (includes teratomatous neoplasms)
6. Trophectoderm (cells of extraembryonic origin)

The six major classes of neoplasms and their major subclasses can be depicted in a simple radial graph (Figure 15.1).

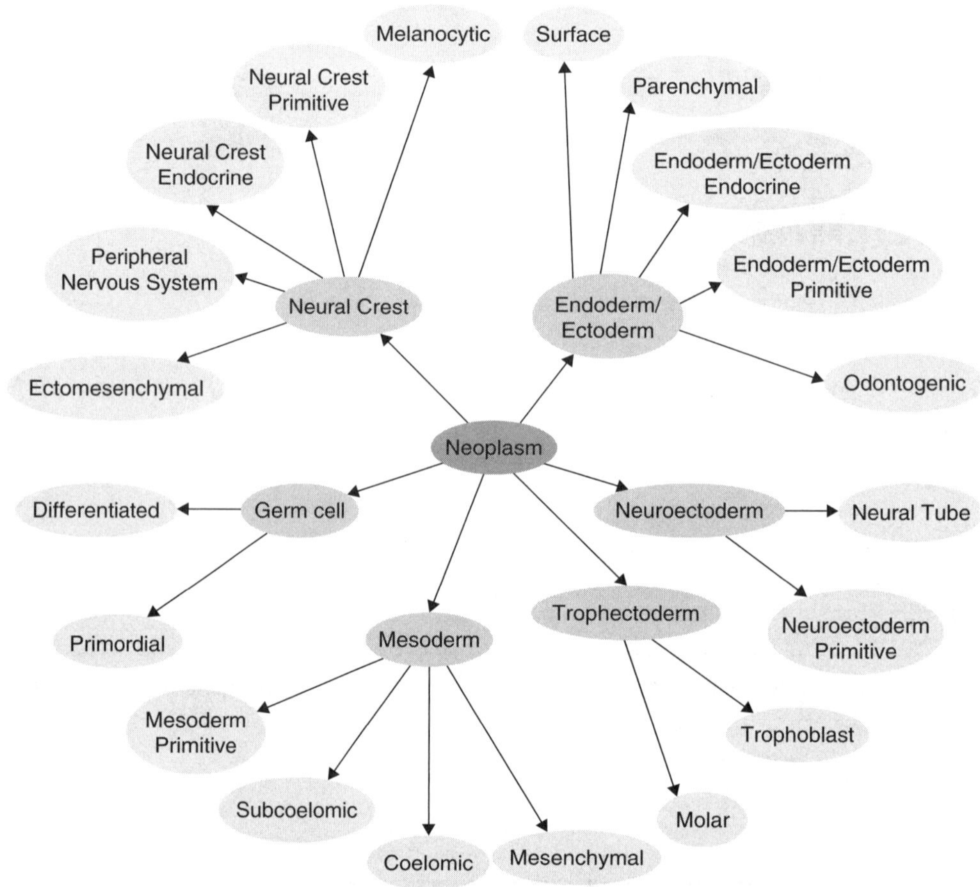

FIGURE 15.1 Graphic representation of the first three class tiers of the Developmental Lineage Classification of Neoplasms.

The purpose of this chapter is to provide a quick overview of the six classes of neoplasms. A detailed explanation of each major class is included in subsequent chapters.

15.2 Class Endoderm/Ectoderm Neoplasms

In humans, gastrulation occurs after implantation on about gestation day 16 and creates the two outer germ layers (ectoderm and endoderm) as well as the mesoderm (Figure 15.2).

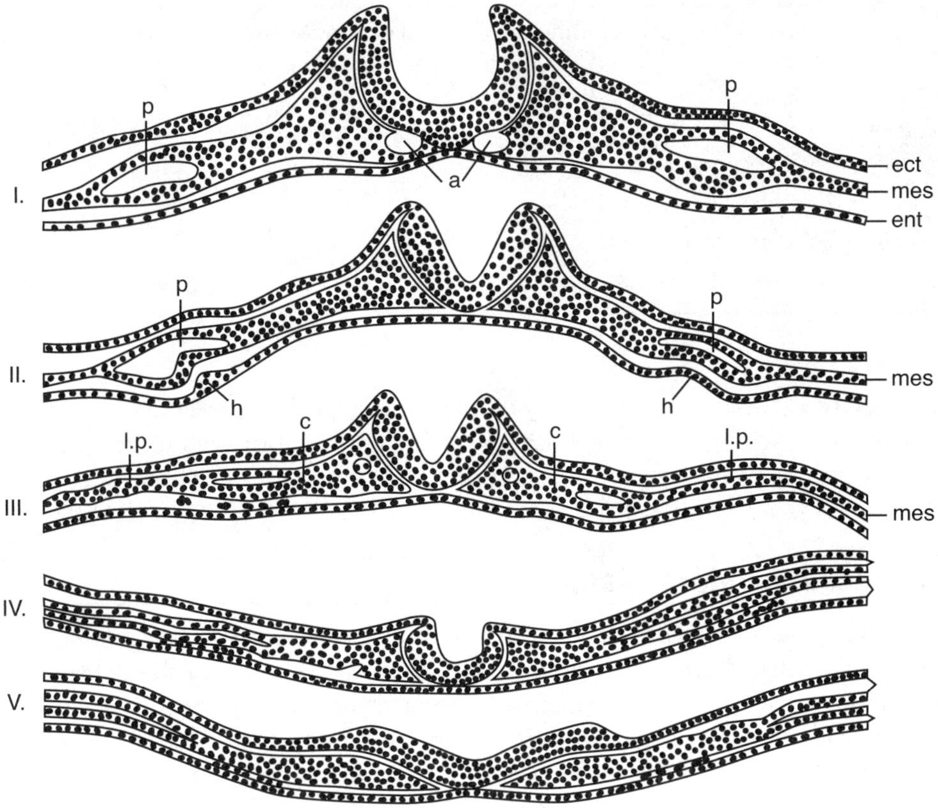

FIGURE 15.2 Transverse sections through a dog embryonic disk. The mesoderm (mes) lies between the ectoderm (ect) and the endoderm (ent). From Gray's Anatomy.

The human body can be envisioned as a topological donut. It is covered by lining cells, with the donut hole lined by endoderm and the donut outer surface lined by ectoderm. The donut pastry would be the mesoderm. There is no anatomic border that separates endoderm from ectoderm, and they can both be perceived as the cells that line the outside of the body. After birth, the endoderm and ectoderm derivatives are directly exposed to exogenous chemicals delivered through the air or through the diet. Both endoderm and ectoderm produce the same basic types of cells (squamous cells and glandular cells). There does not seem to be any benefit in separating the tumors that derive from either of these classes.

The subclassification of tumors of class endoderm/ectoderm is shown in List 15.2.1.

The class of endoderm/ectoderm tumors contains most (but not all) of the neoplasms that are morphologically epithelial. We will see that many epithelial tumors derive from *mesoderm* (e.g., renal cell carcinoma, adrenal cortical tumors, and ovarian carcinoma) or

List 15.2.1 Subclassification of the endoderm/ectoderm class.

Endoderm/ectoderm
- Primitive
- Endoderm/ectoderm surface
- Endoderm/ectoderm endocrine
- Endoderm/ectoderm parenchymal
- Endoderm/ectoderm odontogenic

neural crest (e.g., melanoma, medullary carcinoma of thyroid). A basic assumption in the design of the Developmental Classification is that epithelial tumors in different classes have class-specific phenotypic properties unrelated to their epithelial morphology.

15.3 Class Mesoderm Neoplasms

The mesoderm is a germ layer present in all animals descending from about the level of Coelomata. Mesoderm forms during gastrulation when some of the cells migrating inward to form the endoderm form an additional layer between the endoderm and the ectoderm. This inner layer gives rise to the internal body spaces (cell-lined coelomic cavities) and to a variety of structures that develop from the coelomic folds (i.e., subcoelomic derivatives) and to the connective tissues of the body (List 15.3.1).

15.4 Class Neuroectoderm Neoplasms

The neuroectoderm develops from the ectoderm and produces the central nervous system (List 15.4.1). Tumors of Class Neuroectoderm account for most of the so-called brain tumors but exclude tumors that derive from all the cranial nerves with the exception of the optic nerve. The optic nerve is a purely central nervous system structure. The other cranial nerves derive from the neural crest.

15.5 Class Neural Crest Neoplasms

The neural crest derives from a specialized compartment of cells in the mesoderm between the ectoderm and the neural tube. The neural crest gives rise to the peripheral nervous system, to the connective tissue of the head, to several endocrine glands, and to the connective tissue component of the teeth (List 15.5.1).

List 15.3.1 Subclassification of the mesoderm class.

- Mesoderm
  - Primitive
  - Mesenchyme
    - Connective tissue
    - Muscle
    - Fibrous tissue
    - Vascular
    - Adipose tissue
    - Bone cartilage
    - Heme lymphoid
  - Coelomic
    - Coelomic ductal
    - Coelomic cavities
    - Coelomic gonadal
    - Subcoelomic
      - Subcoelomic gonadal
      - Subcoelomic endocrine
      - Subcoelomic nephric

List 15.4.1 Subclassification of the neuroectoderm class.

- Neuroectoderm
  - Primitive
  - Neural tube
    - Neural tube parenchyma
    - Neural tube lining

15.6 Class Germ Cell (Includes Teratomatous) Neoplasms

Germ cells are produced during gastrulation as specialized cells derived from endoderm. They give rise to differentiated **gametes**, the ova, and the spermatocytes. Early-stage germ cells are called "primordial" and are multipotent (List 15.6.1).

Totipotent cells of germ cell origin (teratomas and embryonal carcinomas and related tumors) can arise by one of two different mechanisms: by a malignant transformation of a primordial germ cell or through a transformative process

List 15.5.1 Subclassification of the neural crest class.

Neural crest
- Primitive
- Peripheral nervous system
- Neural crest endocrine
- Neural crest melanocytic
- Neural crest odontogenic

List 15.6.1 Subclassification of the germ cell class.

Germ cell
- Differentiated
- Primordial (totipotent)
- Includes mixed differentiated germ cell and primordial germ cell

occurring in neoplastic differentiated germ cells (dysgerminoma cells or seminoma cells).

15.7 Class Trophectoderm (Cells of Extraembryonic Origin) Neoplasms

In humans, at about the third day of gestation, the morula divides into two parts: the embryoblast from which the embryo develops, and the trophoblast from which the non-embryonic parts, including the placenta, develop (List 15.7.1). The germ layer containing the tissues developing from the trophoblast is the trophectoderm. Tumors of Class Trophectoderm are somatic neoplasms that contain differentiated cells somewhat misleadingly named trophoblasts (i.e., the cells are not undifferentiated and "blastic"). These cells include cytotrophoblasts and syncytiotrophoblasts. Tumors of Class Trophectoderm occur in mothers, during or soon after gestation.

Some germ cell tumors produce subpopulations of cells that include trophoblasts, and these cells can be the dominant cell-type in a germ cell tumor. Although composed

List 15.7.1 Subclassification of the trophectoderm class.

Molar
Trophoblast

largely of trophoblasts, these germ-cell–derived tumors occur in the offspring, not the mother (i.e., are non-gestational). The trophoblast-dominant tumors of Class Germ cell behave differently than similarly named tumors of Class Trophectoderm and have very different genetic and epigenetic properties. Trophoblastic neoplasia is often divided into gestational and nongestational types, referring to trophoblastic tumors that occur in mothers during or soon after gestation and trophoblastic tumors that occur in offspring. In the Developmental Lineage Classification, the two types of tumors are distinguished by their lineage: Class Trophectoderm (gestational), and Class Germ cell (non-gestational).

15.8 Primitive Tumors Are Somatic Neoplasms that Occur in Different Classes and that Are Unrelated to Germ Cell Tumors

In the Developmental Classification, all primitive tumors are somatic and derive from one the four embryonic layers (endoderm/ectoderm, mesoderm, neuroectoderm, or neural crest). Examples of primitive tumors are neuroblastoma (Class Neural Crest), medulloblastoma (Class Neuroectoderm), Ewing sarcoma (Class Neural Crest), hepatoblastoma (Class Endoderm/Ectoderm), and nephroblastoma (Class Mesoderm). All of these tumors derive from somatic stem cells with phenotypes that are determined by the phenotypes of the primitive cells of their lineage.

Primitive tumors should not be confused with **undifferentiated tumors** or poorly differentiated tumors, which have a completely different biological origin. Undifferentiated tumors are equivalent to high-grade tumors. They are called "undifferentiated" because they have few of the morphologic features that are present in the cells from which they derive. For example, an undifferentiated colon carcinoma will not have well-formed glands and will not produce much mucin. Most undifferentiated carcinomas arise through the process of tumor progression in which accumulating genetic defects lead to an increasingly aggressive phenotype and a worsened prognosis.

A primitive tumor derives from a cell that is not committed to a differentiated phenotype. Its genome and epigenome are similar to the genome and epigenome of a normal stem cell of the same lineage as the tumor.

15.9 Organs with Multiple Anlagen

The ovary and the testis derive from mesoderm and from germ cells. The mesodermal components of the ovary and the testis derive from four different subclasses (List 15.9.1).

The coelomic cavity mesoderm gives rise to the specialized epithelial tumors that grow from the surface of the ovary, including the common papillary serous carcinomas of the ovary.

None of the ovary is derived from endoderm or ectoderm, the lineage of origin of the commonly occurring cancers of humans.

List 15.9.1 Multiple lineages of the ovary and testis.

- Mesodermal coelomic cavity cells—epithelial covering of ovary and testis.
- Mesodermal subcoelomic gonadal cells—specialized stromal cells of the ovary and testis, consisting primarily of granulosa cells, theca cells in the ovary and Sertoli cells and Leydig cells in the testis.
- Mesodermal mesenchymal cells—nonspecialized connective tissue cells in the ovary and testis, including fibrovascular tissue.
- Mesodermal coelomic ductal cells—paramesonephric and mesonephric duct tissues contributing to vestigial tissues attached to the ovary/testis.
- Germ cells—ova in ovary and spermatocytes in testis.

The strange and complex developmental origin of the ovary and the testis explains why these organs share a diverse array of developmentally homologous tumors that are not encountered in other organs.

15.10 Rhabdoid Tumor: The Neoplasm that Breaks All the Rules

Life would be simple if we all followed the same rules. Unfortunately, life is complex, and complex systems have a tendency to produce exceptions to the general rules of behavior. Before closing this chapter, it would be worthwhile to discuss a very rare cancer that occurs in childhood and infancy and that seems to have several properties that are different from all other tumors.

Rhabdoid tumors arise in just a few dozen children each year in the United States. These aggressive tumors arise from brain (class Neuroectoderm) and from kidney (class Mesoderm) and contain morphologically distinctive cells (so-called rhabdoid cells, named for their superficial similarity to muscle cells). Not long ago, the rhabdoid tumors that arose in the kidney were thought to be different from the rhabdoid tumors that arose from the brain. The common rhabdoid cell was considered to be a peculiar morphologic variant that just happened to occur in both types of tumors. The kidney tumor was called MRT (malignant rhabdoid tumor). MRT was considered, by many pathologists, to be a variant form of nephroblastoma. The rhabdoid brain tumor was called AT/RT (atypical teratoid rhabdoid tumor), and some pathologists classified AT/RT among the PNETs.

The perceived distinction between MRT and AT/RT started to disappear when it was found that about 10% of patients with MRT also developed AT/RT. Recently, a characteristic genetic abnormality has been found in rhabdoid tumors of CNS or renal origin: biallelic loss of INI1 gene expression.

Rhabdoid cells are large cells with an eosinophilic cytoplasm. Ultrastructural examination of rhabdoid cells shows characteristic whorls of intermediate filaments. Intermediate filaments are proteins that contribute to the structure of cells and provide resistance to deformity. Normal cells contain intermediate filaments that are specific for their developmental lineage. Cells of endodermal or ectodermal origin contain cytokeratin filaments. Cell of neuroectodermal origin contain neurofilaments filaments. Cells of mesenchymal origin contain desmin filaments. Rhabdoid cells contain all these different types of intermediate filaments.

Immunostaining for the protein produced by the INI1 gene is absent in almost all reported cases of rhabdoid tumor cells. Staining for INI1 gene is positive in all normal cells and in the cells of almost every other examined type of neoplasm. Recently, rhabdoid cells have been reported as a subpopulation arising in tumors of adults including mucinous carcinoma of pancreas (235) and uterine carcinosarcomas (236). In both these cases, INI1 immunostaining was negative in rhabdoid cells and positive in the nonrhabdoid tumor cells.

Rhabdoid tumors apparently disobey many of the rules of neoplastic development (List 15.10.1).

How is this possible? How can a tumor suppressor gene mutation in a nongerm cell produce tumors that arise from tissues of different developmental lineage and contain a common tumor cell that contains intermediate filaments specific for multiple cell lineages?

The INI1 gene codes a subunit of the SWI/SNF chromatin remodeling complex. The study of SWI/SNF complexes in mammals, flies, and plants suggests that they strongly influence many developmental pathways (238). This suggests two possible mechanisms for the action of the INI1 tumor suppressor gene in rhabdoid tumors (List 15.10.2).

Of these two possibilities, the first seems unlikely because subpopulations of INI1-negative rhabdoid cells occur secondarily within adult tumors, and this would not be expected to occur if the INI1 mutation only produced rhabdoid tumors derived from primitive cells. Also, if there were a very primitive cell (with a lineage that precedes the development of the germ layers), why would the INI1 mutation be the only mutation that could produce tumors of these cells?

The second possibility, if true, might explain the odd biological features of rhabdoid tumors. The idea of mutations in differentiated cells conferring pluripotentiality is not unprecedented. In recent work, Yamanaka and coworkers successfully created **pluripotent stem cells** from differentiated fibroblasts by introducing (with retroviruses) several chosen genes (Oct3/4, Sox2, c-Myc, and Klf4) and subsequent selection for Fbx15 (239). These experiments demonstrated that differentiated cells could be altered to become stem cells and provide researchers with a source of stem cells other than embryos.

At this time, I favor the hypothesis that loss of INI1 gene expression can create a pluripotent malignant cell with a specific (rhabdoid) phenotype. Regardless of the lineage of origin, INI1 biallelic loss creates tumors of a unique phenotype.

Rhabdoid tumors open several research opportunities (List 15.10.3).

List 15.10.1 Summary of rhabdoid tumor properties that disobey the rules of neoplasia.

Rhabdoid tumors are lineage nonspecific and can arise from neuroectodermal cells or mesodermal cells. All other tumors of somatic cells (nongerm cells) arise from a single germ lineage.

Cells within a single rhabdoid tumor seem to have differentiated along several developmental lineages (ectodermal, endodermal, neuroectodermal, and mesodermal). This phenomenon is otherwise restricted to totipotent germ cell tumors.

Cells within a single rhabdoid tumor may include primitive cells indistinguishable from PNET tumors (primitive neuroectodermal tumors). PNET tumors are typically monomorphic tumors.

Rhabdoid tumors are all associated with a specific phenotypic cell (the rhabdoid cell) regardless of the developmental origin of the tumor (neuroectoderm or mesoderm).

The rhabdoid cells contain several different intermediate filaments. In normal cells, only one type of intermediate filaments is found in any single cell, and that filament is specific for the lineage of the cell.

Almost all tumors arise from cells that resemble an observable normal cell. For example, a squamous cell carcinoma is composed of cells that resemble normal squamous cells biochemically, ultrastructurally, and by light microscopic examination. The rhabdoid cell has no known counterpart in any adult tissue or in any stage of development.

Rhabdoid tumors seem to be caused by biallelic loss of the tumor suppressor gene, INI1. Tumor suppressor genes are normally associated with different types of tumors. INI1 gene loss seems to be exclusively associated with rhabdoid tumors or with rhabdoid tumor cell subpopulations developing within other types of tumors. Currently, there is no other known genetic mutation that produces a specific phenotype akin to the association between INI1 and rhabdoid cells.

INI1 loss produces rhabdoid tumors in mice. Eight of 125 mice with germline haploid complement of INI (Snf5+/– mice) developed INI1-negative tumors of soft tissue origin and rhabdoid cell morphology (237). The mouse tumor is morphologically and genetically identical to the human tumor. Despite the phenotype and genotypic similarities between murine and human rhabdoid tumors, the mouse tumor arises from the branchial arch soft tissue, a tissue of origin not observed in human rhabdoid tumors.

List 15.10.2 Possible explanations for the peculiar biology of rhabdoid tumors.

1. The cell of origin of rhabdoid tumors is a primitive, pluripotent somatic stem cell with a lineage position that precedes the development of germ layers.
2. Loss of INI1 gene expression produces a pluripotent stem cell when it occurs in cells of several different lineages (e.g., neuroectodermal or endodermal or mesodermal cells).

List 15.10.3 Research challenges and opportunities provided by rhabdoid tumors.

1. Because these tumors seem to be caused by loss of INI1 gene products, inhibition of downstream pathway molecules may have therapeutic value (240).
2. If biallelic loss of INI1 gene expression is sufficient to yield a rhabdoid tumor cell subpopulation in differentiated tumors occurring in adults, this would inform us that major alterations in tumor cell biology can result from a single gene loss and that tumor progression does not always result from multiple small genetic alterations occurring over time.
3. It may be possible to produce a pluripotent stem cell from somatic cells through a single gene mutation.

Chapter 15 Summary

A tumor classification can be designed along purely developmental lines, encompassed by six major classes: endoderm/ectoderm, mesoderm, neuroectoderm, neural crest, germ cell, and trophectoderm. Every cell in the body falls into exactly one of these six classes, and every type of neoplasm (cancer, benign tumor, precancer and hamartoma) can be assigned to exactly one class. Every specific tumor can be assigned to a class, and the tumor will not transit from one class to another during the life of the tumor. Because tumors acquire their phenotype through normal pathways inherited through developmental class, all of the tumors of a class will share a set of fundamental properties. In the next five chapters, we describe some of the specific tumor entities that occupy each of the major neoplasm classes.

CHAPTER

16 Ectodermal and Endodermal Neoplasms

Our similarities are different.

—Yogi Berra

16.1 Background

In the Developmental Classification, the tumors derived from the endoderm and the ectoderm are combined (Figure 16.1). This is done simply because there seems to be no biologic, clinical, morphologic, or molecular differences among the tumors derived from either of these germ layers. This combined class contains most of the commonly occurring tumors of humans. Subclasses of Class Endoderm/Ectoderm include epithelial tumors that arise from endodermal or ectodermal surface epithelium (which would include squamous cell carcinoma of skin, bronchus, or esophagus and adenocarcinoma of colon), epithelial tumors that arise from the endoderm/ectoderm organ epithelium (breast carcinoma, salivary gland carcinoma, pancreatic carcinoma, hepatocellular carcinoma), and epithelial tumors arising from the endodermal or ectodermal endocrine epithelium (thyroid papillary and follicular carcinoma and pituitary adenoma).

Six tumors of the endoderm and ectoderm account for approximately 60% of all cancers that kill humans (38) (List 16.1.1).

The purpose of this chapter is to review the special properties of Class Endoderm/Ectoderm and to explain why this class accounts for the majority of cancers occurring in adults.

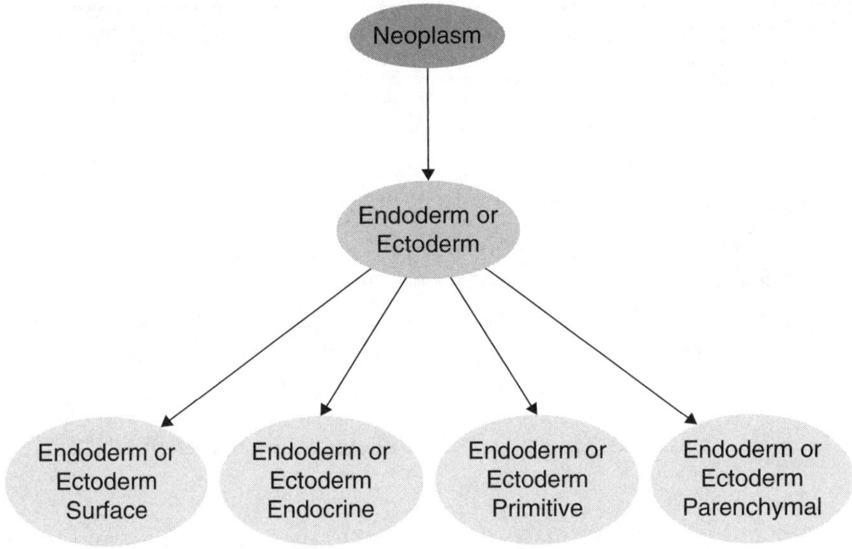

FIGURE 16.1 Hierarchy of Class Endoderm/Ectoderm tumors.

List 16.1.1 Global cancer deaths, 2000 (38).

Trachea, bronchus and lung—1,211,000 (17.2% of total)
Stomach—835,000 (11.9% of total)
Liver—611,000 (8.7% of total)
Colon and rectum—608,000 (8.6% of total)
Breast—473,000 (6.7% of total)
Esophagus—430,000 (6.1% of total)

16.2 Properties of Class Endoderm/Ectoderm

Four general properties of endodermal and ectodermal cells (and absent from most mesodermal, neuroectodermal, and neural crest cells) explain why most tumors of mankind arise from these two developmental layers (List 16.2.1).

16.3 Neoplasms that Are Too Numerous to Count

In addition to causing most of the cancer deaths in humans, Class Endoderm/Ectoderm contains the tumors that occur most frequently in man but which seldom cause death.

The two most common cancers of humans are basal cell carcinoma and squamous cell carcinoma of the skin. The exact number of occurrences of these lesions is

List 16.2.1 Four reasons why most tumors in man are of ectodermal or endodermal origin.

1. Endodermal and ectodermal cells are located on the outside (topologically) of the body and are thus the cell populations that are directly exposed to the exogenous chemicals delivered by air (pollution), food (food-borne toxins), or space (ultraviolet radiation).
2. The endoderm and ectoderm happen to be the greatest metabolic activators of chemical carcinogens received in food (particularly the liver).
3. The endodermal and ectodermal cells have the greatest rates of cell division and are thus much more susceptible to mutation than nondividing cells of mesoderm, neurectoderm, or even neural crest. After a mutation has been replicated into the genome of a dividing daughter cell, the cell can produce a clone of cells that propagate the mutation.

unknown, but conservative estimates suggest that more than a million of these tumors occur each year in the United States. This is roughly equal to the occurrences of all other tumors combined.

For various reasons, counts of surgical pathology reports for excised basal cell carcinoma and squamous cell carcinoma of skin grossly underestimate the incidence of these tumors (List 16.3.1).

Hospital-based Cancer Registries, charged with recording new cases of cancer, do not collect data on new cases of basal cell carcinoma or squamous cell carcinoma of skin. There are just too many of these lesions.

16.4 Tumors of Class Endoderm/Ectoderm in Childhood

Neoplasms rarely occur in children. In 1995, there were 2,275 cancer deaths in children in the United States (184). In a single year in the United States, about 550,000 men and women will die of cancer. Childhood cancer deaths account for just 0.4% of all cancer deaths.

Children almost never develop the most common tumors of adults: squamous cell carcinoma of skin, basal cell carcinoma of skin, lung carcinoma, and colon carcinoma. The cause of these tumors is repeated long-term exposures to environmental carcinogens. Most skin cancer is caused by sunlight exposure. Most lung cancer is caused by cigarette smoke exposure. The cause of colon carcinoma is likely to be carcinogens found in food. Colon tumors take many years to develop and cannot be expected to occur in young children.

The brain and the hematopoietic system are rapidly growing, large organs in the fetus, and it is not surprising that these two tissues account for about half of the tumors occurring in children (184).

List 16.3.1 Reasons why counting surgical pathology reports cannot account for all occurrences of basal cell carcinoma and squamous cell carcinoma of skin.

Basal cell carcinomas are small and subtle skin lesions. Many people have these tumors and never seek treatment.

A single individual may have multiple basal cell carcinomas, all located in the same anatomic region (e.g., face and shoulder). Multiple biopsies submitted from a single office visit may be counted as a single case of basal cell carcinoma when they actually represent several distinct occurrences.

Most squamous cell carcinomas of skin develop from actinic keratoses. Small squamous cell carcinomas and large actinic keratoses are easily mistaken for one another. It is not uncommon to treat keratotic lesions of skin, which may include actinic keratoses and small squamous cell carcinomas, with ablative therapy (such as liquid nitrogen). When skin lesions are ablated, without a biopsy to confirm the diagnosis, it is impossible to determine the number of occurrences of the different types of lesions that were removed.

Some patients in the advanced geriatric population and some patients who are suffering from terminal illnesses may choose to defer treatment for conditions that are unlikely to do them harm prior to their death. In these instances, basal cell carcinomas and squamous cell carcinomas of skin will not be treated. Untreated lesions cannot be recorded by tumor surveillance efforts.

16.5 Why Are the Major Embryonic Layers of Ectoderm and Endoderm Combined in the Developmental Classification?

Although the ectodermal and endodermal layers are quite distinct, the tumors that arise from these two layers are similar and even share the same terminology. For example, there is a squamous cell carcinoma of skin (ectodermal origin) and a squamous cell carcinoma of lung (foregut endoderm origin). There is a ductal carcinoma of breast (ectodermal origin) and a ductal carcinoma of pancreas (endodermal origin). The reason for this is that animals have only a few different ways of building epithelial linings, pavemented epithelium using squamous cells, acinar epithelium using glandular cells, or tubular epithelium using ductal cells. We learn in histology that there are anatomic variations for all of these structures (stratified epithelium, pseudostratified epithelium, ciliated lining cells, brush-bordered lining cells, keratinizing squamous epithelium, parakeratinizing squamous epithelium, and so on). Despite these minor variations, the vast majority of tumors derived from endoderm and ectoderm fall into a few cell types distinguishable principally by their anatomic site of origin.

16.6 The Differences Between Endoderm and Ectoderm

A few shared features unite the endoderm and the ectoderm. Both are characterized by a large epithelial lining that yield squamous cell carcinomas and adenocarcinomas. All glandular organs and glandular mucosal surfaces can be thought of as infolded layers of epithelial cells, not much different from the infolded epithelial lining of ducts. Squamous cell carcinoma and adenocarcinoma account for almost all of the tumors that arise from either embryonic layer.

There are, however, a few significant differences between ectoderm and endoderm that bear explanation. The ectoderm gives rise to another embryonic layer, the neuroectoderm. A strict phylogenist would assign all Neuroectodermal tumors (including all tumors of glial cells) as a sub-sub-class of ectoderm. In the Developmental Classification, neuroectoderm is placed alongside endoderm/ectoderm as another embryonic layer with no descent relationship to ectoderm. Neuroectoderm has no known inherited properties of ectoderm; thus, despite its embryologic origin, it does not qualify as a useful subclass of ectoderm.

The endoderm gives rise to germ cells. Germ cells have no features in common with other ectodermal derivatives and have no known properties that descend from endoderm. In the Developmental Classification, germ cells occupy a separate class from endoderm/ectoderm and are not assigned a descendent category.

A few histological features distinguish ectoderm from endoderm. Only ectoderm gives rise to ameloblasts, specialized odontogenic epithelium. An ameloblastoma is always of ectodermal origin.

Myoepithelial cells to occur almost exclusively around ducts and glands of ectodermal origin. Myoepitheliomas of endodermal origin are rare.

Pseudostratified epithelium seems to be an endodermal feature.

16.7 Polyphyly and the Strange Origin of Transitional Cells

A class is polyphyletic if it does not contain the most recent ancestor of all of its subclasses. The urogenital transitional cell tumors, when subclassed under tumors of ectodermal/ectodermal origin, yield a polyphyletic classification.

Transitional cells have an unusual embryologic origin. Several different germ layers contribute to the urogenital epithelium. Any transitional tumor may derive from any one of several embryonic layers, dependent on their precise anatomic origin (List 16.7.1).

There seems to be no biological difference between transitional cell neoplasms that derive from differing embryonic anlagen. The Developmental Classification lumps all transitional tumors under Class Endoderm/Ectoderm.

List 16.7.1 Origins of transitional cell carcinoma.

Mesonephros
Urogenital sinus
Metanephros
Paramesonephric duct and mesonephric ducts

16.8 Squamous Cell Carcinomas

Squamous cell cancers are one of the most common cancers of skin. Squamous cell carcinoma is also the most common type of bronchogenic carcinoma. The mouth (Figure 16.2), larynx, and esophagus are also common sources of squamous cell carcinoma. The mucosa of the nasal cavity produces neoplasms of cells that have a phenotype that straddles squamous and glandular morphology. Nasal papillary neoplasms are lined by nonkeratinizing squamous cells (alternatively called transitional cells). As far as anyone knows, all squamous cell cancers, regardless of their anatomic origin, arise from squamous cell precancers. A variant of squamous cell precancer is called Bowen's disease (Figure 16.3) and is characterized by atypical squamous cells occupying the full thickness of the epidermis.

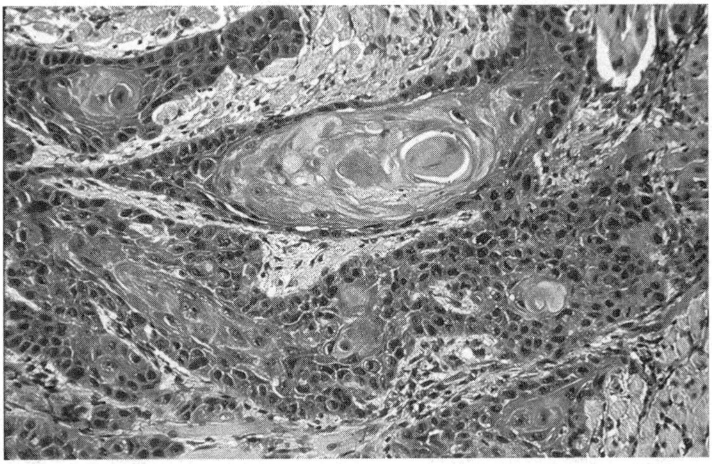

FIGURE 16.2 Oral squamous cell carcinoma. This is a high-power view of keratinized malignant squamous cells.

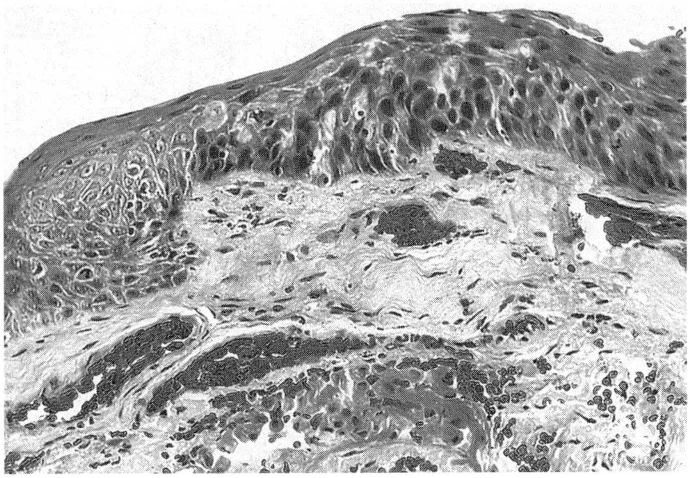

FIGURE 16.3 Bowen's disease. Bowen's disease is a variant of squamous cell carcinoma in situ. Atypical squamous cells are present throughout the epidermis.

16.9 Adenocarcinoma

Adenocarcinomas comprise the second largest type of neoplasm (second to squamous cell carcinomas). All adenocarcinomas arise from glands. Glands are tubes, open on one and closed on the other end, of epithelial cells. The end part of the tube is called the acinus and sits as a bulb on the tube. The open end of the tubular part of the gland opens into another tube or a duct or a major lumen, or in the case of sweat glands, the gland opens into the outside of the organism (Figure 16.4). The closed end of the tube (the acinus) is often lined by epithelial cells that have a different morphology than the epithelial cells that line the straight-tube section (the duct).

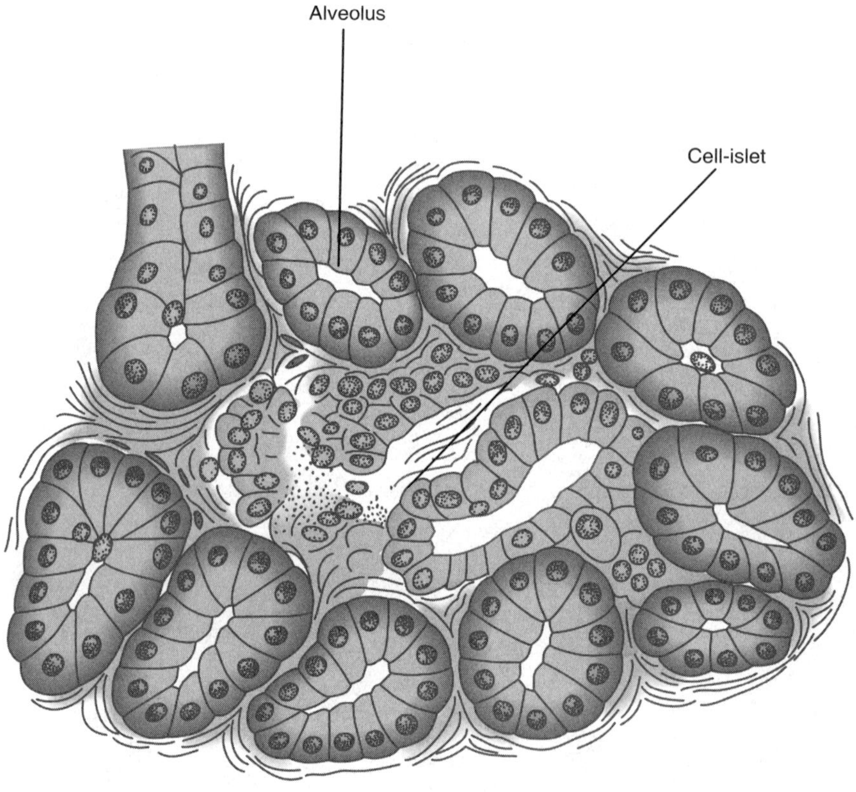

FIGURE 16.4 A section of pancreas, with several types of glands.

Adenocarcinomas contain glands, but the glands are often poorly formed, irregular, complex, and may not open into a lumen (Figure 16.5).

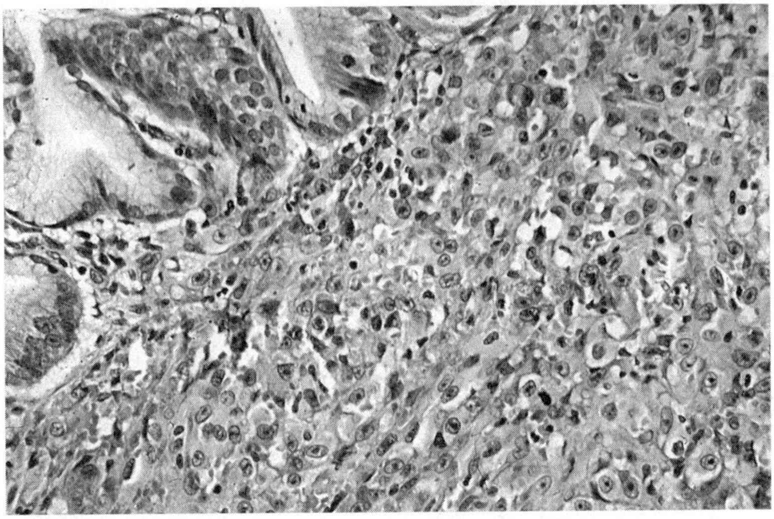

FIGURE 16.5 Adenocarcinoma of stomach. Slit-like glands lined by highly atypical epithelial cells invade the stomach wall. A few non-cancerous glands are present in the top, left corner.

The prostate gland is composed of fibromuscular tissues and glands that empty into ducts that connect to the urethra (Figure 16.6).

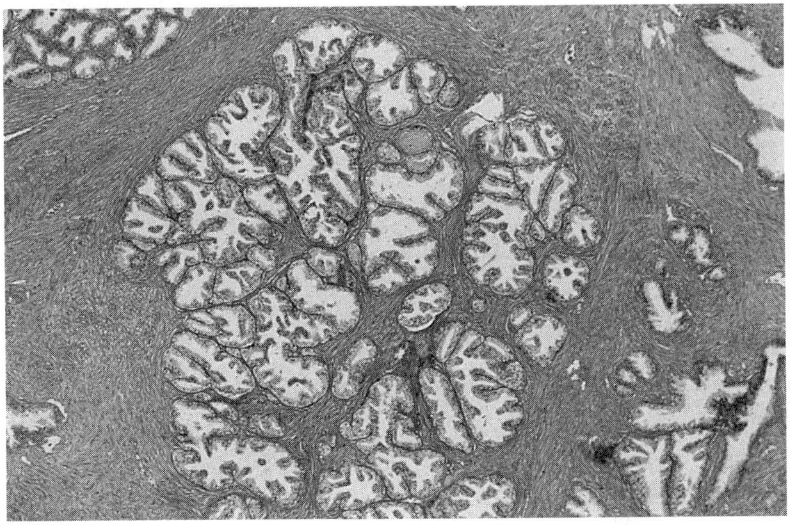

FIGURE 16.6 Normal prostatic tissue.

Each year in the U.S. about 30,000 men die of adenocarcinoma of the prostate (Figure 16.7).

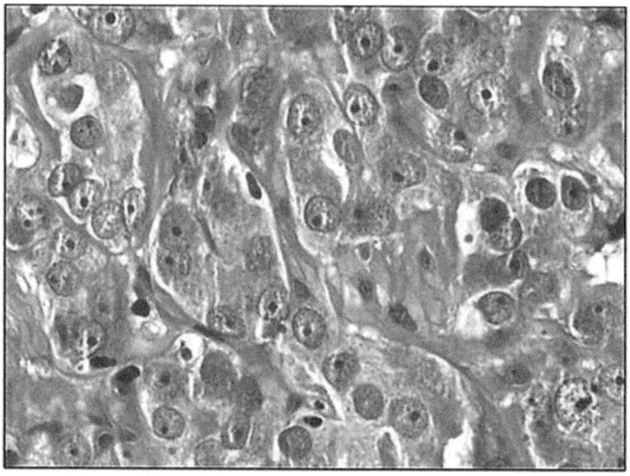

FIGURE 16.7 Adenocarcinoma of prostate. Glands are closely packed. Glands are lined by nearly identical atypical cells. Most cells have a prominent nucleus.

Like other adenocarcinomas, prostate cancer is preceded by a morphologically recognizable precancer lesion. The precancer of prostatic adenocarcinoma is prostatic intraepithelial neoplasia (PIN) (Figure 16.8).

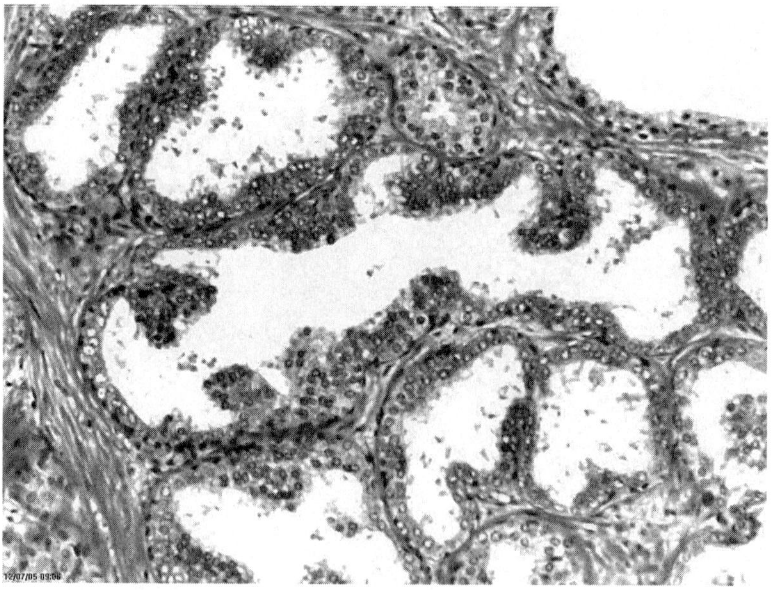

FIGURE 16.8 Prostatic intraepithelial neoplasia (PIN). Glands have a near-normal structure (glands lined by epithelial cells overlying smaller, darker basal cells), but the lining epithelial cells are dysplastic. The nuclei of the epithelial cells are enlarged, and the texture of the chromatin in these lining cells is abnormal and variable from cell to cell.

16.10 Endodermal and Ectodermal Tumors that Do Not Seem to Follow the Rules

Metaplasia is the process whereby tissues acquire cell types that are not normally present in those tissues. Examples of metaplasia are squamous metaplasia of bronchus and Barrett's esophagus of the esophagus. Squamous metaplasia of the bronchus is a common, almost universal, finding in smokers. The normal ciliated epithelial cells that line the bronchus are slowly replaced by squamous cells. Presumably, the squamous cells derive from the same basally located reserve cells that normally give rise to ciliated cells. For reasons unknown, chronic exposure to toxic cigarette fumes causes the reserve cells to differentiate toward squamous epithelium rather than ciliated epithelium. Both squamous cells and ciliated epithelial cells are among the epithelial cells normally derived from the endodermal germ layer, although squamous cells are not a normal constituent of bronchial mucosa.

Barrett's esophagus is a metaplastic process whereby the normal squamous cells lining the esophagus are replaced by foci of mucus-producing glandular epithelium. The process is caused by the chronic effects of a toxin (stomach acid in most cases) on the esophageal mucosa. The esophageal, endodermally derived reserve cells begin to produce mucin-containing cells rather than squamous cells. The mucin-containing cells are apparently more resistant to stomach acid than are squamous cells.

It is not unusual to find tumors of ectoderm or endoderm origin that have a mixed squamous and glandular morphology.

Occasionally, a tumor of ectodermal or endodermal origin will develop foci that appear to be mesenchymal. The tumor may contain fibrous tissue, cartilage or bone. If the tumor is malignant, the mesenchymal components are likely to be malignant as well.

In the case of breast carcinoma, foci of chondrosarcoma, fibrosarcoma, or osteosarcoma may arise. When this occurs, the tumor is called a metaplastic breast carcinoma, indicating that the tumor contains cells that have a different origin from the malignant ductal cells that comprise a typical breast cancer.

How is it possible for a monoclonal tumor of ectodermal origin to contain mesenchymal foci? The trick is that mesenchymal differentiation is not the exclusive property of mesoderm. The ectoderm and the endoderm have a very special cell that can produce mesenchymal components, and this cell is the *myoepithelial cell.*

The myoepithelial cell is located adjacent to ductal cells and plays a role in ductal motility. As its name implies, it has both epithelial and myoid (muscular) morphologic features. Pure tumors of myoepithelial cells (myoepitheliomas) occur rarely. More often, myoepithelial cells are found as a component of ectodermal tumors (commonly) or endodermal tumors (less commonly). When myoepithelial cells are admixed in ductal tumors, the myoepithelial component is usually observed as loose myxoid or chondromyxoid tissue. In *metaplastic breast carcinoma*, mixed ductal and myoepithelial differentiation yield an ectodermal tumor with sarcomatous elements.

Examples of mixed tumors (ductal and myoepithelial) are pleomorphic adenoma of salivary gland and chondroid syringoma of eccrine (sweat gland) origin. These benign mixed tumors are monoclonal, just like most other neoplasms (241). The bipartite feature of the tumor results from a clonal or subclonal progenitor cell that is capable of ductal or myoepithelial differentiation.

16.11 The Genetic Causes of Class Endoderm/Ectoderm Tumors

As a general rule, tumors of endoderm/ectoderm arise rather late in life, often from the mucosal linings most exposed to environmental carcinogens (epidermis, lung, colon, esophagus, and stomach).

Tumors of endoderm/ectoderm origin tend to have innumerable cytogenetic abnormalities. A recent manuscript by Vogelstein and coworkers analyzed every well-annotated gene that codes for a human protein in a series of breast and colon cancers.

Of the 13,023 genes analyzed, it was shown that individual tumors accumulate an average of 90 mutant genes (23). Tumors of Class Endoderm/Ectoderm accumulate many different gene mutations, consistent with a process that occurs over time, possibly due to repeated exposures to carcinogens, and that retains the mutations in growing tumor cell populations.

16.12 Can a Single Gene Alteration Characterize a Tumor of Class Endoderm/Ectoderm?

Tumors of Class Mesoderm are often characterized by simple cytogenetic translocations, many of which result in fusion genes (composed of a translocated gene fragments fused to a gene residing in the recipient chromosome). The simplicity of genetic markers in the tumors of Class Mesoderm contrasts sharply with the complexity of genetic changes associated with most carcinomas of Class Endoderm/Ectoderm. Nonetheless, there are examples of benign (see List 16.12.1) and malignant (see List 16.12.2) tumors of Class Endoderm/Ectoderm that have simple gene alterations.

List 16.12.1 Benign tumors of class ectoderm/endoderm with simple gene alterations.

Cylindroma (52)
Pilomatrixoma, beta-catenin (54)
Pleomorphic adenoma, PLAG1 (55)
Clear cell hidradenoma of the skin, TORC1-MAML2 gene fusion (57)

List 16.12.2 Malignant tumors of class endoderm/ectoderm with simple gene alterations.

Secretory carcinoma of breast in young women, with ETV6-NTRK3 fusion gene in secretory (135)
Midline carcinoma of children and young adults with NUT rearrangement (124)

Both secretory carcinoma of breast and midline carcinoma of the young are extraordinary examples of Class Endoderm/Ectoderm tumors. They occur in children and young adults and are not associated with any known exogenous carcinogens.

Mitelman has noted that fusion genes and rearranged genes, the kinds of gene alterations that characterize tumors of Class Mesoderm, are also found in tumors of Class Endoderm/Ectoderm (242). The key cytogenetic differences between tumors of Class Mesoderm and tumors of Class Endoderm/Ectoderm seems to be that tumors of Class Endoderm/Ectoderm usually have many additional cytogenetic abnormalities superimposed on genetic hodgepodge of fusion genes, re-arranged genes, and various other gene mutations.

Chapter 16 Summary

Tumors of Class Endoderm/Ectoderm account for more than 95% of all human cancers. The reasons for this high frequency are threefold: cells of endoderm/ectoderm line the topological outside of the body (skin and gut) and are directly exposed to carcinogens received from air, water, and food; cells of the endoderm/ectoderm are capable of metabolizing carcinogens to their proximate (active) forms; cells of the endoderm/ectoderm are dividing cells and therefore targets for carcinogens.

Tumors of class endoderm/ectoderm also tend to be the most aneuploid, heterogeneous, and genetically unstable tumors of the body. The reason for this is that these tumors form over a long period of time, during which altered cells are continuously exposed to carcinogens. This results in cumulative alterations in initiated cell populations. In addition, some of the most common epithelial tumors undergo a long period of very slow expansion during which tumor cells simply slough into a lumen (gastrointestinal tumors) or the air (skin tumors) without appreciably contributing to the growth of the primary tumor mass. During this period, the selection of increasingly aggressive tumor cells proceeds unabated.

The same factors responsible for the high incidence of tumors of class endoderm/ectoderm are also responsible for remarkable aneuploidy, genetic instability, and heterogeneity of these tumors. It is unlikely that we will find cures for the metastatic tumors of Class Endoderm/Ectoderm before we find cures for the tumors with simpler genetic alterations, especially tumors of Class Mesoderm and Class Germ cell. The best way to reduce deaths from tumors of Class Endoderm/Ectoderm is to prevent tumor initiation (the earliest step in carcinogenesis) or to eradicate precancers (the last step of cancer development before the occurrence of an invasive neoplasm).

CHAPTER

17

Mesodermal Neoplasms

If you don't know where you're going, you'll wind up somewhere else.

—Yogi Berra

17.1 Background

Tumors of Class Mesoderm include the sarcomas, as well as some carcinomas and other tumors of mixed sarcoma and carcinoma morphologies. The purpose of this chapter is to provide a simple way of understanding tumors of mesodermal origin (Figure 17.1).

17.2 The Mesenchyme

In human development, the mesoderm is the embryonic layer separating the ectoderm and the endoderm. The mesoderm gives rise to all of the connective tissue of the body (i.e., the mesenchyme). The mesenchymal tissues are usually divided into the soft tissues (deriving from muscle, fibrous tissue, and vascular tissue), hard tissue (i.e., bone and cartilage lesions), and hematopoietic (which would include all lymphomas and leukemias).

17.3 Coelomic Tissues

In addition to the mesenchyme, the mesoderm gives rise to cavities lined by specialized mesodermal epithelium (coelom). The coelomic cavities become the pleura, peritoneum, pericardium, and joint spaces. All of these cavities are lined by mesothelium (body cavities), or by a specialized lining closely related to mesothelium, the synovium

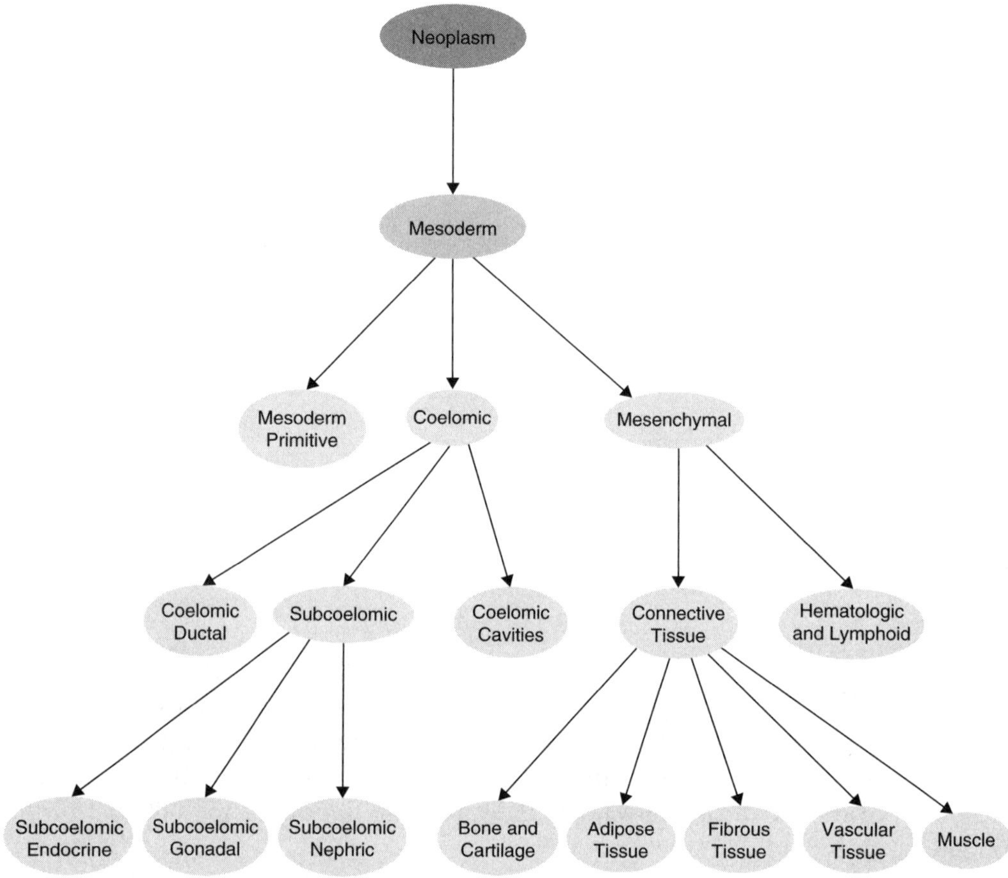

FIGURE 17.1 Hierarchy of Class Mesoderm tumors.

of joint cavities. These give rise to mesothelioma and synovial sarcoma, two tumors that are morphologically similar, composed of both epithelial cells and spindle cells (64).

A very specialized coelomic lining cell covers the gonads. These cells are morphologically, clinically, and genetically distinct from the other coelomic lining cells and are assigned their own subclass, which includes the papillary carcinomas of ovary, the most common ovarian cancer.

The embryonic coelom is also capable of forming epithelial-lined ducts. These ducts (such as the paramesonephric duct) give rise to the fallopian tubes and uterus. All uterine and cervical cancers fall in the class of mesodermal, coelomic-ductal tumors. The mesodermal lineage of these tumors is consistent with the variety of epithelial, mixed epithelial and nonepithelial, and sarcomatous tumors arising from the uterus, including endometrial carcinoma, carcinosarcoma of endometrium, heterologous mixed mesodermal tumor of uterus, and leiomyosarcoma.

17.4 The Subcoelomic Mesodermal Tissues

A specialized mesoderm develops subjacent to coelomic cavities. Specialized subcoelomic mesoderm develops into the stroma of the gonads, the adrenal cortex, and the parenchyma (tubules) of the kidneys. Tumors derived from these specialized mesodermal tissues are assigned their own subclasses.

The mesoderm that develops as gonadal stroma gives rise to the sex cord-stromal tumors of the ovary and testis (e.g., granulosa cell tumor, thecal tumors, Sertoli-Leydig tumors, and sex cord tumor with annular tubules). The mesoderm that develops as adrenal cortex gives rise to cortical adenomas and carcinomas. The mesoderm that develops as kidney gives rise to renal cell carcinoma and all of its variants.

17.5 Generalizations about Mesodermal Tumors

A collection of morphologic, biologic, and clinical features distinguish the mesodermal tumors from other classes of tumors (List 17.5.1).

17.6 Why Do Soft Tissue Tumors Have Simple Genetic Markers?

The first five items in List 17.5.1 have been discussed in other sections of this book. Item six is an oft-ignored property that may explain many of the observed cytogenetic differences between epithelial tumors and sarcomas. Most epithelial tumors

List 17.5.1 General features of Class Mesoderm tumors.

1. They are rare. Mesodermally derived tumors account for only a small percentage of all of the tumors that occur in humans.
2. They tend to be composed of spindle-shaped cells. This is especially true of the mesenchyme-derived subclass of mesodermal tumors.
3. They tend to metastasize via blood vessels (hematogenous seeding). Tumors of endoderm/ectoderm origin tend to metastasize via lymphatics.
4. As a group, they tend to occur among all ages (although individual types of tumors may target a narrow age range). Class Endoderm/Ectoderm tumors tend to occur largely in adults in a late stage of life (over 50 years of age).
5. They tend to be characterized by simple cytogenetic markers.
6. They have no physiologic mechanism to slough off tumor cells. This is stark distinction from tumors of endoderm/ectoderm origin, which slough tumor cells into gut (endoderm) or air.

develop from intraepithelial dysplasia. The dysplastic cells mimic the normal cells of the surrounding mucosa and differentiate upward as they traverse the mucosal width. Eventually, they slough into the lumen (of the gut or duct) or into the environment (external to the skin). During the precancerous phase, the size of the nascent neoplasm scarcely changes so long as the rate of extrusion from the tumor equals the rate of cell replacement. In effect, epithelial neoplasms have an opportunity to experiment with new subclones of cells that occur continuously as genetic instability adds additional genetic variants to the population of dysplastic cells. Eventually, a population of cells with a higher proliferation rate or a lower apoptotic rate will emerge. The growth of new cells will exceed the extrusion of old cells and a tumor mass will appear. The combined influences of genetic instability, selection of rapidly growing subclones, and extrusion of most of the tumor population will result in a tumor filled with highly atypical cells with multiple genetic abnormalities.

Most mesenchymal neoplasms cannot grow like epithelial neoplasms because they have no mechanism to slough their cell population. When a neoplastic cell grows from the center of a muscle, a lump of adipose tissue, or a fascia, its offspring have no place to go, and a soft-tissue neoplasm retains tumor cells from every cell generation of tumor growth. This is an oversimplification, of course. Many of the tumor cells in a soft-tissue neoplasm will die, and some of these cells will be engulfed by reticuloendothelial cells (macrophages), to be slowly removed from the tumor. The basic difference between soft-tissue tumor growth and epithelial tumor growth is that of garbage collection, but this difference may explain the cytogenetic differences between epithelial tumors and sarcomas. Epithelial tumors may undergo hundreds or thousands of generations before a tumor mass develops. During this time, epithelial tumors accumulate genetic abnormalities. Soft-tissue tumors undergo a smaller number of growth cycles before a clinically detectable mass emerges (because they cannot extrude cells as they grow). With fewer growth cycles in which to accumulate genetic abnormalities, you might expect that soft-tissue tumors would have simpler genetic markers (and they do).

17.7 Why Do Tumors of Mesodermal Origin Occur Infrequently?

The mesoderm is sandwiched between the ectoderm and the endoderm. At this location, it is protected from direct exposure to carcinogens in the food we eat and the air we breathe. The mesenchyme is relatively protected from ultraviolet light, which must pass through the epidermis and the melanocytic layer before reaching the dermal mesenchyme. When carcinogens manage to stray into the mesenchyme, they encounter cells that are very poor metabolizers and that seldom divide. It is not surprising that the mesoderm, which is spared the brunt of carcinogen exposure, accounts for just a small percentage of the neoplasms of humans.

17.8 Renal Cancer and Uterine Cancer, So Misunderstood

The most common tumors of mesodermal origin are those of the kidney and the uterus. Because most tumors of the kidneys and the uterus have an epithelial morphology, tumor biologists may forget that these tumors have a mesodermal origin and are much more closely related to sarcomas than to any tumors arising from endoderm or ectoderm. The kidney is an example of an epithelial organ that does not derive from endoderm/ectoderm. The tubules of the kidney derive from mesodermal cells (sometimes referred to as mesenchymal cells because they arise from loose mesodermal tissue that otherwise contributes to the connective tissue of the retroperitoneum). During kidney organogenesis, the mesodermal cells condense around the (metanephric) ureteric bud and acquire an epithelial morphology. They form convoluted tubes that merge into the ureteric bud. Tumors of the adult kidney usually have a purely epithelial morphology.

Despite having an epithelial morphology, renal tumors share biological properties with the sarcomas. A recent article compared the gene expression profile of renal cell carcinomas (RCCs) with the gene expression profiles of other tumors. The authors of the article found that "a RCC-specific set of oncogenesis-related genes was identified and surprisingly shared by sarcomas" (243). The finding that surprised the authors was that the gene expression profile of RCCs was not similar to the gene expression profile of epithelial tumors. They did not expect that RCCs would cluster with sarcomas, but we know that an epithelial morphology is not a biological class and does not correlate to a tumor's developmental lineage. RCCs are mesodermal tumors, just as all sarcomas are mesodermal tumors. The shared expression profile of RCCs and sarcomas illustrates that the basic structure and pathways of a cell are inherited through its developmental lineage.

The renal epithelial cancers consist of a collection of tumors that fall under the common name RCC, and these include clear-cell (conventional) renal carcinoma, papillary renal carcinoma, and chromophobe renal carcinoma. These tumors have phenotypes that approximate the anatomic parts of the proximal tubule. Collecting duct carcinomas also occur. Benign renal tumors include papillary adenoma, renal oncocytoma, and metanephric nephradenoma (adenofibroma).

Monoclonal mixed epithelial/stromal tumors would be expected to arise from mesodermal organs, because epithelial and connective tissues derive from mesoderm. Mixed tumors arising from the kidney include rare entities such as primary renal synovial sarcoma, low-grade myxoid renal epithelial neoplasm with distal nephron differentiation, epithelioid angiomyolipoma, and of course, nephroblastoma (Wilms tumor).

The uterus, like the kidney, is derived entirely from mesoderm. The uterus is formed from a duct that forms within the mesoderm adjacent to the coelomic cavity (the paramesonephric duct). This paramesonephric duct gives rise to all of the uterus, including the endometrial epithelium and the underlying specialized stroma, and all of the fibromuscular tissue composing the uterine wall. Consequently, tumors of endometrial epithelial cells and endometrial stromal cells share the same lineage in the developmental classification (subcoelomic ductal).

List 17.7.1 Theories of uterine carcinosarcoma histogenesis (244).

Collision tumor theory: Synchronous biclonal tumors produce a mixed carcinoma/sarcoma histology.

Composition theory: The sarcomatous component is an exuberant, non-neoplastic overgrowth of connective tissue in reaction to the presence of the neoplastic carcinoma that invades the uterine musculature.

Combination tumor theory: Both carcinoma and sarcoma elements are subclones of a single progenitor stem cell.

Because pathologists have been "stuck" on the epithelial/spindle cell dichotomy, the histogenesis of mixed epithelial/spindle tumors in the uterus has puzzled generations of morphologists. Theories for the dual nature of uterine tumors have been proposed through the years and summarized by Fuji and coworkers (List 17.7.1) (244).

Fuji and coworkers demonstrated that carcinosarcomas have a monoclonal origin and must have arisen from a progenitor stem cell that is capable of differentiating as a carcinoma and as a sarcoma. In the uterus, the line between epithelial tumors and spindle tumors is blurred because every tumor arising in the uterus is of mesodermal origin, and mesodermal cells can be epithelial (like endometrial mucosal cells) or spindled (like myometrium). This is another example wherein the Developmental Classification accommodates morphologic class ambiguity.

The uterus gives rise to adenocarcinomas, adenocarcinoma precancers, sarcomas, and variously named mixed tumors, including carcinosarcomas (Figure 17.2), adenosarcomas, and mixed müllerian tumors with heterologous components.

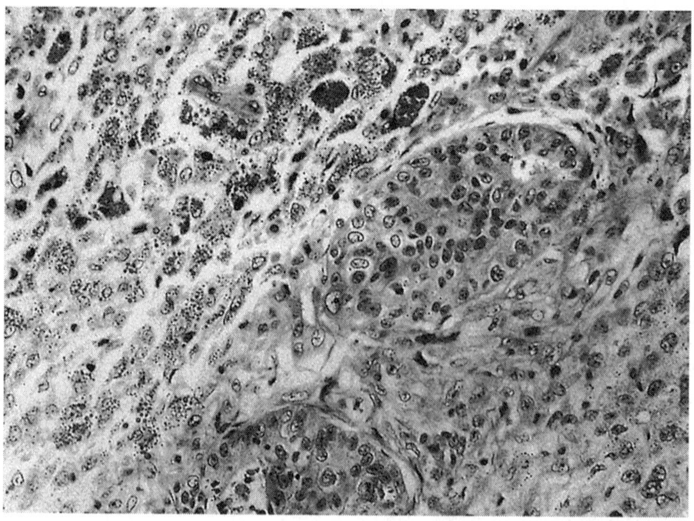

FIGURE 17.2 Carcinosarcoma of uterus. A loosely coherent collection of highly atypical cells is seen. Some cells are round. Some cells are spindle shaped.

Both epithelial and nonepithelial tumors of the uterus seem to have an association with tamoxifen therapy (245), suggesting the possibility that a single stem cell targeted by tamoxifen can give rise to epithelial and stromal tumors of the uterus.

17.9 Leukemias and Lymphomas

Both leukemias and lymphomas derive from mesenchymal cells. The hematopoietic stem cells give rise to red blood cells, granulocytes (neutrophils, eosinophils, and basophils), monocytes, and lymphocytes. There are a variety of well-described preleukemic lesions (the myelodysplasias and myeloproliferative disorders) and less well described **lymphoproliferative lesions** that precede the development of leukemias and lymphomas.

The lymphomas derive from lymphoid cells, which develop from lymph nodes and from the bone marrow and from lymphoid aggregates found in skin, mucosal surfaces, and other body sites. Lymphomas often have a phenotype that is fixed at a specific stage of lymphoid development, corresponding to any of the normally occurring lymphocytic cells at any of their developmental stages.

There are two major lineages of lymphoid cells: T and B. Each lineage is functionally distinct, and each derives from a different fetal organ (B cells derive from the bone marrow in humans or the bursa of Fabricius in birds, and T cells derive from the thymus). Lymphomas are divided into tumors of B cells or tumors of T cells. The lymphoid type occurring in any tissue is likely to be of the cell type that is present in greatest number in the tissue. In the gut, B cells predominate over T cells, and most gut-derived lymphomas are B-cell lymphomas. In the skin, T cells are more numerous than B cells, and T-cell lymphomas are common in the skin. In most cases, these tumors are intransitive (a B-cell tumor does not become a T-cell lymphoma or vice versa).

Sometimes lymphocytic leukemias evolve during the blast crises of myeloid leukemias, when a large portion of the leukemic population consists of early myeloid cells capable of division. When this happens, the resulting lymphocytic leukemias are usually B-cell type, consistent with lymphoid cells of bone marrow origin and indicating the common developmental origin of lymphocyte and myeloid cells.

Occasionally, a tumor of T cells will develop B-cell markers or vice versa. This is referred to as **lineage infidelity**. The mechanism for this transition is not well understood.

Burkitt lymphoma is a B-cell lymphoma that occurs most often in children (Figure 17.3). It is remarkable for its constitutive dependence on c-MYC. The c-MYC protein is a transcriptional regulator that regulates cell proliferation, differentiation, and death. The expression of c-MYC can be deregulated by any of several different mechanisms: the juxtaposition of c-MYC to **regulatory elements** of the Ig loci that enhance c-MYC expression, mutations in c-MYC regulatory regions, and amino acid substitutions in exon 2 enhancing the c-MYC transactivation domain.

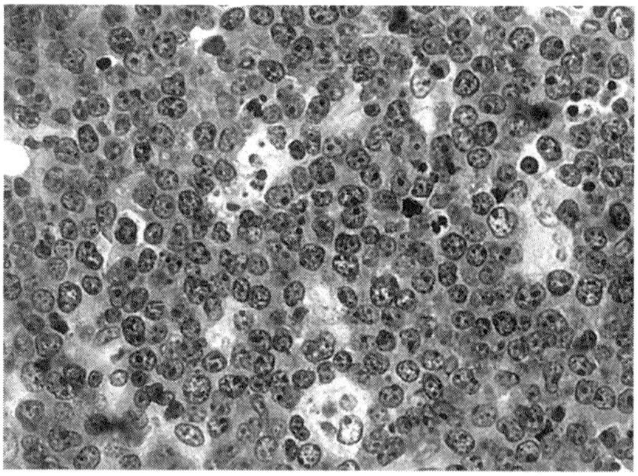

FIGURE 17.3 Burkitt lymphoma. A dense population of uniform atypical, round cells with scant cytoplasm is seen. Some cells are dying, and these cells are characterized by shrunken (pyknotic) or broken nuclei. A few large histiocytes, containing fragments of phagocytized cells, are noted. These large phagocytizing histiocytes are called tingible body macrophages.

17.10 *Helicobacter*-Induced Prelymphoma

Most neoplasms have an identified precancerous lesion that precedes the appearance of the neoplasm and that contains a clone of progenitor cells from which the neoplasm will arise. Surprisingly, the lymphomas have no designated prelymphoma lesions. We can infer, however, that prelymphomas must exist. *Helicobacter pylori* are bacteria that cause acute and chronic gastritis. People with long-standing *H. pylori* infections have an increased likelihood of developing a particular type of stomach tumor known as a MALToma (mucosa-associated lymphoid tissue lymphoma). After treatment with antibiotics to eliminate the *H. pylori* infection, the MALToma can regress (246). Precancers commonly regress. Cancers almost never regress. The facile regression of MALTomas after antibiotic therapy is consistent with the biology of a precancer.

Rosenberg and Horning have shown that up to 25% of non-Hodgkin's lymphomas may regress spontaneously (47). This suggests that many of the clonal lymphoid growths that are commonly called lymphomas would best be called prelymphomas. At this time, we have no criteria to distinguish prelymphomas from lymphomas.

Chapter 17 Summary

The mesodermal tumors include a wide array morphologically disparate neoplasms. The most familiar mesodermal tumors are tumors of mesenchyme (connective tissue, vascular tissue, hematopoietic cells and lymphoid tissues). The mesoderm also gives rise to epithelial populations that derive from coelom (mesothelial cells), subcoelomic ducts (including the uterus), and periductal mesenchyme (kidney epithelia). Much of the confusion over tumors of "mixed" epithelial and mesenchymal origin disappears when we acknowledge that populations of neoplastic cells of coelomic/subcoelomic origin can have spindle morphology or epithelial morphology.

Tumors of the mesoderm are rare, and this may simply be a consequence of the topologic organization of the body. Cells of the ectoderm receive the brunt of exogenous carcinogens. Mesodermal cells are relatively spared. In neonates and children, where tumors are not the result of long-term exposure to exogenous carcinogens, tumors of mesodermal origin occur as frequently as tumors of other developmental lineages.

CHAPTER

18 Tumors of Neuroectoderm (Brain and Central Nervous System)

Discovery is ". . . seeing what others have seen, but thinking what others have not."

—Albert Szent-Gyorgyi

18.1 Background

The neuroectoderm gives rise to the neural tube, which forms the brain, the spinal cord, and virtually nothing else. Years ago embryologists considered the neural crest to be an offshoot of the neural tube, and tumors of neural crest origin were grouped along with brain tumors as neuroectodermal neoplasms. Today we still encounter terminology that includes neuroblastoma among the neuroectodermal neoplasms. For example, PNET (primitive neuroectodermal tumor) refers to a host of peripheral tumors with no anatomic relationship to brain or spinal cord and no proven derivation from neuroectoderm. In this chapter, we argue that PNETs are tumors of class neural crest.

The current *Nomina Embryologica Veterinaria* and the *Nomina Anatomica* provide an authoritative listing of current opinion regarding the histogenesis of the neuroectoderm (List 18.1.1) (223, 224). This simple histogenetic list makes it very easy for us to classify neuroectodermal tumors (Figure 18.1). The purpose of this chapter is to describe the general subclasses of neuroectodermal tumors and to provide some indication of the special features that distinguish these tumors from other classes of tumors.

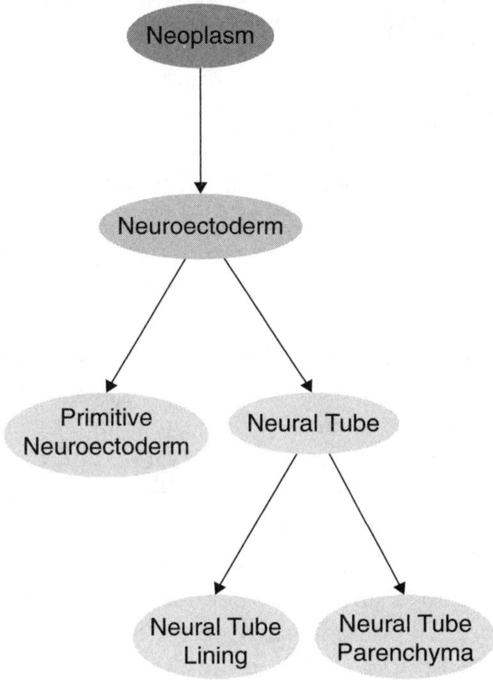

FIGURE 18.1 Hierarchy of Class Neuroectoderm tumors.

List 18.1.1 Histogenesis of epithelium tubi neuralis (neuroectoderm).

- Ependymoblasti
  - Spongioblasti
- Glioblasti
  - Myelinisatio
- Neuroblasti
  - Neuroblasti apolares
  - Neuroblasti unipolaries
  - Neuroblasti bipolares
    - Processificatio
    - Dendrificatio
    - Coni augmentales
- Neuroni

18.2 The Types of Neuroectodermal Tumors

Neuroectodermal tumors include all the so-called brain tumors, which include the variants of ependymomas, gliomas (including astrocytomas, oligodendrogliomas, and glioblastoma multiforme), and tumors of central nervous system (CNS) neuronal origin (including gangliomas, gangliogliomas, and medulloblastomas).

Through an accident of anatomy, brain tumors are malignant even though they almost never metastasize. They are malignant because they occur in the brain, a vital organ in which even small tumors can cause death. The skull restricts the outward growth of brain tumors. Any tumor that grows persistently inside the brain, even though it may not invade locally or metastasize distantly, will eventually kill its host. Eliminating the tumor is the only treatment option that has curative potential.

Class Neuroectoderm excludes all peripheral tumors of neuronal origin (because these tumors derive from neural crest neurons). Tumors of Class Neuroectoderm also exclude tumors of Schwann cells (because Schwann cells derive exclusively from neural crest). Neuroectodermal tumors also exclude meningiomas (which have neural crest origin), and they exclude all mesenchymal tumors that occur in the brain (such as lymphomas, angiosarcomas, and osteosarcomas). Teratomas and germ cell tumors occurring in the brain are not classed as neuroectodermal tumors. These are tumors of Class Germ Cell. Atypical teratoid rhabdoid tumors are a special case and are discussed separately in Chapter 15.

18.3 PNET Tumors Do Not Belong in Class Neuroectoderm

PNETs, primitive neuroectodermal tumors, do not belong in Class Neuroectoderm. This assertion seems as though it cannot be correct, but please bear with me. There is a rational explanation.

The neural crest and the neuroectoderm both produce nervous systems. The neuroectoderm produces the central nervous system, which is chiefly composed of neurons and specialized supportive cells called glia. The neural crest produces the peripheral nervous system, which consists chiefly of neurons and specialized supportive cells called Schwann cells. During embryologic development, the neural crest develops directly adjacent to the neuroectoderm. Many embryologists think of the neural crest as a derivative of neuroectoderm. This would justify labeling tumors of neural crest origin as neuroectodermal tumors. In the Developmental Classification, the neural crest is considered a separate germ layer, distinct from the neuroectoderm. The neural crest is not a subclass of neuroectoderm, just as neuroectoderm is not a subclass of ectoderm. Because Class Neural Crest is different from Class Neuroectoderm, some tumors of neural crest origin that are traditionally called "neuroectodermal" must be reclassified.

The PNETs are a subset of the Ewing family of tumors, which include Ewing sarcoma, primitive neuroectodermal tumors (PNETs), and Askin tumor. The Ewing family of tumors derive from the neural crest and are characterized by a translocation, usually t(11;22)(q24;q12), producing a fusion gene involving EWS. The primitive tumors of CNS origin (medulloblastoma, retinoblastoma, and pineoblastoma) lack this characteristic genetic marker.

In the Developmental Classification, the PNETs are all considered to be tumors of Class Neural Crest. The primitive tumors of Class Neuroectoderm (CNS) are a diverse set of lesions that are not unified by a group-characteristic marker, and they should not be included as PNETs.

18.4 The So-Called Blastomas of Neuroectodermal Origin (Glioblastoma, Ependymoblastoma, Medulloblastoma, Retinoblastoma, and Pineoblastoma)

18.4.1 Glioblastoma

Glioblastoma is a poorly differentiated astrocytoma. There is a known progression of astrocytoma, through worsening grades of differentiation, leading eventually to glioblastoma. Despite its name, glioblastoma is not a tumor composed of a primitive cell type "locked" into an undifferentiated state (i.e., not a true blastoma). Glioblastomas commonly occur in adults and uncommonly occur in children. All the true "blastomas" are rare tumors that occur predominantly in children.

18.4.2 Ependymoblastoma

The neural tube is lined by ependymal cells from which ependymomas derive. In children, a rare tumor of primitive round cells may derive from the ependymal lining, and this tumor is called ependymoblastoma. Most ependymomas arise from the ventricular system of the brain.

18.4.3 Medulloblastoma

Medulloblastomas are sometimes classified as a type of PNET, for reasons that have some merit. First, medulloblastomas are primitive, and they are neuroectodermal tumors. Medulloblastomas are tumors of small cells that often occur in children, like the other PNETs. The problem is that the PNETs belong to Class Neural Crest (247). Medulloblastoma is a CNS tumor of Class Neuroectoderm. Gene expression profiles of medulloblastomas are nothing like those of PNETs or of CNS atypical teratoid/rhabdoid tumors (Class Germ Cell) (248). Putting each of these three entities into separate tumor classes fits our classification.

Medulloblastomas occur in the posterior fossa, usually within the cerebellum and often along the midline (central axis of the brain). High levels of genes encoding transcription factors specific for cerebellar granule cells are found in medulloblastomas (248). Medulloblastomas are one of the few malignant brain tumors that can metastasize outside the brain.

18.4.4 Retinoblastoma and Pineoblastoma

Retinoblastoma is a tumor of Class Neuroectoderm. The optic nerve is an extension of the brain, and the nerve cells of the retina are specialized cells of neuroectodermal origin. The primitive nerve cells that produce retinoblastomas have a neuroectodermal origin.

In Chapter 5, we discussed pineoblastoma in the context of its role as the trilateral tumor in familial retinoblastoma syndrome. Like the retinoblastoma, the pineoblastoma is a tumor derived from primitive neuroectodermal cells.

18.5 Phakomatoses

The phakomatoses are inherited syndromes that produce hamartomas of neuroectodermal tissues (i.e., phakomas). The prototypical phakomatosis is tuberous sclerosis (also known as tuberous sclerosis complex [TSC]), which is characterized by a variety of rare hamartomas and neoplasms (249) (List 18.5.1).

The genetic basis of tuberous sclerosis involves biallelic inactivation of either of two unlinked genes that seem to have equivalent pathogenic roles. The genes are TSC1 (coding for hamartin) and TSC2 (coding for tuberin). The hamartomas of tuberous sclerosis can be found in tissues of several different developmental lineages. This fits with our previously discussed observation that tumor suppressor gene inactivation is associated with multiple tumor types derived from more than one developmental lineage.

List 18.5.1 Hamartomas and neoplasms associated with tuberous sclerosis complex (81).

Cortical tubers of brain
Retinal astrocytoma
Cardiac rhabdomyoma
Lymphangiomyomatosis (very rarely)
Facial angiofibroma
White ash leaf-shaped macules
Subcutaneous nodules
Cafe-au-lait spots
Subungual fibromata
Myocardial rhabdomyoma
Multiple bilateral renal angiomyolipoma
Ependymoma
Renal carcinoma
Subependymal giant cell astrocytoma

Cortical tubers of the brain are pathognomonic for tuberous sclerosis and consist of localized but poorly demarcated malformations of neuronal and glial cells. The neurons are typically large and misshapen.

Like other hamartoma syndromes, the germline mutation in tuberous sclerosis produces benign hamartomas and carcinomas. As previously observed, the occurrence of both hamartomas and carcinomas in inherited syndromes supports the inclusion of the hamartomas among the different kinds of neoplasms of animals.

Chapter 18 Summary

Neuroectodermal tumors are limited to tumors of ependymal, glial, or neuronal origin that arise from the brain and the spinal cord. A wide variety of tumors occur in, on, or near the brain that are not of neuroectodermal origin, and these may have an origin from any of the other cell lineages, including germ cells, neural crest, mesoderm, ectoderm, and endoderm.

CHAPTER

19 Neural Crest Tumors

The only interesting thing about vertebrates is the neural crest.

—Thorogood

19.1 Background

As the story goes, a genie grants three wishes. After the first two wishes are squandered (they always are!), we can only wonder why the hero of the story does not ask, as his third wish, for three more wishes. Apparently, recursive wishes are not permitted in fairy tales.

Surprisingly, nature provides us with a recursive opportunity that exceeds the plausibility of fairy tales. In human evolution, our third wish is for a neural crest that recursively employs all of the cell types that previously evolved.

During evolution, just about the time that vertebrates appeared, a new germ layer arose: the neural crest. This new germ layer had the remarkable property of recapitulating cells that would otherwise be found in ectoderm, endoderm, mesoderm, or neuroectoderm (250). In addition, the neural crest offered some fascinating cell types not present in any of the preceding germ layers.

Indeed, the remarkable complexity of the neural crest has made it difficult for us to grasp its biological roles. Even today, developmental biologists have not fully characterized the neural crest, and pathologists have not applied our growing knowledge of the neural crest to questions of neoplasm classification (Figure 19.1).

Let us review the phylogenetic evolution of man to determine the point at which the neural crest was invented by our distant ancestor (List 19.1.1).

Not very far down, tenth in line just after Chordata, are the Craniata. This is the class of organism in which the neural crest appears as a fully functioning germ layer. Order Craniata emerged well before the evolution of the mammals and includes all animals

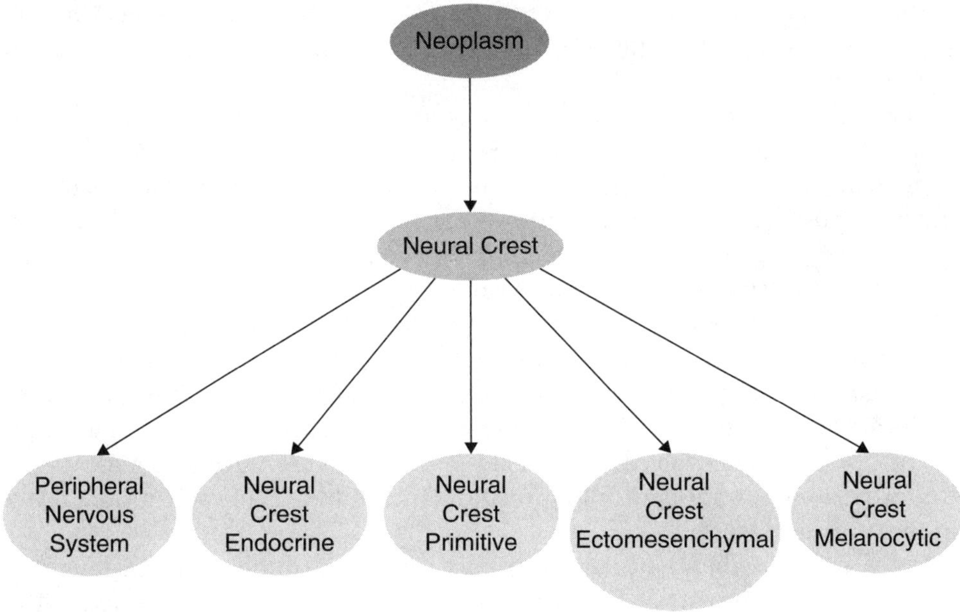

FIGURE 19.1 Hierarchy of Class Neural Crest tumors.

List 19.1.1 Our family tree.

Cellular organisms	Tetrapoda
Eukaryota	Amniota
Fungi/metazoa group	Mammalia
Metazoa	Theria
Eumetazoa	Eutheria
Bilateria	Euarchontoglires
Coelomata	Primates
Deuterostomia	Haplorrhini
Chordata	Simiiformes
Craniata	Catarrhini
Vertebrata	Hominoidea
Gnathostomata	Hominidae
Teleostomi	Homo/Pan/Gorilla group
Euteleostomi	Homo
Sarcopterygii	Homo sapiens

that have a face, because the neural crest is responsible for the development of a skull and of the bony tissues that give animals a face, something a little more distinguished than a mere head.

All Craniata have a neural crest. Humans and fish are distant cousins and both belong to Class Craniata. Because humans and fish both have neural crests, we both develop neural crest tumors. It is easy to find schwannomas and melanomas in fish. It would be impossible to find such tumors in organisms that do not descend from order Craniata. Do not expect to find schwannomas in a tarantula, a crab, a beetle, or in any of the crowd of faceless species that dwell among us.

The neural crest produces cell types not produced by the other germ layers. These would include odontoblasts, melanocytes, Schwann cells, and paraganglial cells. None of these cell types are present in orders outside order Craniata. Multicellular animals first appeared approximately 600 million years ago. In all this time, only a few hundred different cell types had evolved in chordates. The neural crest deserves a great deal of credit for creating several new cell types and a method of recursively producing cells from the other germ layers.

Many years ago, when I began my training in cancer biology, I was taught that the neural crest derived from neuroectoderm and gave rise to the peripheral nervous system, to melanocytes, and to a widely distributed collection of neuroendocrine cells that intermingled with the endodermal lining cells. The endoderm is divided into the foregut (which gives rise to the lungs and esophagus), the midgut (giving rise to the small intestines), and the hindgut (giving rise to the colon). All of these organs are sites where carcinoids may occur. Melanomas, schwannomas, pulmonary carcinoids, gastrointestinal carcinoids, islet tumors, and paragangliomas were all putative neural crest tumors. Neuroblastomas were considered primitive tumors with features consistent with a neuroectoderm origin. In the past, the neural crest was considered a derivative of the neural tube, and it made sense that any primitive tumor of neural type would be consigned to the neuroectoderm. Most of these cherished notions are incorrect and ignore a large body of recent scientific evidence.

Although modern anatomists have verified the close relationship between the neuroectoderm and the developing neural crest, the biologic properties of the neural crest are not derivative of the neuroectoderm. For the purposes of neoplasm classification, tumors of neuroectodermal origin must be distinguished from tumors of neural crest origin. For example, neuroblastomas are almost certainly of neural crest origin and not of neuroectodermal origin. Neuroblastomas share properties with Class Neural Crest tumors and exclude properties of Class Neuroectoderm tumors. Carcinoids and tumors of the pancreatic islets are almost certainly endodermal tumors, not neural crest tumors. Almost everything I learned was wrong.

The purpose of this chapter is to describe our current understanding of the neural crest, to provide a newer and better classification of the neural crest tumors, and to demonstrate the relevance of this classification to cancer prevention and treatment.

19.2 The Neural Crest in Human Development

In embryos of order Craniata, neural crest cells condense along the length of the neural tube (neuroectoderm), in the space between the neural tube and the ectoderm. The cells of the developing neural crest are motile and invest the mesoderm with their derivatives. The neural crest has four major divisions, producing tissues of specific functions (List 19.2.1).

The current authoritative listings of neural crest ontogeny are the *Nomina Embryologica* (human embryologic nomenclature) and the *Nomina Embryologica Veterinaria* for animals (List 19.2.2).

List 19.2.1 Anatomic divisions of the neural crest.

Cranial neural crest—cranial nerves, mesenchyme of the face and skull, including the mesenchyme of several pharyngeal arches and pouches, odontoblasts, and C cells of the thyroid
Trunk neural crest—melanocytes and dorsal root ganglia
Vagal and sacral neural crest—parasympathetic ganglia
Cardiac neural crest—arteries arising from the heart (including the aortopulmonary septum), connective tissue of the several pharyngeal arches, and melanocytes

List 19.2.2 Ontogeny of the neural crest, combined excerpts from the *Nomina Anatomica* and the *Nomina Anatomica Veterinaria* (223, 224).

- Textus cristae neurali (mesectoderma or ectomesenchyma)
- Segmenta cristae neuralis
  - Ganglia craniospinalia
  - Ganglia autonomica (missing in *Nomina Anatomica*)
- Neuroblasti
  - Chromaffinoblasti
  - Corpora para-aortica (missing in *Nomina Anatomica*)
  - Neurolemmoblasti
    - Myelinisatio
  - Glioblasti ganglionici
  - Melanoblasti
  - Mesenchyma capitis
    - Chondroblasti
    - Odontobiasti

The human and veterinary ontogenies for the neural crest differ in only two items. The veterinary edition contains "Ganglia anatomica" and "Corpora para-aortica," both missing from the human ontogeny. Both these neural crest structures are actually present in humans and both give rise to tumors.

19.3 Tumors of the Peripheral Nervous System

The neural crest ganglia are the site of origin of paragangliomas and of extra-adrenal neuroblastomas. The corpora para-aortica are the site of origin of paragangliomas and of chemodectomas. The most prominent para-aortic body in humans is the organ of Zuckerkandl, which is a site of origin of extra-adrenal pheochromocytomas and of paragangliomas.

The peripheral nervous system derives wholly from the neural crest and consists of all of the nerves of the body other than the nerves of the central nervous system. The central nervous system is of neuroectodermal origin and is anatomically confined to the brain and the spinal cord, whereas the peripheral nervous system permeates the body.

The peripheral nervous system contains the so-called autonomic nervous system (the nervous system that functions automatically, without mental instruction), and the motor neurons that synapse from the spinal CNS motor neurons and extend to the effector organs (muscles). All of the cranial nerves, with the exception of the optic nerve (cranial nerve II) are neural crest derivatives.

Because the neural crest produces neurons and the CNS also produces neurons, it is easy to imagine confusion over the correct histogenetic assignment of tumors that have features of primitive neural cells. Many tumors that have neural crest origin are still called neuroectodermal tumors. It is common to read journal articles that refer to neuroblastomas as neuroectodermal tumors. Similarly, the term neuroectodermal is permanently welded into the names of various tumors, such as "neuroectodermal tumor of infancy." In both cases, these tumors are of neural crest (not neuroectoderm) origin.

Because neurons may have either a neuroectodermal origin or a neural crest origin, how can we be sure of the derivation of tumors with neuronal differentiation? The simplest method is by anatomic distribution. As already noted, the neural crest and the neuroectoderm have strict anatomic boundaries. Tumors of neuroectoderm origin almost always arise within the brain or the spinal cord. If a primary neuronal tumor occurs peripherally, it is always of neural crest origin. Neuroblastomas are tumors that usually arise from peripheral ganglia, and thus, we infer that neuroblastomas are of neural crest origin.

In tumors of the head or from the spinal cord, proximity of the tumor to the CNS may be problematic. Often, the presence of subpopulations of cells within the tumor may help assign an accurate histogenesis for the tumor. Neurons of neuroectodermal origin are invested by glial cells, whereas neurons of neural crest origin are invested by Schwann cells. Schwann cells give rise to schwannomas, and glial cells give rise to gliomas. If the tumor has a Schwann cell component, it must be a tumor of neural crest origin (e.g., schwannoma, neurofibroma, and ganglioneurofibroma). If the tumor has a glial component, it must be of neuroectoderm origin (e.g., glioma, glioblastoma, astrocytoma, and ganglioglioma). Luckily, tumors of mixed type (Schwann cell and glial cell) are never observed.

When Schwann cells and glial cells are both absent from a primitive tumor of neural differentiation, it may still be possible to accurately assign tumor lineage. Neuroectodermal tumor of infancy (also known as juvenile progonoma, retinal anlage tumor, and melanotic neuroectodermal tumor of infancy, among others) is a rare, primitive tumor that occurs primarily in infants. Most of these tumors secrete measurable amounts of vanillylmandelic acid, a compound that is characteristically produced by certain endocrine cells of neural crest origin. The production of neural crest-specific hormones by the so-called neuroectodermal tumors of infancy indicate that the tumors derive from neural crest and not from neuroectoderm or retinal anlage.

The neural crest-derived neuron (ganglion cell) is associated with a peculiar round cell called a paraganglion cell. It is convenient to think of paraganglial cells as a postsynaptic (postganglionic) epithelial cell whose normal functions are involved in some way with responses to ganglionic (nervous) stimulation. The paraganglionic cells grow as epithelial nests in the locations of peripheral nervous system ganglia. These nests of paraganglia and ganglia are dispersed throughout the trunk, neck, and head. The largest nest of paraganglia is the adrenal medulla. In this location, the paraganglial cells are called chromaffin cells (so named because they turn a rich brown color when exposed to a dichromate-chromate mixture). The general term for tumors arising from paraganglial cells is paraganglioma, but the special term for paragangliomas arising from the adrenal medulla is pheochromocytoma. All paragangliomas are neural crest-derived tumors.

Chemodectoma is another type of paraganglioma. Chemodectomas arise from specialized chemoreceptive paraganglia that reside in the walls of large vessels. Chemodectomas are sometimes named for the specific vessel in which they occur (e.g., aortic-body tumor, carotid-body tumor, and glomus jugulare tumor).

19.4 Melanocytic Lesions

The neural crest also produces melanocytes and nevus cells. Melanocytes are large, pigmented cells that can be found throughout the body, but are found in highest concentration sprinkled throughout the basal (lowermost) layer of the epidermis (the epithelial covering of skin). Nevus cells are round and are found in the dermis of the skin; they produce melanocytes and nevi. Nevi, also known as moles, are focal collections of nevus cells and melanocytes usually confined to the dermis or to the dermis and the lower epidermis. The number of morphologic and biologic variants of the common nevus is startling. Some are spindled. Some are epithelioid. Some grow deep, penetrating into the subcutaneous tissue. Some are confined to the epidermis (lentigo). Some are highly pigmented, and some are devoid of pigment. Some are congenital (present at birth), and some develop in adults. Nevi are benign, neural crest-derived hamartomas that typically grow to a certain size (under 5 mm) and then stop and never enlarge to any appreciable extent during the human lifespan. It is not unusual for people to have dozens or even hundreds of benign nevi. The clinical importance of nevi comes from their occasional resemblance to melanomas (a highly malignant lesion) and the rare progression from benign nevus to malignant melanoma.

19.5 Odontogenic Tumors

In the stomatodeum (primitive mouth), the neural crest gives rise to highly specialized cells that, together with ectoderm-derived enamel-forming cells (ameloblasts), produce teeth. Tooth formation is a fascinating example of differentiation that requires the mutual, coordinated participation between two different germ layers (ectoderm and neural crest). Buds of epithelial cells of ectoderm origin localize in the primitive gingival mucosa. Each ectodermal bud marks the site of a future tooth. The growing buds induce differentiation in cells located in the underlying mesenchyme. This mesenchyme of the mouth is of neural crest origin (so-called **ectomesenchyme**). The induced cells are the future odontoblasts (the cells that produce dentin and that will compose the largest and hardest portion of the tooth). The odontoblasts, in turn, induce the overlying epithelial cells to transform into ameloblasts, which become the enamel organ of the tooth (the enamel that crowns the tooth). The root of the tooth is composed of fibrous tissue and vessels that are also derived from neural crest.

The developed tooth is a wonder of evolutionary effort. It may appear that evolution has devoted undue attention to developing the humble tooth (a structure that is absent or vestigial in many craniates). Close examination of fish and reptiles reveals that scales are closely related to teeth. Mammals swapped scales for teeth in a few evolutionary steps.

Because the ectoderm-derived ameloblasts induce differentation in the underlying mesenchymal cells, odontoblasts are dependent on ameloblasts for their development. Hence, odontogenic cancers are virtually all of ectodermal origin, and the odontoblasts, when present, are merely there to support the neoplastic ameloblastic population.

19.6 Ameloblastomas

Ameloblastomas are aggressive tumors of purely ectodermal origin and consist morphologically of ameloblastic cells without odontoblasts. Ameloblastomas never contain enamel, although enamel is the substance that ameloblasts exclusively produce during normal tooth development. The reason that enamel is not found in ameloblastomas is that during normal tooth development, enamel production is dependent on the inducing influence of odontoblasts. Because ameloblastomas contain no odontoblasts, the biologic pathway to enamel production is broken.

Odontomas are hamartomatous malformations of teeth. They are benign and contain all of the cells and substances that contribute to tooth formation, including odontoblasts that produce dentin and ameloblasts that produce enamel. As malformations, odontomas may contain multiple cell lineages, and it is not surprising that a malformation of teeth would contain cells of ectoderm origin (ameloblasts) and neural crest origin (odontoblasts).

In rats, both odontomas and ameloblastomas are produced with experimental carcinogens (251, 252). In one study, 29 odontogenic tumors developed in 134 rats injected only once with methylnitrosourea (0.4 millimole/kg body weight) via the portal vein after partial hepatectomy. Twenty-six tumors were odontomas, and three were ameloblastomas. One of the ameloblastomas occurred adjacent to an odontoma. It is tempting to surmise that the carcinogen produced primarily hamartomatous malformations of teeth, and from the hamartomatous growths, occasional clonal populations of neoplastic ameloblasts arose. The occasional development of a malignant neoplasm from a hamartomatous growth is a common theme. It is seen in Cowden syndrome (253) and in neurofibromatosis (254).

19.7 Endocrine Tumors of Neural Crest Origin

The neural crest is the source of several endocrine organs. The adrenal medulla, an endocrine gland composed of specialized paraganglial cells, has already been discussed. Also of neural crest origin are the C cells (calcitonin-secreting cells of the thyroid) that grow as inconspicuous nests among the thyroid follicles (of endodermal origin). Both the thyroid and the adrenal are examples of mixed, multipartite glands containing distinctive and anatomically partitioned cell populations derived from two germ layers. The pituitary gland (adenohypophysis of endodermal origin and neurohypophysis of neural origin) is another example of multipartite endocrine glands.

19.8 Cladistics and the Neural Crest-Derived Mesenchymal Tumors

A cladistic classification is based purely on biologic lineage. Biologic classifications seldom have a strictly cladistic hierarchy. For example, by removing birds from their ancestral lineage (reptiles) and placing them, along with mammals, under Amniota, the classification of organisms lost its cladistic organization. Simply put, organisms are not classified strictly by lineage. Organisms are sometimes classified by weighing the importance of lineage against the commonality of characteristic inherited traits.

The neural crest makes it difficult to construct a strict cladistic classification of neoplasms. Because mesenchymal cells of the head and neck derive from neural crest, not mesoderm, the mesenchymal tumors of the head and neck have a different developmental lineage than the mesenchymal cells of the rest of the body. A cladistic classification, based purely on lineage, would require us to assign mesenchymal tumors of the head and neck (e.g., angiosarcomas, osteosarcomas, chondrosarcomas, and liposarcomas of the head) to a subclass of neural crest-derived tumors. Mesenchymal tumors of the rest of the body would be assigned to a subclass of mesodermal tumors. Whenever we referred to any of these tumors, we would need to know its anatomic location (head and neck or someplace else) before we could determine the proper class of the tumor.

In the Developmental Classification, we place the mesenchymal tumors of the head and neck into Class Mesoderm. This makes Class Neural Crest **paraphyletic** (because it does not include all of its descendants), and it makes Class Mesoderm polyphyletic (because it contains members that did not descend from the mesodermal lineage).

As a general rule, bones of neural crest origin are membranous (sometimes called dermal) bones. A membranous bone forms in situ from a fibrous structure that has the shape of the final bone and that becomes mineralized and ossified over time. In contrast to bone of neural crest origin, bones of mesodermal origin are **enchondral**. Enchondral bones elongate from a cartilaginous growth zone that calcifies as bone forms from one side of the zone.

Other than this superficial phenotypic distinction, there seems to be no difference between the bone tumors that arise from neural crest and the bone tumors that arise from mesoderm. The only biologic difference between bone tumors that derive from neural crest and bone tumors that derive from mesoderm is location: neural crest-derived bone tumors are located in the head. All other bone tumors derive from mesoderm. Consequently, tumors that derive from neoplastic osteocytes and chondrocytes of neural crest-derived bone have no known differences from tumors of the same cell types that derive from mesodermal bone. For this reason, the neural crest bone tumors are combined with the mesenchymal bone tumors in the Developmental Lineage Classification of Neoplasms.

Are there tumors that arise from neural crest bone that are not related to osteocytes or chondrocytes? Yes. Melanotic neuroectodermal tumor of infancy is an extremely rare tumor that usually arises from the mandible. The tumor grows rapidly and aggressively invades adjacent tissues. The cells of the tumor may resemble melanocytes or neuroblasts or primitive neuroepithelium (255). Because this tumor is distinct from other tumors of bone, only arises in facial bones and skull (i.e., neural crest origin), and has a neural crest phenotype, melanotic neuroectodermal tumor of infancy is assigned to Class Neural Crest.

19.9 Meningiomas

Meningiomas are tumors that grow from the coverings of the brain (the meninges). Specifically, meningiomas are presumed to be the neoplastic equivalent of the meningothelial cells of arachnoid villi. Arachnoid villi are specialized structures that project from the arachnoid layer of the meninges into the cerebrospinal space and actively absorb cerebrospinal fluid. The areas of the meninges from which the greatest number of meningiomas are known to arise, are the same areas known to contain the greatest concentration of arachnoid villi.

Meningiomas are not members of Class Neuroectoderm. They belong to Class Neural Crest. Evidence suggests that the meninges develop as neural crest derivatives (256). Meningiomas co-occur with other tumors of neural crest origin and are associated with neurofibromatosis, a neurocristopathy. Warwar and coworkers have reported a case of neurofibroma (Class Neural Crest), uveal melanoma (Class Neural Crest), and meningioma occurring in the same patient (257). More convincingly, meningiomas may occur in parts of the body outside the CNS. Meningiomas of the lung, though rare, have appeared in numerous case reports (258). Tumors of Class Neuroectoderm occur exclusively in the brain and spinal cord. Tumors of Class Neural Crest occur throughout the body.

19.10 Triton Tumors

Triton tumors are malignant schwannomas with a rhabdomyosarcomata (striated muscle cell tumor) component. Because these tumors seem to arise from two distinct lineages (neural crest and mesoderm), some pathologists have classified these tumors as teratomas (i.e., arising from a pluripotent stem cell capable of multilineage differentiation). A more likely explanation is that this tumor derives from a neural crest progenitor that can differentiate toward Schwann cells or rhabdomyoblasts (259). We have previously described that neural crest can develop into any type of mesenchymal cell. The occasional observance of malignant triton tumors arising from pre-existing malignant schwannomas strengthens our argument that the multilineage differentiation of triton tumors has a purely neural crest origin.

Chapter 19 Summary

The neural crest is a fascinating embryonic layer because it gives rise to the peripheral nervous system, melanocytes, several different endocrine organs (adrenal medulla and C cells of thyroid), and specialized cells of odontogenic mesenchyme. Because neurons derive from both neuroectoderm and neural crest, the separate classification of biologically distinctive neurocytic and neuroblastic tumors of Class Neuroectoderm and Class Neural Crest renders the neoplasm classification polyphyletic.

Class Neural Crest is a particularly challenging tumor class because the cranial division of the neural crest recapitulates the mesenchymal cell types (of mesodermal origin). Virtually all of the connective tissue of the head and neck has a neural crest origin, not a mesodermal origin. Sarcomas of head and neck origin are biologically and morphologically equivalent to sarcomas arising from Class Mesoderm. Consequently, in the developmental lineage classification, neoplasms of neural crest origin are grouped in Class Mesoderm, thus introducing paraphyly to the neoplasm classification.

CHAPTER

Tumors of Class Germ Cell and Class Trophectoderm

20

Things that are new are wont to be set forward rudely and formlessly, and then must be polished and perfected in succeeding centuries.

—Pappus, one of the last Hellenistic mathematicians, circa 350–300 BC

20.1 Background

The neoplastic entities described in this chapter are controversial, mysterious, and rare. Even the definitions of the terms used in this chapter are subject to change without notice. The reason for all the mystery is that most tumors of class germ cell and class trophectoderm are not composed of cells that are similar to any identifiable differentiated cells found in the developed human body. The tumors described in this chapter seem to develop from primordial cells whose normal physiologic role is to disappear.

The purpose of this chapter to provide a simple and consistent way to think about tumors that do not derive from any fully differentiated cell found in adult humans.

20.2 Definitions

Germ cell tumors, stem cell tumors, extraembryonic tumors, and their relationships to one another are among the most contentious subjects in pathology. Too often, all of these types of tumors are lumped as "germ cell tumors." If you think of a germ cell tumor as a tumor composed of cells that are normally present at the beginning of life (i.e., the germinal period), you might conclude that all these different types of tumors are "germ cell tumors." This kind of lumping is not acceptable because we now know that the cellular pathogenesis of these different classes of tumors is distinct and has direct relevance to the manner in which these tumors will be clinically managed. The term "germ cell tumor" should be restricted to tumors whose cell of origin is a particular type of cell (the germ cell) (Figure 20.1).

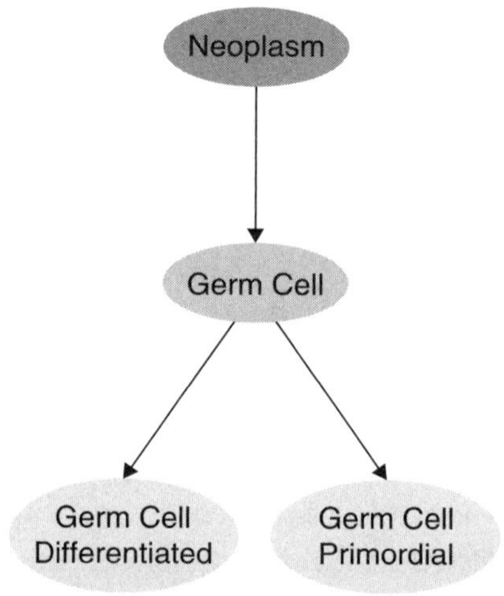

FIGURE 20.1 Hierarchy of Class Germ Cell tumors.

Germ cells are the progenitor cells of ovarian oocytes (eggs) and testicular sperm. In the human, germ cells arise from the endoderm, and in theory, all germ cell tumors could be classified as a subclass of endodermal/ectodermal tumors. This would be a terrible mistake. Germ cells are very special and have no properties exclusively shared with other cells of endodermal origin. For the purposes of classification, germ cell tumors fall into their own class. After germ cells are formed (during gastrulation), they migrate along the central axis of the body until they find the developing ovary (in the case of females) or testes (in the case of males). The ovary and the testes develop from mesoderm and are wrapped by a specialized coelomic epithelium (also of mesodermal origin). Because the ovary and the testis are composed of two different classes of mesodermal tissues as well as germ cells, and because they have no components derived from ectodermal origin or from nongerm cell endodermal origin, both ovary and testis develop a remarkable array of tumors that are not encountered in other organs.

During germ cell migration, some germ cells come to rest in midline structures along the length of the body. For this reason, tumors of germ cell origin occasionally develop in extragonadal locations, almost always along the midline of the body.

As the germ cells migrate to the primitive gonads, they perform a remarkable epigenomic process called "**erasure**." The process of cell lineage development is a process of epigenomic modification. DNA methylation, histone modifications, and possibly many other alterations to the packaging of DNA are required to produce cells that behave in a restricted manner that can be passed along to the progeny of the cell. Epigenomic modification is the normal mechanism through which all cellular differentiation is achieved. The problem for the organism is that there are two specific occasions

List 20.2.1 Things that germ cells can do (sometimes).

1. They can give rise to tumors composed of cells with features similar to those of differentiated germ cells (germinomas), including dysgerminoma (of ovary) and seminoma (of testis).
2. They can give rise to tumors of totipotent cells, including teratomas and embryonal cell carcinomas.
3. They can give rise to mixed tumors (of differentiated germ cells and of embryonic cells).
4. Female germ cells can produce new organisms through a process of *parthenogenesis* in which an unfertilized ovum develops into a viable embryo.
5. Cultured primordial germ cells, supplemented with growth factors, can produce *totipotent stem cell* lines (260, 261).
6. Female germ cells can use their "erasing" equipment to activate a parthenogenetically inserted nucleus taken from a differentiated cell, yielding an organism (as in Dolly, the cloned sheep).

for which cells must be dedifferentiated. In the case of the postfertilized ovum (zygote), the nucleus must be dedifferentiated if it is to serve as the totipotent progenitor for all the cells of the embryo. In the case of germ cells, they must erase their epigenomic pattern so that they can contribute meiotically recombined chromosomes that lack epigenomic changes found in the parent cells. For example, in the female, one x chromosome is inactivated through epigenomic modifications. The progeny of the female must have x chromosomes composed by meiotic recombination of both x chromosomes (even the inactivated x chromosome). To achieve meiotic recombination, the inactivating epigenetic modifications to the x chromosome must be erased before the time that the ovum is formed. To preserve meiotic recombination, the germ cell erases its epigenomic program (the program that told the cell that it was a germ cell of endodermal lineage) before it commits to the development of an ovum or a sperm.

Because germ cells erase their differentiated programming, they achieve a state, for some period of time, of totipotency. For this reason, germ cells are capable of some of the most remarkable phenomena of nature (List 20.2.1).

The germ cell is such a remarkable entity that all other cell types pale in comparison. We see that the germ cell is our key to understanding a range of rare neoplasms for which phenotype (including morphology and cell of origin) is disconnected from genotype.

20.3 Diverse Nature of Germ Cell Tumors

Nowhere is the dissonance between morphology, molecular biology, and developmental biology more striking than in the realm of the germ cell tumors. The germ cell tumors include tumors of diverse morphology, including seminomas (male), dysgerminomas (female), teratomas, embryonal carcinomas, **endodermal sinus tumors**, and some (but not all!) cases of choriocarcinoma (262). Germ cell tumors also have diverse molecular markers with cytogenetic changes correlating with the age of the patient. Pediatric germ cell tumors show imbalances in chromosome 1 and loss of 6q, whereas adult germ cell tumors often have an isochromosome 12p or amplification of 12p (262). Germ cell tumors arise in almost any part of the body (particularly midline locations) and any age (with a pediatric dominance).

Pathologists could not place teratomas of infants, teratomas of adults, seminomas, dysgerminomas, and nongestational choriocarcinomas into a single class of neoplasms until Teilum, in 1965, suggested that all these tumors had the same cell of origin, the primordial germ cell (263). The biologic mechanism by which one cellular progenitor can give rise to these seemingly unrelated tumors is still unfolding as a fascinating saga of developmental biology (262).

Germ cells normally follow a narrow developmental path. Male germ cells give rise to sperm cells. Female germ cells give rise to oocytes. Under normal circumstances, germ cells are not pluripotent and are not related to embryonic stem cells. The only tumors arising directly from neoplastic germ cells are pure seminomas and dysgerminomas (the female equivalent of seminomas). As described earlier, primordial germ cells undergo a complete erasure of epigenetic programming. This phenomenon precedes the transformation of primordial germ cells into totipotent cells. It is the putative mechanism whereby germ cells are the indirect precursors of totipotent tumors, and it provides a biologic rationale for separating germ cell tumors from primitive stem cell tumors (260, 261, 264). It permits us to think of germ cell tumors as a distinct tumor class.

We can think of the pluripotent (not totipotent) tumors as tumors of somatic stem cells (not germ cells). Similarly, gestational choriocarcinomas are somatic tumors that derive from the trophectoderm (not germ cells). Furthermore, new drugs that target the germ cell tumors may exploit differences in the epigenomic state of germ cell tumors and embryonic tumors (i.e., methylation patterns). Finally, a developmental classification that distinguishes germ cell tumors and embryonic tumors reconciles the fundamental differences between the morphologic and cytogenetic incongruities among these unique tumors.

There are several types of tumors of germ cell origin (List 20.3.1).

20.4 Embryonic Stem Cell Tumors

Embryonic stem cell tumors are quite easy to understand. They are simply another somatic tumor (nongerm cell tumor) derived from cells that happen to have a lineage position that precedes the common differentiated cell types that we observe in the fully developed human.

Children seem to have more stem cells than do adults. This is because children are still developing as organisms. Consequently, most embryonic stem cell tumors occur in

List 20.3.1 World Health Organization classification of germ cell tumors (47).

Choriocarcinoma (nongestational)
Embryonal carcinoma
Germinoma (includes dysgerminoma and seminoma)
Mixed germ cell tumor
Teratoma (mature, immature, teratoma with malignant transformation)
Yolk sac tumor

children. Stem cell tumors can be nondifferentiating (i.e., composed of monomorphic cells that have features of one stem cell at a particular stage of development), or they may be differentiating (composed of polymorphic cells with features of all the differentiated cells that descend from the neoplastic stem cell). When an embryonic stem cell tumor shows virtually no differentiation and consists of small, round, blue cells, it is usually called a primitive tumor.

Most of the embryonic stem cell tumors fall into the category of blastomas. Without exception, every embryonic stem cell tumor will fall into Class Endoderm/Ectoderm, Mesoderm, Neuroectoderm or Neural Crest. One of the most important things to remember about embryonic stem cell tumors is that they are very rare.

20.5 Germ Cell Tumors Are Not Equivalent to Stem Cell Tumors

Germ cell tumors are biologically distinct from embryonic stem cell tumors (List 20.5.1).

20.6 What Is the Stem Cell Theory of Cancer?

Some researchers believe that the phenomenon of dedifferentiation in cancers supports the stem cell theory of cancer. The stem cell theory of cancer holds that during carcinogenesis, tumor cells arise from the stem cell compartment, and that every tumor contains neoplastic stem cells that renew themselves and that produce differentiated cells. In the stem cell theory, the differentiated tumor cells form the major population of cells in tumors.

Proponents of the stem cell theory believe that chemotherapeutic regimens that target the predominant cell population in tumors (the differentiated cells) may not be effective against the stem cell subpopulation. If the stem cell theory of tumor biology is correct, tumors will continue to grow so long as a viable subpopulation of tumor stem cells remains.

List 20.5.1 Some expected biological differences between germ cell tumors and embryonic (pluripotent) stem cell tumors.

Germ cell tumors would include extraembryonic structures (yolk sac, choriocarcinoma), and embryonic stem cell tumors would not. In other words, germ cell tumors are potentially totipotent, able to differentiate to all of the embryonic layers, including extraembryonic components, and additional germ cells.

Embryonic stem cell tumors would tend to occur almost exclusively in children because stem cells seem to disappear in adults. Germ cell tumors would occur in children and adults because both children and adults contain germ cells.

Stem cell tumors do not contain germ cells. Stem cell tumors can be pluripotent but not totipotent.

Embryonic stem cell tumors are somatic tumors and would not have an epigenomic pattern indicative of germ cell tumors (265).

Germ cell tumors may respond to a totally different range of targeted drug therapies than embryonic stem cell tumors (such as agents that target epigenomic modifications found in germ cell tumors).

Germ cell tumors might be expected to have a completely different etiology than embryonic stem cell tumors.

Extragonadal germ cell tumors are almost all midline. Round cell tumors of childhood are usually lateral.

The tumor stem cell theory holds that classic response measurements to tumor chemotherapy (e.g., tumor shrinkage) may mislead us into thinking that a treatment is effective, even when there is no useful reduction in the tumor stem cell population. Several observations lend support to the tumor stem cell theory. The first is the low level of success in curing cancers with toxic chemotherapeutic regimens. We can kill a large percentage of a patient's tumor cells, often achieving remission of the cancer, but we seldom achieve a lasting cure. In many instances, a subpopulation of cells survives, leading to tumor recurrences. The stem cell theory would suggest that a small population of stem cells persists after therapy and renews the tumor population.

Another observation that supports the stem cell theory is that tumors tend to have a phenotype that is less differentiated than the phenotype of mature cells that the tumor resembles morphologically. For example, hepatocellular carcinomas are often well-differentiated tumors that bear a striking similarity to mature hepatocytes. Despite appearances, hepatocellular carcinomas often produce high amounts of alpha-fetoprotein. Alpha-fetoprotein is the fetal equivalent of albumin. It is produced normally for a period in fetal development, to be replaced forever by albumin production in the postnatal organism. The production of alpha-fetoprotein, often in high concentration by hepatocellular carcinoma, suggests that an early precursor cell of the hepatocyte has joined the tumor population.

Many tumors are composed of cells that are, under normal circumstances, postmitotic (unable to divide further). A good example is squamous cell carcinoma, the most frequently occurring tumor of man. Squamous cells compose the epidermis (the outer layer) of our skin. They are flat cells with tight intercellular connections (desmosomes) that render our skins impervious to water, bacteria, and other detritus encountered at the interface between us and the rest of the universe. The nucleus of a mature squamous cell is small and dark-staining. The most mature squamous cells of skin (the so-called orthokeratotic squamous cells) have no nucleus altogether. Squamous cells flatten and lose their nuclei as they rise through the levels of the epidermis, pushed upward by younger squamous cells produced at the basal layer. Squamous cells never divide. When the top layer of squamous cells dies, it simply sloughs into the air we breathe. Most of the house dust we see dancing under a ray of sun is composed of sloughed squamous cells. Clumps of sloughed squamous cells account for dandruff. If squamous cells cannot divide (and they cannot), then the squamous cells we see could not have arisen from other squamous cells. They must have been produced by stem cells.

At the bottom layer of the epidermis (at the so-called dermoepidermal junction) are the basal cells of the skin. Basal cells are the direct precursors of all of the postmitotic squamous cells of the skin, and they have a morphology that is distinctly different from mature squamous cells, being round with a relatively large nucleus with loosely dispersed chromatin. The basal cells of skin can produce other basal cells, or they can produce squamous cells. It is reasonable to think of squamous cell carcinomas as tumors of neoplastic skin stem cells, with non-dividing tumor squamous cells representing the products of the neoplastic stem cells.

Stem cells account for non-neoplastic growth disorders. Polycythemia vera is a blood disorder that is characterized by an increase in the number of circulating mature red blood cells. Mature red blood cells are anucleate. They obviously cannot divide; therefore, the increased number of red blood cells found in polycythemia must result from a growth abnormality within an erythrocytic stem cell and not of the mature red blood cell.

In most cases, we can infer that the stem cells that produce neoplasms occupy the penultimate lineage for the predominant cell type in the tumor. For example, the stem cell that causes squamous cell carcinoma is likely to be the stem cell that produces a squamous cell and another stem cell when it divides. It is not likely to be a cell that is two, three, four, or more ancestors distant from the dominant cell type of the tumor. How do we know this? When we study the morphology of squamous cell carcinomas, we see dividing cells that morphologically resemble differentiated squamous cells, with keratin-filled cytoplasm and prominent intercellular desmosomes (connections). Since the dividing tumor cells have morphologic features of squamous cells, we can infer that the population of tumor stem cells must include a nearly-differentiated precursor.

Sometimes tumors display properties that clarify certain mysteries of their genesis. **Multiple myeloma** is a neoplasm in which increased numbers of neoplastic plasma cells infiltrate the bone marrow and other tissues. Plasma cells are a type of B lymphocyte that can secrete immunoglobulins. An immunoglobulin-producing plasma cell produces replicate copies of one unique immunoglobulin molecule. In multiple myeloma, a unique immunoglobulin molecule is secreted by all the cells of the tumor. If the stem cell progenitor for a multiple myeloma preceded the stage at which the unique immunoglobulin molecule is determined, you would expect to find many different immunoglobulin molecules produced by the tumor. This never occurs, and thus, we infer that the multiple myeloma stem cell can produce a unique immunoglobulin or descends from a cell that can produce a unique immunoglobulin.

How is the uniqueness of an immunoglobulin molecule determined, and at what stage in development does it occur? The creation of a unique immunoglobulin molecule is an elaborate process, involving specialized B-cell components that are acquired at different stages of B-cell development. Immunoglobulin gene selection starts with recombination of several regions of the gene that are responsible for the basic protein structure of the molecule (V(D)J recombination). Two additional processes follow recombination: gene conversion and somatic hypermutation. In gene conversion, sequences of DNA are copied and patched into variable regions. In somatic hypermutation, single point mutations occur in a special region of the immunoglobulin gene and are shielded from repair, ensuring the uniqueness of the resulting immunoglobulin molecule (266). These specialized cellular processes occur in lymphocytic B cells (not in ancestral cells); therefore, the stem cell for multiple myeloma cannot arise any earlier than the committed B cell stage.

The hope among proponents of the stem cell theory is that agents that specifically target tumor stem cells (and not the fully-differentiated cells in tumors) may succeed in curing cancers where other regimens have failed. In addition, agents that specifically target stem cells may have lower toxicity than standard chemotherapies that are known to kill normal, differentiated cells.

Do all neoplasms arise from stem cells? Maybe not. The EWS-ERG oncogene has been found in several sarcomas, and its target is thought to be an uncommitted mesenchymal stem cell. In a recent study, Codrington and coworkers have shown that the EWS-ERG fusion protein confers the malignant phenotype on mature lymphoid target cells. Their work demonstrates that mature cells (not stem cells exclusively) can propagate tumor populations (267).

There are other limitations to the stem cell theory of tumor biology. We know that in some tissues, there seems to be little or no difference between differentiated cells and stem cells. Hepatocytes (the differentiated epithelial cells of the liver) do not divide under normal circumstances, but every nondividing hepatocyte retains the lifelong ability to divide following liver injury. After two thirds of the liver is resected, the full weight of the liver is regenerated in a very short time (a week in the case of rats) from the remaining hepatocytes. Studies in rodents have shown that virtually every remaining hepatocyte behaves like a stem cell, dividing to enlarge existing liver lobules with new cells. If every differentiated cell can produce more differentiated cells, then, in the case of liver, there is no reason to separate tumor cells into stem cell and non-stem cell compartments.

The stem cell theory raises fundamental questions related to carcinogenesis. It is generally believed that most tumors are caused by carcinogens that require metabolic activation. The ability of a cell to metabolize drugs is a feature acquired through differentiation. Undifferentiated stem cells cannot metabolize many of the carcinogens known to produce tumors in animals; therefore, the target cell of carcinogens must be different from the stem cell of the resulting tumor. How does a stem cell emerge as the progenitor of a tumor when the initiation process occurs in a differentiated cell (that metabolized the initiating carcinogen)?

Although the stem cell theory of carcinogenesis has weaknesses, we observe that tumors express properties determined by their cellular lineages. Within any tumor, there may be neoplastic cells that closely approximate normal differentiated cells. There may be other tumor cells with properties that approximate those stem cells that would normally produce the differentiated cells of the tissue of origin of the tumor. In either case, all of the cells of the tumor share a common developmental lineage and express their properties through lineage-specific cellular pathways.

20.7 Tumors of Class Trophectoderm (Gestational Trophoblastic Tumors)

Trophoblast cells are the primary cellular constituent of the **extraembryonic** component of the embryo. This oxymoron ("extraembryonic component of the embryo") indicates that the cellular tissues (placenta, umbilical cord, yolk sac, and amniotic membranes) are not part of the developed embryo; hence, they are extraembryonic. The embryonic layer that produces the extraembryonic components is called the trophectoderm (Figure 20.2). Gestational trophoblastic tumors are equivalent to tumors of Class Trophectoderm (and both terms are used in this book).

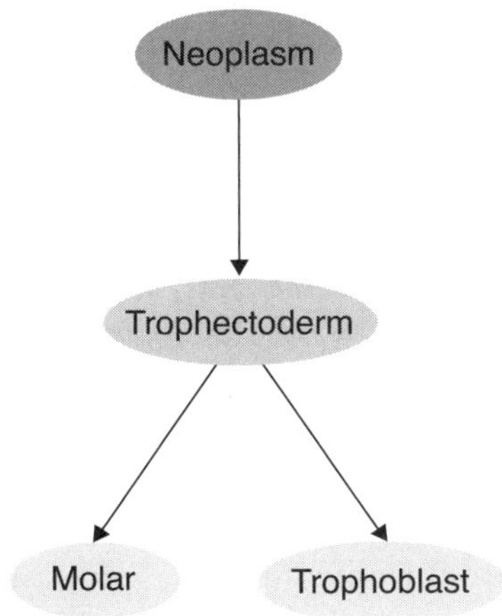

FIGURE 20.2 Hierarchy of Class Trophectoderm tumors.

As discussed in a prior section, some trophoblastic tumors arise from germ cells that become pluripotent and from which cells of the trophectoderm lineage predominate. This mechanism accounts for every trophoblastic tumor arising in children and males. Women may also develop trophoblastic tumors that arise from germ cells, but more often, trophoblastic tumors occurring in women of child-bearing age develop from a conceptus (a fertilized ovum).

Gestational trophoblastic disease is a somewhat euphemistic term that covers a variety of neoplasms and proliferative lesions of trophoblast cells. The lesions included as gestational trophoblastic diseases are: **hydatidiform mole** (complete and partial types), **invasive mole**, **choriocarcinoma**, **placental site trophoblastic tumor**, and epithelioid trophoblastic tumor (268). These lesions all arise from the extraembryonic lineage of a conceptus and have no relation to germ cell tumors that may also yield a neoplastic trophoblastic component.

20.8 Complete and Partial Hydatidiform Moles

About half of all gestational choriocarcinomas (malignant trophoblastic tumors) arise with or after normal-term pregnancies. About one quarter of gestational carcinomas arise after miscarriages, ectopic pregnancies, or other abortive pregnancies, and about 25% occur with gestational moles (269).

A gestational mole, also called hydatidiform mole, is a malformed conceptus that has no chance of developing into a normal organism. All hydatidiform moles consist primarily of placental villi with varying amounts of disorganized fetal parts. Although only one quarter of choriocarcinomas occur with hydatidiform moles, each occurrence of a hydatidiform mole is associated with a significant risk of trophoblastic neoplasia.

There are two types of hydatidiform mole: complete mole and partial mole. A complete mole is characterized by edematous placental villi and little or no embryonic tissue. Complete moles can develop into **invasive moles** or choriocarcinoma. An invasive mole grows directly from preexisting villi of the hydatidiform mole, which invade into the myometrium (wall of the uterus). A choriocarcinoma that derives from a hydatidiform mole is a monoclonal neoplasm that grows from a progenitor trophoblastic cell within the hydatidiform mole. Choriocarcinomas differentiate into two types of trophoblastic cells, the cytotrophoblast and the syncytiotrophoblast. Complete moles occur after an anucleate ovum is fertilized by two sperm. The cells of a complete mole are diploid, with both sets of chromosomes having a paternal origin.

A partial mole is characterized by nonedematous placental villi and a variable amount of malformed embryonic tissue. A partial mole, like a complete mole, can develop into an invasive mole or a choriocarcinoma, but the incidence of conversion is much smaller (about 20% for complete mole and about 3% for partial mole). Partial moles are triploid and formed by the dispermic fertilization of an ovum, which produces two sets of paternal chromosomes and one set of maternal chromosomes.

20.9 Why Does Paternal Disomy Lead to Hydatidiform Moles?

In both complete and partial mole, the ovum receives a double dose of paternal chromosomes. Studies of mouse embryogenesis have demonstrated that both maternal and paternal chromosomes are necessary for normal embryonic development. When there is a deficiency of maternal chromosomes and a redundancy of paternal chromosomes in a mouse conceptus, growth of extraembryonic tissues (trophectoderm) is exaggerated, and growth of embryonic tissue is stunted (270, 271). Presumably, the same biological effect accounts for the overgrowth of trophoblastic tissues in human gestational moles.

20.10 The Totipotent Fascination of the 129 Mouse

In 1954, Leroy Stevens and coworkers published observations that had a profound and lasting influence in the fields of cancer research and developmental biology (272). Stevens and coworkers found that testicular teratomas arose spontaneously in strain 129 male mice. Moreover, teratomas could be induced in these mice by several different methods: by transplanting fetal genital ridges into the testes of adult male mice, by transplanting eggs taken from the tubes of females and inserted into the testes of adult male mice, and by transplanting postimplantation mouse embryos directly into the testes of adult male mice.

The tumors that resulted were totipotent (i.e., differentiated into cells of every embryonic lineage) and were used to establish numerous embryonic cell tissue culture lines that were used for decades as sources of mouse **embryonic stem cells**.

The most fascinating application of these teratoma cells was demonstrated by Brinster, Illmensee, Mintz, Papaioannou, and colleagues (273–275). Brinster showed that a cell from a malignant teratoma in a 129 mouse could be inserted into the blastocyst of a gestating mouse of the same strain. Amazingly, the offspring were seemingly normal mice containing a subpopulation of cells that had developed from the inserted malignant cell: a chimeric animal composed of "tamed" malignant cells and normal blastocyst-derived cells (273).

This area of research became even stranger with a publication of Mintz and coworkers provocatively titled, "Normal Genetically Mosaic Mice Produced From Malignant Teratocarcinoma Cells" (274). Occasionally, a teratoma-blastocyst chimeric mouse will develop with oocytes derived from a malignant teratoma cell. Likewise, another chimeric mouse may develop with spermatocytes derived from a malignant teratoma cell. If these two mice mate, their offspring would be a mouse derived exclusively from malignant teratoma cells. This, in fact, was shown to be possible (274, 275).

The modern-day techniques for producing transgenic and knockout mice were developed from the stunning experiments and observations made on spontaneous malignant teratomas occurring in mouse strain 129. When these early studies proved

List 20.10.1 Exceptional properties of the mouse 129 teratoma model.

1. The teratomas occurred spontaneously and abruptly and did not undergo the usual process of carcinogenesis and tumor progression. The mouse 129 teratomas simply grew out of transplants of normal germ cells. Most malignant tumors develop through a multistage process whereby malignant phenotypic properties are acquired as mutations accumulate. This fundamental difference between mouse 129 teratomas and other malignancies suggests that the mouse teratoma results from a peculiar developmental abnormality in strain 129 mice.
2. The teratomas used to produce chimeras were euploid. Most cancers are aneuploid and simply lack the genetic material to yield normal cells and normal organisms. There is no way to "normalize" a dividing cell that lacks a normal DNA complement.
3. When mouse teratoma cell lines become aneuploid, they lose its pluripotent potential. When mouse 129 teratoma cells acquire the hallmark of malignancy (aneuploidy) they lose the ability to produce a chimera (276).
4. Nobody seems able to transmit a mutation to the germ line of chimeras using teratomas. The modern-day technique of transmitting mutations via the germ line of chimeric mice is done using embryonic stem lines obtained from normal blastocysts, not teratomas.
5. Nobody has repeated these observations (spontaneous teratomas and tumor chimeras) in mice of other strains or in other animals, suggesting that there was something very odd about the germ cells of mouse 129 that predisposed to teratomatous growth of euploid cells.

that malignant cells could be tamed to differentiate into normal tissues, there was enormous enthusiasm for the idea that cancer was simply a mistake in normal differentiation that could be reversed by modifying the tumor's environment. Like alchemists searching for a way to change lead to gold, cancer researchers sought a method to change cancer cells to normal cells.

A few observations of the mouse 129 teratoma model could have dispelled this fanciful notion (List 20.10.1).

The **field theory** of carcinogenesis holds that cancer and all of the genetic changes observed in cancer cells are the result (not the cause) of disordered tissue growth. The theory suggests that carcinogens act by poisoning the stroma underlying epithelial cells, resulting in the loss of normal epithelial development, and the subsequent growth of a tumor.

The theory is based in no small part on the observations of Mintz and colleagues on mouse 129 teratomas. The central idea of the theory is that tumor cells lack normal developmental control (supplied by a normal stroma). Tumor cells can be "normalized" by placing them in a normal developmental environment, as observed in a few instances with primitive or blastemal tumors (277–279). An excellent discussion of this theory, also called the **tissue organization field theory** has been provided by Sonnenschein and Soto (280).

My own opinion is that scientists should be wary of extrapolating too much from one or several tumor models. Very few present-day researchers use the mouse 129 teratoma model. Although the model has taught us many secrets about development and transgenic biology, cancer researchers learned a sobering lesson: there are many pathways that lead to neoplastic development. Neoplasms are biologically diverse, highly complex biological entities that have not been successfully modeled (to everyone's liking) by any single theory of carcinogenesis.

Chapter 20 Summary

The tumors of man that differentiate along all of the embryonic and extraembryonic cell lineages (i.e., totipotent differentiation) are included in Class Germ Cell. The mechanism whereby germ cells give rise to neoplastic totipotent cells seems to occur after epigenomic erasure, leading to a "blank" primordial cell of no predetermined lineage. This uncommitted cell is the progenitor cell for teratomas, embryonal carcinomas, non-gestational choriocarcinomas, and their many variants.

Under normal circumstances, germ cells follow a restricted path of differentiation towards gametocytes (ovum and sperm). In the ovary, tumors of differentiated germ cells are called dysgerminomas. In the testis, the homologous tumors are known as seminomas. Neoplastic germ cells can undergo transformation into totipotent primordial cells; therefore, in addition to dysgerminomas and seminomas, a variety of teratomatous, embryonal, and extraembryonal neoplastic cell types can emerge from neoplastic germ cells.

Gestational trophoblastic tumors are neoplasms that arise from the extraembryonic component of an embryo. The tumors that arise from extraembryonic tissues contain neoplastic trophoblasts (the predominant cell of the trophectoderm). Sometimes gestational trophoblastic tumors arise in conjunction with an otherwise normal pregnancy (i.e., the embryonic portion of the conceptus is normal). Some gestational trophoblastic tumors arise from hydatidiform moles. Gestational trophoblastic tumors have no relation to germ cell tumors. Nongestational trophoblastic tumors are germ cell tumors with a predominating component of trophoblast cells. Not surprisingly, gestational germ cell tumors arising from hydatidiform moles (all of which have a paternal gene component) and nongestational germ cell tumors arising from hydatidiform moles (formed after epigenomic erasure of germ cells) have distinctly different biologies and clinical courses.

Tumors that arise from early lineage somatic cells within the embryo comprise many of the so-called blastomas and primitive tumors (e.g., nephroblastomas, hepatoblastomas, and primitive neuroectodermal tumors) and are examples of the so-called primitive tumors of childhood. These tumors do not derive from germ cells.

The tumors that are called stem cell tumors are simply tumors that contain cells of an earlier lineage stage than the most differentiated cell type in the tumor population. In stem cell tumors, the clonal progenitor of the tumor is an early-stage cell, and the tumor consists of a mixture of stem cells and descendant cells all within the same developmental lineage. The stem cell theory holds well for neoplasms and clonal proliferative disorders composed largely of nondividing cells (such as chronic myelogenous leukemia and polycythemia vera). Currently, many tumor biologists believe that every tumor contains a subpopulation of stem cells (and is thus a stem cell tumor). This belief may be incorrect. Every cell type in the body that is capable of dividing is a potential clonal progenitor for a tumor. When a fully differentiated dividing cell (such as a hepatocyte or a renal tubular epithelial cell) is the progenitor cell for a tumor, there is no compelling reason to think that the resulting tumor is a stem cell tumor.

CHAPTER

21

Specialized Cancer Nomenclatures and the Developmental Lineage Classification and Taxonomy of Neoplasms

Definition of a lexicographer: "harmless drudge"
—Samuel Johnson

21.1 Background

Just a few decades ago, uses of a biological classification were simple (List 21.1.1).

Today, classifications are expected to serve many additional purposes (List 21.1.2).

The purpose of this chapter is to explain the technical, social, and legal aspects of cancer classification and how they impact on modern cancer research and on our current perceptions of neoplasms.

List 21.1.1 Traditional purposes of a biological classification.

1. Organize and simplify the natural world.
2. Provide a comprehensive conceptual structure into which every part of the world has a unique place.
3. Provide relationships among all of the classes of natural things.
4. Develop a general way of looking at nature that encourages the generation of new questions and new ways of testing the structure of the classification.

List 21.1.2 Modern purposes of a biological classification.

1. Serve as the key to data retrieval and data organization within the taxonomy.
2. Conform to a standard file format that permits the classification to be distributed, opened, and parsed by any computer.
3. Conform to computational standards for identifying data elements and for establishing the relationships among the data elements in the classification.
4. Support the annotation of other biological data sets with related terms from the classification.
5. Enable the retrieval of data contained in other data sets through terms and relationships established within the classification.
6. Employ modern data extraction methods to find new taxonomic terminology and to add these terms to the classification.
7. Support an endless variety of research and health care activities that depend on evaluation of biological objects that can be grouped under a common name and included within a biological class.
8. Support the continual re-examination and improvement of the classification through informatics technology.
9. Provide all the other traditional functions of biological classifications.

21.2 Nomenclatures, Taxonomies, Classifications

Modern nomenclatures connect **heterogeneous data** and are the key to all efforts towards data organization, data retrieval, and the development of new data relationships within and among medical knowledge domains (281–283).

The tools of modern informatics are nomenclatures, taxonomies, classifications, and ontologies (List 21.2.1).

21.3 Inconsistent Meanings Within Standardized Medical Nomenclatures

There is a movement to standardize medical nomenclatures. The idea behind nomenclature standardization is that if everyone used the same terms, it would be feasible to index and retrieve information collected from many different medical data sources. A report from the National Committee on Vital and Health Statistics states that "without national standard vocabularies, precise clinical data collection and accurate interpretation of such data is difficult to achieve. Further, this lack of standard vocabularies makes it difficult to study best practices and develop clinical decision support" (284).

There are several problems with relying on a single, standard nomenclature. One problem relates to competing concepts in different biological domains. Single words have

List 21.2.1 The seven basic tools of cancer informatics.

Nomenclature. A list of all of the significant words used within a knowledge domain. A nomenclature is equivalent to a vocabulary list, but differs from a dictionary. A dictionary is a *nomenclature* (or vocabulary), with a provided definition for each entry.

Controlled vocabulary. A controlled vocabulary contains edited terminology ensuring that only approved terms are included. Deprecated terms are excluded. Controlled vocabularies can be used as an authoritative source of "correct" synonyms for any given concept. The problem with a controlled vocabulary is that it cannot be used to retrieve information related to concepts occurring as deprecated terms.

Taxonomy. The list of every instance in an entire classification, along with its assigned class. You can think of the classification as a scaffold to hold organized objects and the taxonomy as the collection of named objects that fills the scaffold.

Concept. A concept is an abstract representation of a defined entity that is known by many different terms. Most vocabularies assign concepts a unique identifying number, which applies to a set of equivalent terms (i.e., synonyms).

Term. In a nomenclature, a term is a name for a concept. Typically, a medical concept will have many alternate names, all synonymous with one another.

Classification. We use the biologist's definition of a classification, explained in Chapter 7 (119). A classification is a hierarchical and comprehensive grouping of all of the members of a well-defined knowledge domain. In a classification, a member can only be assigned one class. Each class can be the subclass of exactly one immediate ancestral class (although one class can have more than one subclass).

Ontology. An ontology is a rule-based grouping of members of a knowledge domain. Ontologies support queries and logical inferences pertaining to the (ontologic) group members. They can be used to test data for logical consistency and can be designed to discover relationships between different classes of data. Ontologies, like classifications, assign members to classes.

many different meanings. The term "dysplasia" means precancerous atypia to a cancer pathologist, but indicates a growth dysmorphism to a neonatologist.

Even multiword terms have different meanings to different health care workers. To a pulmonary specialist, chronic obstructive pulmonary disease may represent a collection of diseases that produce a similar set of clinical complaints. To a health care administrator, chronic obstructive pulmonary disease is one of many long-term illnesses that are high cost and that may require special support services not usually supplied within an acute-care setting. To pathologists, chronic obstructive pulmonary disease is a lay term that they would not use in their surgical pathology reports.

As discussed in Chapter 4, within the small realm of neoplasm terminology, there is ample room for disagreement. For every lesion listed below as a cancer, a benign

List 21.3.1 Some neoplasms whose categories can be disputed.

- Cancer
  - Transitional cell carcinoma of bladder
  - Astrocytoma
  - Basal cell carcinoma
- Benign tumor
  - Colonic villous adenoma
  - Renal tubular adenoma
  - Keratoacanthoma
- Hamartoma
  - Dermal nevus
  - Seborrheic keratosis
- Precancer
  - In situ carcinoma of bladder

tumor, hamartoma, choristoma, or precancer, another pathologist might object and provide evidence supporting an alternate category (List 21.3.1). Transitional cell carcinoma of bladder may occur as a noninvasive papillary growth, and some pathologists do not consider these lesions "cancers." Other pathologists call these lesions cancers to indicate that their natural biology typically progresses to invasion. Astrocytoma is a tumor of the brain that infiltrates through the brain parenchyma. Some pathologists believe that this tumor is benign because it almost never metastasizes and lacks the full malignant phenotype. Others believe that the astrocytoma is malignant because it kills people. Basal cell carcinoma is a common, invasive tumor of the skin. It almost never metastasizes. You can infer from its name that some pathologists consider it to be a malignant lesion. In England, the common term for this tumor, "basal cell epithelioma," is noncommittal, implying neither malignancy nor benignity. The villous adenoma of the colon is usually considered a benign tumor, but it commonly progresses to cancer, and some pathologists might call this lesion a precancer. Renal tubular adenomas are considered benign if they are smaller than 2 cm in diameter. The same lesion, after it exceeds 2 cm, is usually considered malignant because a sizable percentage of "adenomas" this size will metastasize. Keratoacanthoma is a skin lesion that grows rapidly, contains aneuploid cells, and looks very much like a squamous cell carcinoma. After a few months growth, it regresses spontaneously, in most cases. Regression of keratoacanthomas results when the dividing, atypical squamous cells differentiate into keratin, enlarging the so-called keratin cup that sits atop the formerly dividing population. Some pathologists do not consider a keratoacanthoma to be a true neoplasm because it lacks progressive, sustained growth. Others consider it to be a variant of squamous cell carcinoma because occasional cases fail to regress.

A dermal nevus is a common mole. They grow to a certain size and stop growing. Some pathologists believe these lesions to be **focal hyperplasias**, not true neoplasms. Other pathologists consider these lesions to be precancers for melanomas because some melanomas are known to emerge from preexisting nevi. Seborrheic keratosis is a skin lesion, common among seniors, that often resembles a drop of flattened tar sticking to the skin. Some pathologists consider these lesions to be focal hyperplasias of small, round squamous cells. Others consider these lesions to be neoplasms because they often have a clonal morphology (i.e., every cell morphologically alike and distinct from normal squamous cells). In situ carcinoma of the bladder is widespread dysplasia of transitional cells lining the entire surface of the bladder. Some pathologists consider this a precancer because the cells do not form a mass, and the lesion is superficial and noninvasive. Other pathologists consider this lesion to be a fully malignant neoplasm because it often leads to invasion and ultimately to the death of the patient.

Even when an assemblage of prominent scientists reach consensus on the meaning of terms included in a nomenclature, their consensus may be dead wrong, or their consensus might make sense under some conditions but not under other conditions.

21.4 Incompleteness in Standard Nomenclatures

Another limitation of a standardized medical nomenclature relates to incompleteness. Omnibus nomenclatures attempt to provide a controlled vocabulary that covers many specialized fields (e.g., pathology, internal medicine, and surgery). These general medical nomenclatures have enormous value, but they may lack terms and concepts found in specialized nomenclatures.

By far the largest medical nomenclature is the U.S. National Library of Medicine's Unified Medical Language System **Metathesaurus** (UMLS Metathesaurus), which contains several million terms grouped under about a million concepts (285). The UMLS is comprised of over 100 individual thesauruses, many of which are contributed by efforts funded by the U.S. government. The UMLS metathesaurus can be acquired at no cost from the National Library of Medicine. UMLS users must agree to abide by the terms of the UMLS license.

SNOMEDCT (Systematized Nomenclature of Medicine Clinical Terms) is a designated standard used in U.S. Federal Government systems for the electronic exchange of clinical health information and is included as a **thesaurus** within the UMLS (286). SNOMEDCT is also being implemented throughout the National Health Service in the United Kingdom. Special usage restrictions, available at the UMLS website, apply to the SNOMED thesaurus.

An extraction of neoplasm terms included in the 2005 version of SNOMEDCT found about 2,000 distinct neoplasm concepts (i.e., types of neoplasms) comprising about 7,100 different names of neoplasms.

The UMLS metathesaurus and the SNOMEDCT nomenclature contain fewer than one third of the neoplastic entities listed in the Developmental Lineage Classification and Taxonomy of Neoplasms (also called the Developmental Classification), a nonstandard taxonomy developed by the author of this book, and available at no cost to the public under an open source license. The current version of the Developmental Classification contains over 5,800 neoplasm concepts and over 130,000 different terms. This demonstrates that specialized (nonstandard) cancer nomenclatures can supplement standard vocabularies with terms and concepts relevant to narrow knowledge domains.

The Developmental Classification attempts to include the name of every neoplasm that occurs in humans. The Developmental Classification is available at no cost from this book's listing at the Jones and Bartlett web site: http://www.jbpub.com/ and from the author's web site: http://www.julesberman.info/.

21.5 Striving for Completeness

A specialized nomenclature strives for completeness. A complete nomenclature contains every concept within a domain, as well as every synonym for every concept. A nomenclature cannot be used to retrieve a term from a corpus of text, unless the term is included in the text and in the nomenclature.

In the Developmental Classification, prostate cancer is an example of a neoplasms concept. It is assigned the unique concept code C486300, and all of the term variants for this concept are attached to the same concept code (287) (List 21.5.1). The Developmental Classification contains triple the number of neoplasm concepts as SNOMEDCT and nearly 20-fold the number of neoplasm terms. In general, the concepts that are present in the Developmental Classification and absent from SNOMEDCT are rare tumors, rare subtypes of common tumors, benign tumors, and precancers. The relative lack of benign neoplasms and precancers in SNOMEDCT compared with the Developmental Classification may be due to SNOMEDCT's historical roots in ICD-O (International Classification of Diseases—Oncology). The ICD-O was created to provide Cancer Registrars with codes for the major cancers treated and followed by the Cancer Registry (288). Benign tumors and precancers are generally not collected by cancer registrars.

List 21.5.1 Synonyms for prostate cancer.

C4863000 prostate with adenoca
C4863000 adenoca arising in prostate
C4863000 adenoca involving prostate
C4863000 adenoca arising from prostate
C4863000 adenoca of prostate
C4863000 adenoca of the prostate
C4863000 prostate with adenocarcinoma
C4863000 adenocarcinoma arising in prostate
C4863000 adenocarcinoma involving prostate
C4863000 adenocarcinoma arising from prostate
C4863000 adenocarcinoma of prostate
C4863000 adenocarcinoma of the prostate
C4863000 adenocarcinoma arising in the prostate
C4863000 adenocarcinoma involving the prostate
C4863000 adenocarcinoma arising from the prostate
C4863000 prostate with ca
C4863000 ca arising in prostate
C4863000 ca involving prostate
C4863000 ca arising from prostate
C4863000 ca of prostate
C4863000 ca of the prostate
C4863000 prostate with cancer
C4863000 cancer arising in prostate
C4863000 cancer involving prostate
C4863000 cancer arising from prostate
C4863000 cancer of prostate
C4863000 cancer of the prostate
C4863000 cancer arising in the prostate
C4863000 cancer involving the prostate
C4863000 cancer arising from the prostate
C4863000 prostate with carcinoma
C4863000 carcinoma arising in prostate
C4863000 carcinoma involving prostate
C4863000 carcinoma arising from prostate
C4863000 carcinoma of prostate

(continues)

List 21.5.1 (continued)

C4863000 carcinoma of the prostate
C4863000 carcinoma arising in the prostate
C4863000 carcinoma involving the prostate
C4863000 carcinoma arising from the prostate
C4863000 prostate adenoca
C4863000 prostate adenocarcinoma
C4863000 prostate ca
C4863000 prostate cancer
C4863000 prostate carcinoma
C4863000 prostatic cancer
C4863000 prostatic carcinoma
C4863000 prostatic adenocarcinoma
C4863000 prostate gland adenocarcinoma
C4863000 adenocarcinoma of the prostate gland
C4863000 adenocarcinoma of prostate gland
C4863000 prostate gland carcinoma
C4863000 carcinoma of the prostate gland
C4863000 carcinoma of prostate gland

21.6 Informatics Support Within the Developmental Classification

A modern classification serves as the key to information contained in any digital document located anywhere on earth. In Chapter 15, we provided an overview of the Developmental Classification. The features of the classification most relevant to biomedical informaticians are summarized here (List 21.6.1).

21.7 Coding Medical Information

Medical nomenclatures are implemented by coding: assigning a concept code from the medical nomenclature to each occurrence of a concept found in a medical text (List 21.7.1).

An encounter with a clinician (a clinic visit, a therapeutic procedure, a laboratory rest, a dispensed pharmaceutical) is provided with codes from a medical nomenclature. The coded data can be retrieved, merged, shared, analyzed, summarized, and used in

List 21.6.1 General features of the Developmental Classification relevant to biomedical informatics.

Instance uniqueness. Each tumor entity occurs only once within the Developmental Classification.

Comprehensive. The Developmental Classification ensures that every tumor of man can be placed somewhere within the classification. The curator must constantly review the medical literature to add new concepts and terms. For the Developmental Classification, software was developed to parse (electronic) cancer literature, automatically extracting candidate new terms for any knowledge domain (289).

Consistent. Ensuring consistency involves determining that no term occurs redundantly, either under one concept or under multiple concepts, and that no redundant concepts are included in the nomenclature (290).

Simplicity. One of the purposes of a classification is to drive down the complexity that exists when the domain taxonomy is large. The entire Developmental Classification is contained in about 40 classes. The neoplasms in each class share common biological properties, and these common properties may lead us to common treatments for all of the tumors of a class.

Principled. The Developmental Classification is based on known principles of developmental biology, not on political or artifactual distinctions between tumors. A counterexample would be a tumor classification based on medical specialty (e.g., dermatologic neoplasms, hematologic neoplasms, and head and neck tumors).

The classification has "competence." In the field of informatics, competence is the ability to answer questions related to the instances of a data group.

Standard method of organization. The Developmental Classification is represented in a standard, computer-parsable syntax. Equivalent versions of the Classification are available as XML (eXtensible Markup Language), RDF (Resource Description Framework), and plain text formats.

Scalability. It is easy to expand the Developmental Classification with new subclasses. This is important, as the molecular analysis of tumors is likely to provide new taxa (ways of distinguishing one class from another).

Modifiable. It is easy to move subdivisions of the Developmental Classification. Classifications are hypothetical re-creations of reality and must be changed as information is accrued.

Understandable. The classification is easily understood by developmental biologists. Developmental biologists are major participants in postgenomic science and need to have tools to relate basic research with clinical exigencies.

Credible. The Developmental Classification complements modern theories of the "stem cell" origin of tumors.

(continues)

List 21.6.1 (continued)

Compatible with other visions of reality. The Developmental Classification does not invalidate existing diagnoses found in pathology reports. The medicolegal importance of this feature cannot be exaggerated. This relieves anatomic pathologists from reviewing all of their prior cases and rediagnosing them in conformance with a new classification.

Open access. The Developmental Classification is an open access, easily accessed document that can be used or criticized freely by the biomedical community.

List 21.7.1 Example of an automatically coded title of a published research paper (291).

Mucinous carcinoid as an unusual manifestation of endodermal differentiation in ovarian yolk sac tumor

<autocode term="*carcinoid*" code="C4139100" />

<autocode term="ovarian yolk sac tumor" code="C8107000" />

any number of ways to help the patient, health care workers, agencies that provide care, and agencies that coordinate or regulate care.

Cancer-related data cannot be fully coded by humans. Modern hospitals produce terabytes of information each week. Research laboratories and medical literature flood us with data. All of these data need to be coded, and the only feasible way to code large amounts of data is to use computer programs to extract concepts from the literature and to match the concepts with concept codes found in large, complete medical nomenclatures. The Developmental Classification can be autocoded with free, publicly available algorithms and software implementations (292). Text can be autocoded at the rate of nearly 1 megabyte of text per second on a desktop computer using a modest 1.6 GHz processor (292). There are about 25 million surgical pathology reports generated each year in the United States. Each report contains a diagnosis that has an average length of 1 kilobyte. A slow computer (1.5 GHz) could autocode every surgical pathology report generated in the United States each year in about 25,000 seconds. Supercomputers complete hundreds of trillions of operations per second. This translates to a speed of about 34,000 times the speed of a 1.5 GHz computer. A supercomputer could perform the chore of autocoding the same number of surgical pathology reports in under a second.

21.8 Re-Coding Data

For decades, the approach to coding medical data has been to provide a persistent and unchanging code for medical concepts that will be used for current and future data retrieval needs. This approach provides impetus for having national medical nomen-

List 21.8.1 Limitations of autocoding using a standard nomenclature.

Even standard nomenclatures change. Records autocoded with a 1995 version of any nomenclature may have a different set of terms and codes than records coded with the 2009 version of the same nomenclature.

Different autocoding strategies may produce remarkably different coding results using the same nomenclature and the same data set. For instance, one autocoder may use a parsimonious autocoding strategy, producing the single best nomenclature code for a diagnosis; whereas another autocoder may prefer a comprehensive strategy, providing codes for all terms in the text that match any terms in the vocabulary.

Data sets may have markedly different content and structure, making it impossible to rationally combine autocoding results, even when the same nomenclatures and autocoding strategies are used. For instance, one data set may have poor orthography or may represent many terms as nonstandard abbreviations, making it impossible for an autocoder to match terms from the data record with terms from the nomenclature. If the other data set contains terms that are correctly spelled and eschews abbreviated forms, the coding results will vary greatly, even when the same terms are included in both data sets.

List 21.8.2 Instances when cancer data should be re-coded.

1. Whenever you need to change from one nomenclature to another (eliminates problem of brand-name loyalty)
2. Whenever you need a new version of a nomenclature
3. Whenever you need a new coding algorithm (e.g., parsimonious versus comprehensive)
4. Whenever you need to add legacy data
5. Whenever you need to merge different surgical pathology data sets

clatures. If everyone were to use the same nomenclature to code their data, data accumulated over the decades could be electronically exchanged and integrated. The problems with this approach are several (List 21.8.1). After a text corpus is coded, future uses of the text will almost certainly require re-coding. Automatically re-coding large data sets is practical and necessary under many conditions (List 21.8.2).

When data are being re-coded, the opportunity arises to change nomenclatures, coding algorithms, and even data sets. For instance, if a data set is poorly constructed, some effort may be expended to transform the data into a syntax that is consistent with the data set with which it will be merged. An autocoding algorithm suited to an anticipated study can be designed that will apply to all the data sets. Autocoding large data sets can now be done quickly, using any nomenclature that associates terms with concept codes (281). When data sets are re-coded, they can be re-coded with more than one nomenclature as long as the concepts are fully identified by their source nomenclature. Anticipated mandates to implement standard nomenclatures will accommodate dual-coding strategies that permit data retrieval by the most suitable nomenclatures (standard and nonstandard).

Chapter 21 Summary

Modern biological classifications are the key to relating (and making sense of) the seemingly infinite amount of data collected by cancer researchers, oncologists, and everyone who helps cancer patients and who wants to find a cure for cancer.

The Developmental Classification is by far the largest *taxonomy* of cancer, holding over 130,000 different names of neoplasms categorized under about 6,000 concepts, all related under about 40 classes. This document is available as a free, open source file.

It is now feasible to use a variety of general and special-purposed nomenclatures to retrieve all data records containing synonymous terms associated with any unique concept included in any chosen nomenclature (293). In the next chapter, we learn that an entirely new way of organizing data, the ontology, now challenges the traditional classification approach to biological organization.

CHAPTER

22 Cancer Ontologies

I can see plainly that there are two sides to the matter but unfortunately I can occupy only one of them.

—James Joyce

22.1 Background

The major drawback in any classification (some would say the major strength of any classification) is that they impose a specific worldview of the data domain that cannot simultaneously include alternate views.

Consider this example: a zoologist searches a large, authoritative, and comprehensive taxonomy to obtain a listing of all of the animals that live in North America and that weigh over 130 pounds. The zoologist's search yields no useful results. The taxonomy provides ancestral relationships among the different organisms, but it does not provide properties for the organisms. Because location and weight are not organizing principles for the animal classification, animals of a particular weight and location cannot be extracted.

Ontologies, briefly mentioned in Chapter 21, are much more complex than classifications, but modern information specialists have computer-based methods to parse through complex ontologies. Ontologies are currently gaining popularity in the biomedical field. The purpose of this chapter is to describe how ontologies, unlike standard taxonomic classifications, allow objects (such as neoplasms) to be organized many different ways and all at once.

22.2 Introduction to Ontologies

Ontologies are rule-based systems operating on classes of instance objects. All of the members of a domain that obey a particular rule or set of rules may be collected into a class. A class is a subclass of another class if every member of the class obeys the rules that apply to the superclass.

22.3 Ontologies and Complexity

One of the main functions of a classification is to reduce the complexity of a knowledge domain. This is simply not the purpose of an ontology. The purpose of an ontology is to allow us to compute logical inferences about the members of the ontology. Ontologies differ from classifications on several key points (List 22.3.1).

In a sense, an ontology is a classification with a lot of freedom. Multiclass inheritance and class-specific logic rules (instead of class-specific properties) are the chief features that distinguish ontologies from classifications. A classification constructed in a manner that restates its class features as rules would qualify as a simple ontology. For example, the classification of all living organisms can be converted into an ontology by creating a general rule applied to all classes. The operational rule for the classification of organisms is that each class evolves from its parent class.

There are several advantages of ontologies over classifications (List 22.3.2).

With the good comes the bad. Ontologies create problems that are not encountered with classifications (List 22.3.3).

List 22.3.1 Distinctions between ontologies and classifications.

1. An ontology does not need to provide a theoretic embodiment of a data domain. In fact, an ontology need not be comprehensive (i.e., an ontology does not need to include all of the instances of a knowledge domain) and may extend over several different knowledge domains.
2. Ontologic classes are characterized by one or more logical rules and include object instances that behave in conformity to the class rules. The classes in classifications are determined by a set of features that are shared among the member of the class. The features that define a class within a classification are usually not logical rules.
3. Ontologies permit multiclass inheritance. Any ontologic class can inherit from any number of father classes. A class within a classification can have at most one father class.
4. An ontologic object instance may belong to more than one class, just as long as the object obeys the rules of the class in which it is a member. An object in a classification can belong to only one class.
5. An ontologic class inherits the rules of its superclass; however, an ontologic class is not required to have a superclass (i.e., an ancestor class) or descendant classes (i.e., subclass).

List 22.3.2 Good things about ontologies.

Ontologies are computable.
Ontologies are semantically sensible.
Ontologies have competence, meaning that they can be used to draw a variety of inferences about class members based on class rules.
Ontologies are extensible and can be integrated with other ontologies.
When the same ontology is used by different researchers, concepts in common use will have the same meaning and properties.

List 22.3.3 Bad things about ontologies.

Ontologies place no constraints on internal complexity and can quickly become incomprehensible to humans.
Ontologic complexity may lead to unanticipated consequences (including paradoxes of self-referral).
Ontologies are relatively new, and there are very few examples where they have shown to be of any biomedical value. Classifications have proven their value over millennia.
It is easy to design a flawed ontology, and it is hard to find the flaws after they have been introduced.
Ontologies work on the assumption that medically relevant domains have an intrinsic logic that can be described by rules. This assumption may be false.

Rule-based systems are easily victimized by the law of unintended consequences. An example from the realm of clinical laboratory testing will illustrate. The laboratory creates a rule that applies to all differential blood cell counts (an ontologic subclass of laboratory tests). When a white cell count falls below a certain number (e.g., 4,500, the lower range of normal in adults) the laboratory information system issues a reflexive alert to the patient's physician. An error in the rule-based system arises when a neonate presents with a white blood cell count of 5,500. The normal range of white blood cells in a neonate is 9,000 to 30,000. The normal white cell count in an adult is 4,500 to 10,000. A white cell count of 5,500 in a neonate is leukopenic and may be signal a life-threatening condition. The same white cell count in an adult would be normal. The rule system would have failed to make the proper inference (and would have failed the patient) in this specific instance. In an ontology with many classes, many rules, and complex relationships between classes, it becomes extremely difficult to assess describe, evaluate, or even fathom the system.

Managing the complexity of cancer-related data is an important task. Cancer researchers who have a keen sense of how software and data can be simplified will be in the best position to contribute to their field.

22.4 The National Cancer Institute's Thesaurus

The National Cancer Institute (NCI) has produced an ontology of cancer-related terms, including the names of many neoplasms. This vocabulary has been curated for over a decade by in-house ontologists (NCI employees) and contractors and through the use of domain consultants (including some pathologists). It is updated monthly. A great deal of time and money has been invested in the development of the NCI Thesaurus, and it is one of the most worked-on vocabularies in the medical field. The NCI Thesaurus is freely available for download from an NCI ftp site—ftp://ftp1.nci.nih.gov/pub/cacore/EVS/NCI_Thesaurus/. The NCI Thesaurus is a formal ontology. Every name of a disease in the NCI thesaurus is a subclass of one or more classes, and the relationships among the terms can be represented graphically. As an example, the ontologic class ancestry of the term "hairy nevus" is shown (Figure 22.1).

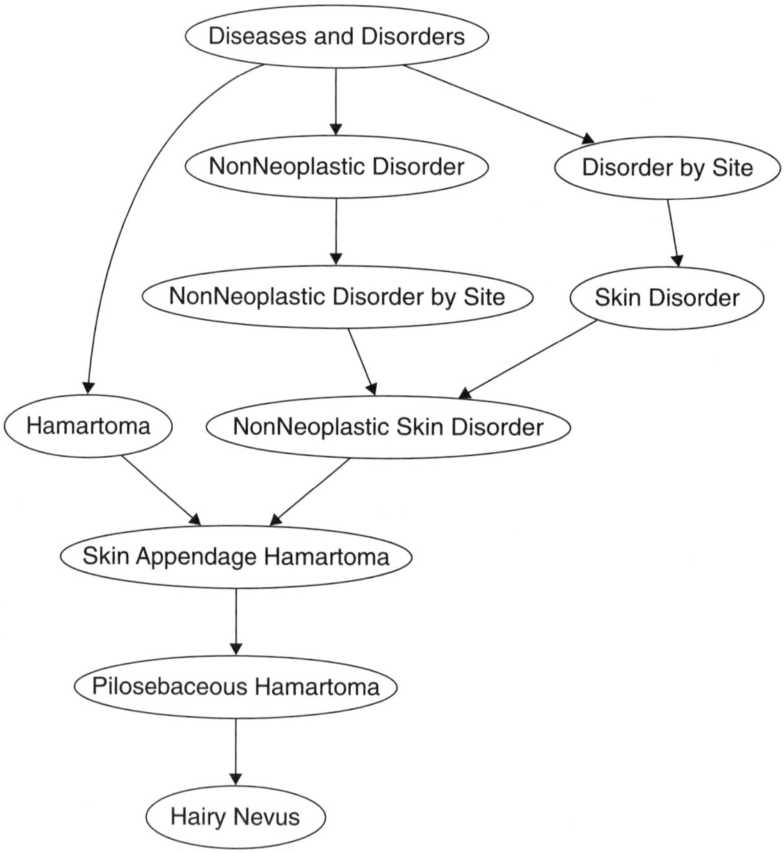

FIGURE 22.1 Ontologic ancestry of Hairy Nevus in the NCI thesaurus.

List 22.4.1 Partial contents of the NCI Thesaurus entry for the object "Skin Appendage Hamartoma."

```
owl:Class rdf:ID="Skin_Appendage_Hamartoma">
<rdfs:label>Skin Appendage Hamartoma</rdfs:label>
<code>C5562</code>
<rdfs:subClassOf rdf:resource="#Hamartoma"/>
<rdfs:subClassOf rdf:resource="#Non-Neoplastic_Skin_Disorder"/>
```

The syntax used in the NCI Thesaurus is OWL (Web Ontology Language), a formal language used by ontologists to describe objects in an ontology. An excerpt of an entry term is shown (List 22.4.1) (294).

The entry is simple to understand. The object "Skin Appendage *Hamartoma*" is a class. Its identifier string is "Skin_Appendage_Hamartoma," and its unique code is "C5562." These alternate forms of the object name can be used by computers to store and retrieve information. Class "Skin Appendage Hamartoma" is a subclass of class "Hamartoma" and class "Non-Neoplastic_Skin_Disorder" and inherits relationships conferred by these ancestor classes.

The problem with ontologies, such as the NCI Thesaurus, is that multiclass inheritance is always confusing for humans. In the case of the ancestral graph for "Hairy Nevus," we see that "Hairy Nevus" inherits from "Hamartoma" but that "Hamartoma" does not inherit from "Neoplasm." In the NCI Thesaurus, hamartomas are considered types of diseases with no inherited relationship to neoplasms. Consequently, insights on the neoplastic process based on properties found among the hamartomas would not be discovered using the NCI Thesaurus.

Because multiclass inheritance is permitted in an ontology, we see that the hairy nevus inherits from three lineages under "Diseases and Disorders," including "Hamartomas," "Non-neoplastic Disorder," and "Disorder by Site." A single term has many ontologic connections. We also see that "Non-Neoplastic_Disorder_by_Site" inherits from "Non-Neoplastic_Disorder" but, strangely enough, does not inherit from "Disorder_by_Site."

In an ontology, there is no single organizing principle for the neoplasm domain. Some neoplasms are subclassed by their anatomic site (e.g., urinary tract neoplasm). Others are subclassed by their tissue type (e.g., soft tissue neoplasm), and so on. This is allowable under an ontology as long as the ontology maintains consistency and competence (ability to answer questions about the members of classes); but I wonder whether this is the best way of organizing tumors.

The NCI Thesaurus has been reviewed by Ceusters and co-workers, who reached the following conclusion: "We found many mistakes and inconsistencies with respect to the term-formation principles used, the underlying knowledge representation system, and missing or inappropriately assigned verbal and formal definitions." (295). The NCI Thesaurus is an impressive piece of work that provides a rich set of relationships that can integrate data from many different biomedical sources. Inadequacies of the NCI Thesaurus may reflect the impossibility of curating an ontology consisting of hundreds of thousands of terms, each with multiple interdependencies and codependent rules of logic.

The biggest difference between an ontology and a classification is that the purpose of an ontology is to achieve "competency," the ability to reach inferences about members in the ontology based on the logical rules that apply to classes. An ontology is not created for the purpose of providing a conceptual understanding of the biologic domain; that is the job of a classification.

Both classifications and ontologies can be described with the same formal languages (such as OWL). The dichotomy between classifications and ontologies are almost entirely accounted for by one logical constraint: the inclusion of multiclass inheritance in ontologies!

22.5 GO, the Gene Ontology that Is Not an Ontology

Certainly, the most famous biologic ontology is the Gene Ontology (GO) (296). No discussion of nomenclatures would be complete without a description of GO and an explanation of the potential uses of this remarkable open access resource; however, GO, despite its name, in not an ontology.

Most of the comments to follow in this section were inspired by Barry Smith and his colleagues who have disregarded the world's ennui and have persevered to search for meaning and purpose in medical ontologies. Those who plan to use ontologies in their work are strongly advised to read Barry Smith's superbly written and indispensable manuscripts (297–299).

GO is an amalgam of three classifications relevant to cell biology: cellular components, biological processes, and molecular functions. GO is not an ontology because it is not a logical construction of rules applied to classes. If GO had simply been constructed as three separate classifications, it would not occupy its unique and perplexing status. The creators of GO decided to combine all three classifications into a single structure. Because a single biologic entity can reside in three GO classes at once (i.e., an entity can be a cellular component, a biological process, and a molecular function), it violates the rules for a classification. Thus, if GO is neither an ontology nor a classification, what is it? GO is a hierarchical, controlled vocabulary. This designation, however, does not do justice to GO's remarkably useful structure.

Having hierarchical biologic domains for uniquely identified objects enables researchers to drive down the complexity of large datasets. Guo and coworkers have used GO to assign gene array profiles to functional domains determined by GO relationships (300). A gene expression value was computed that combined the gene expression measurements collected on all the members of a functional domain was computed. The gene expression signature of a biological sample was transformed into a profile of the activities of functional domains in GO. Instead of having thousands of seemingly unrelated gene expression measurements, the array data were classified under a few functional domains. Likewise, within the domains, gene expression activities of descendant groups could be determined. This novel approach demonstrates how GO can be used to simplify complex data and to represent observations as shared properties of biological classes.

Innovative semantic structures, such as GO, provide cancer researchers with the opportunity of merging neoplastic datasets with molecular datasets and drawing inferences from the combined data. It remains to be seen whether these innovations will lead to breakthroughs in the treatment of cancer.

22.6 Converting Classifications into Ontologies

Any classification can be converted into an ontology if it is rule based. The Developmental Classification has a single rule that determines every class and subclass of neoplasms. The rule underlying the Developmental Classification is that every class of neoplasm is a subclass of its developmental ancestor and thus inherits the properties of the classes in its developmental lineage.

It is technically possible to represent the Developmental Classification as an ontology using RDF (Resource Description Framework), a standard ontology syntax (283). The Developmental Classification is available in RDF syntax at: http://www.julesberman.info/neordf.txt.

An excerpt of the top schema (the part of the RDF document that specifies the relationships among the different classes included in the ontology) is shown in List 22.6.1.

The ontology document for the Developmental Classification continues for 10 megabytes (the equivalent of ten times the length of this book).

List 22.6.1 Excerpts from Ontology for the Developmental Classification.

```
<rdfs:Class rdf:ID="Tumor_classification"
  <rdfs:subClassOf
      rdfs:resource="http://www.w3.org/2000/01/rdf-schema#Class"/>
  </rdfs:Class>

  <rdfs:Class rdf:ID="Neoplasm">
    <rdfs:subClassOf
      rdfs:resource="#Tumor_classification"/>
  </rdfs:Class>

<rdfs:Class rdf:ID="Neural_crest">
<rdfs:subClassOf
      neo:resource="#Neoplasm"/>
  </rdfs:Class>

<rdfs:Class rdf:ID="Peripheral_nervous_system">
    <rdfs:subClassOf
      neo:resource="#Neural_crest"/>
  </rdfs:Class>

  <rdfs:Class rdf:ID="Germ_cell">
    <rdfs:subClassOf
      neo:resource="#Neoplasm"/>
  </rdfs:Class>
```

Chapter 22 Summary

Ontologists have a big job ahead of them. They need to organize the flood of data coming from medical centers and research laboratories and centers for disease statistics. A purely ontologic approach to data organization will allow us to understand the relationships that exist among data of diverse origin—if we can somehow manage the complexity of the ontologies that we create.

If the ontologic approach to data organization is abandoned in favor of the traditional method of strictly hierarchical classification, with each biological object assigned one ancestor (and thus one ancestral lineage), we can achieve a simplified world view of cancer-related information, but we may miss discovering relationships that lie outside the taxonomic principles upon which the classification is based.

Issues of ontology and data classification seem irrelevant and silly to patients struggling to survive cancer or to physicians who are trying provide their patients with hope. When we incorrectly classify a type of neoplasm, we effectively ruin any chance of curing the tumor with a class-specific treatment. When we test new drugs on collected classes of tumor, we reduce the chances of validating drug effectiveness if the tumors in the test population have been misclassified. Building a classification for neoplasms that correctly assigns each instance and type of tumor to a biologically meaningful class is an important, clinically relevant task. The lives of millions of patients will depend on our choices, because the most effective approach to curing cancers is to develop class-based cures.

In the past decade, for better or for worse, the fields of biology and of medicine have been restructured as data-driven, computationally intensive, interrelated disciplines. At this moment in the history of medical science, we must choose the way we think of about complex diseases, such as cancer, and the way we organize the terabytes of data that are produced each week in major medical centers. Researchers with a background in computer science are apt to prefer an ontological approach to organizing data. Developmental biologists are apt to choose classification. Everyone should remember that a rule-based classification can always be converted into an ontology, but an ontology cannot generally be converted into a classification. It may be worthwhile to develop new ontologies under the constraining conditions of a classification to derive the best of both worlds.

PART

III

Cancer Research and the End of Neoplasms

CHAPTER

23

Class-Dependent Cancer Prevention, Diagnosis, and Treatment

Despite the efforts of many talented leaders and dedicated professionals, the last quarter of the 20th century might best be described as the 'era of Brownian motion in health care.'

—Institute of Medicine

23.1 Background

Cancer is one of the most important diseases of humans. The Centers for Disease Control publishes summary data on the number of deaths in the U.S. attributed to leading causes of death (List 23.1.1) (301).

In the United States, cancer kills over half a million persons each year (301). Worldwide, cancer kills about 7 million persons each year and accounts for about 13% of all human mortality (38). This book has no value whatsoever if it cannot be used to decrease the number of cancer deaths. The purpose of this chapter is to describe how we might start to think about the uses of a neoplasm classification to develop and test new hypotheses related to the diagnosis and treatment of neoplasms.

23.2 The Malignancy Detector

A malignancy detector is a mythic device that can scan your body and tell you whether you have a malignancy. What would be the clinical value of such a device? Not much. To be of any value, you would need to know the location of the tumor, the type of tumor, the extent of local invasion, and whether the tumor has metastasized. Simply knowing that a patient has cancer has no real benefit (List 23.2.1).

List 23.1.1 Centers for Disease Control report of annual number of deaths in the U.S. from leading diseases.

Heart disease: 696,947
Cancer: 557,271
Stroke (cerebrovascular diseases): 162,672
Chronic lower respiratory diseases: 124,816
Accidents (unintentional injuries): 106,742
Diabetes: 73,249
Influenza/pneumonia: 65,681
Alzheimer's disease: 58,866
Nephritis, nephrotic syndrome, and nephrosis: 40,974
Septicemia: 33,865

List 23.2.1 Limitations of a malignancy detector.

It does not tell you where to start looking for a primary tumor.
It does not tell you whether the tumor has reached a size that permits it to be found with radiographic imaging techniques.
It does not tell you whether the patient has a type of malignancy that might benefit from diagnosis or treatment.
It does not tell you whether the patient may have several different primary cancers that happen to exist concurrently.

Let us step back a moment to examine a real-life situation. A long-time smoker has a productive cough. Lung cancer can lead to a productive cough, as malignant cells from the cancer slough into a bronchus to be expectorated in sputum. His doctor has the patient capture a sample of his sputum in a specimen container. The sputum sample is transported to the pathology department where a cytologist renders the following diagnosis: "Positive for malignant cells. Cytologic features are consistent with squamous cell carcinoma."

The sputum diagnosis prompts a chest x-ray, which shows a 2-centimeter density in the left lower lung. The patient is advised to have surgery to remove the cancer. At surgery, the left lower lobe of the lung is removed and sent to the pathologist to evaluate the extent of the cancer.

The pathologist examines the specimen and finds no squamous cell carcinoma. The 2-centimeter nodule detected on chest x-ray was diagnosed as a benign chondroid hamartoma, a harmless lesion that could have remained in the patient indefinitely without negative consequence.

Review of the original sputum specimen confirmed the presence of malignant squamous cells. What went wrong? The positive sputum indicated that a malignant squamous cell carcinoma was present. It was a false assumption that the 2-centimeter lung nodule found on the chest x-ray was the source of the malignant squamous cells found in the sputum specimen.

Chastened, the doctors re-examine the original lung x-ray, searching for a small lesion that they might have missed during the initial evaluation. No lesion is found. The doctors recommend spiral computed tomography (CT) imaging. Spiral CT is more sensitive than conventional x-ray. Sure enough, the spiral CT showed a small, 5-mm lesion in the right upper lung. This time the clinicians have learned their lesson and arrange for a bronchoscopic biopsy of the lesion, sparing the patient another round of lung surgery

Bronchoscopy is a procedure that is highly uncomfortable for the patient. The bronchoscopist passes a tube down the patient's mouth, larynx, trachea, and bronchus, inching ever closer to the elusive lesion. A lens located at the end of the tube sends an image of the bronchus back to the bronchoscopist. If a lesion is visualized, a biopsy forceps, cached within the tube, can grab a tissue sample. The purpose of the bronchoscopic procedure is to determine if the visualized lesion is cancer; and if so, to determine the type of cancer. In this instance, bronchoscopy failed to find any endobronchial lesion. Bronchial washings and brushings (additional specimens produced by bronchoscopic manipulation) are obtained, but pathological examination fails to find malignant cells.

The clinicians are contemplating ordering a percutaneous biopsy of the lung lesion, wherein a radiologist inserts a needle through the chest wall and into the lung mass via fluoroscopic guidance. The radiologist doubts that it would be possible to accurately guide a biopsy to a small, 5 millimeter lesion.

Nearly out of options, a medical student suggests sending the patient to an ear, nose, and throat (ENT) surgeon. Maybe the malignant squamous cells in the patient's sputum originated in the mouth and not the lungs. The ENT examination reveals a small focus of leukoplakia just under the tongue. Leukoplakia means "white plaque." In the mouth, leukoplakia may have a number of causes, including mechanical keratosis (benign), candidiasis (benign), squamous cell precancer or squamous cell cancer. Biopsy of the mouth lesion showed squamous cell carcinoma in situ, a precancerous lesion that was easily treated with surgical excision. After further questioning, it seems that the patient has been a long-time tobacco chewer.

The story is not over. The presence of a lesion that may account for the malignant cells in the sputum does not necessarily mean that the lesion is the certain source of the malignant cells. The patient may have multiple precancers and cancers of the mouth, oropharynx, or larynx. The patient may even harbor a small squamous cell carcinoma of the lung.

List 23.2.2 Causes for a "positive" result for a malignant cell detector.

1. The patient has a very small tumor that would likely take many years to develop into a clinically detectable mass.
2. The patient has a precancer with a negligible chance of developing into a malignant cancer.
3. The patient has multiple metastatic cancers, all in a clinically uncurable stage.
4. The patient has a so-called occult cancer that cannot be found with currently available imaging techniques.

In the described case, the patient had a very specific type of malignant cell detected on a sputum examination: malignant squamous cells. With this specific cytologic diagnosis, the clinician and patient were led on a long chase that, after many procedures, finally led to a small mouth tumor.

Imagine the events that would ensue if clinicians were given a universal malignant cell detector: a device that could determine whether the patient harbored any malignant cells anywhere in his or her body (List 23.2.2).

If we had a pill that could cure cancer, we could pop it in our mouths whenever the malignant cell detector turned positive. A universal malignant cell detector has limited utility when we do not have a universal cancer cure.

23.3 Why Are There So Few Tumors that We Can Now Cure?

Currently, there are just a handful of cancers that can often be cured when the cancer is discovered at an advanced stage (List 23.3.1).

Today, all potential cancer cures are tested in **clinical trials**. The term clinical trial is virtually synonymous with prospective randomized clinical trial. Randomization refers to the random assignment of patients to the treated or to the untreated (or standard

List 23.3.1 Tumors that can be cured with chemotherapy (47).

Choriocarcinoma
Acute lymphocytic leukemia of childhood
Burkitt tumor
Hodgkin disease
Acute promyelocytic leukemia
Large follicular center cell (diffuse histiocytic) lymphoma
Embryonal carcinoma of testis
Hairy cell leukemia (probable)]

List 23.3.2 The six classes of neoplasms.

1. Tumors of Class Endoderm/Ectoderm
2. Tumors of Class Mesoderm
3. Tumors of Class Neuroectoderm
4. Tumors of Class Neural Crest
5. Tumors of Class Germ Cell
6. Tumors of Class Trophectoderm

treatment) group. In addition to being randomized, many clinical trials are blinded so that neither the treating physician nor the patients know the assignment group during the trial.

Modern clinical trials are long and expensive. The process of testing a prospective new drug can take many years. For example, the Prostate, Lung, Colorectal, and Ovarian Cancer Screening Trial (PLCO, NIH/NCI trial NO1 CN25512) serves as an example. The PLCO is a randomized controlled cancer trial. Between 1992, when the trial opened, and 2001, when enrollment ended, 155,000 women and men between the ages of 55 and 74 joined PLCO. Screening of participants and the collection of follow-up data will end around 2016. The purpose of the study is to determine whether screening will reduce mortality from prostate, lung, colon, or ovarian cancers (302).

There are thousands of different tumors, and it is unlikely (actually impossible) that we will ever come up with effective, tested treatments for each and every type of cancer. Although prospective trials are considered the only way of determining the efficacy and safety of new treatments, the public may ask whether society has the time, money, and patience to continue our current course (303).

In Chapter 15, we surveyed the six major classes of neoplasms (List 23.3.2).

The neoplasms within each class of tumor use inherited cellular pathways to achieve the neoplastic phenotype, and many of these inherited pathways are shared among the members of a tumor class. Controlling class-dependent pathways may lead to methods of cancer treatment that will be effective for most or all of the tumors within a class.

We can start to think about approaches to preventing, diagnosing and treating neoplasms that are based on the range of neoplasms that occur within a neoplasm classification (List 23.3.3).

23.4 First Priority: Drugs that Induce Precancer Regression

Precancer treatments will likely be class specific. For example, a treatment for precancers of Class Endoderm/Ectoderm may be effective against some or all precancers of Endoderm/Ectoderm but may not be effective against precancers of Class Mesoderm or Class Neuroectoderm.

List 23.3.3 To-do list for conquering cancer.

1. Develop relatively nontoxic drugs that will enhance the natural tendency of precancers to regress.
2. Continue to develop strategies to prevent cancers.
3. Continue to develop effective treatments for stem cell tumors, almost all of which are tumors of childhood and almost all of which operate through simple pathways that can be targeted by relatively nontoxic drugs.
4. Continue to develop vaccines against viral carcinogens.
5. Develop epigenomic drugs that will target germ cell tumors.
6. Develop effective treatments for tumors of mesoderm, neuroectoderm, and neural crest tumors.
7. Develop effective nonsurgical treatment for benign neoplasms.
8. Develop new diagnostic and predictive markers for cancers.
9. Use experience obtained from items 1, 3, 6, and 7 to develop targeted drug treatments for endoderm/ectoderm tumors.

List 23.4.1 Precancer therapies.

Antibiotics cure gastrointestinal *H. pylori* and cause regression of MALToma pre-lymphomas (246).
Sulindac causes regression of colon adenomas (304).
Progestins are active against endometrial precancer (16)

The current experimental drugs that are effective against precancers are relatively nontoxic (List 23.4.1), suggesting that precancers can be treated more easily than cancers.

Precancers and early cancers can be thought of as neoplasms that have not yet stepped through the progression phase, during which multiple genetic and epigenetic alterations accumulate and many subclones emerge, producing a heterogeneous population of tumor cells.

If a tumor is clonal and homogeneous, as is the case of some early cancers, then an event that causes one tumor cell to die is likely to cause every other cell in the tumor to die. If the cells in the tumor depend on exogenous factors to support their continual growth (e.g., growth factors, oxygen levels, pH levels, and calcium levels), then changes that are deleterious to any of the cells will be deleterious to all of the cells in the same clone. Likewise, therapeutic agents are unlikely to encounter resistant subpopulations in a homogeneous and clonal tumor.

It is no wonder that precancers are susceptible to treatment and likely to spontaneously regress (305).

It is hard to understand why precancers have not been the subject of a major effort by the cancer research community. Arguments against developing precancer treatments seem weak. "Precancers are small, so they are very difficult to find, and if you cannot find the precancer, you cannot treat the precancer." If precancers can be treated with nontoxic agents, then you do not need to find and diagnose the precancer. You can treat the entire population of humans at-risk for the precancer. In the United States, people are accustomed to treating diseases that they do not have. Millions of Americans who have never had a heart attack and have never had angina, gladly swallow a daily statin and a daily aspirin, just to reduce the likelihood of a heart attack. Americans would be happy to take an effective precancer medication if they thought it would protect them from cancer.

A legislative stimulus for precancer research has come from the Food and Drug Administration's Modernization Act of 1997 (FDAMA) (306). The FDAMA contains provisions for "surrogate endpoints" in drug evaluation when the endpoints seem clinically useful. This opens the door to using precancers as surrogates for cancers. Foods or drugs that cause precancers to regress are likely to reduce the incidence of cancer, whereas substances that increase the incidence of precancers are likely to increase the incidence of cancers.

The FDAMA provides relief from the creeping fear that neither the pharmaceutical industry, nor the National Cancer Institute, nor society has the resources or the patience to conduct randomized clinical trials for all of the promising new anticancer drugs, particularly if the endpoints are remote (e.g., incidence or long-term survival). We can hope that shifting cancer tests to the precancers will result in a rapid method to evaluate new treatments that will decrease the incidence of invasive cancers.

One of the most interesting biological features of precancers is regression. The prevalence of precancers in the population (e.g., colon adenomas, actinic keratoses, and squamous intraepithelial lesions of the cervix) is much higher than the incidence of cancers that arise from these lesions. Precancers may be amenable to treatment with agents that enhance the natural tendency of regression. It may take less time to determine whether an agent can induce precancer regression (e.g., if a COX-2 inhibitor reduces the number of preexisting colon adenomas) than to determine whether the same agent reduces the number of colon cancer deaths.

Several precancer lesions seem particularly appropriate for drug development (List 23.4.2).

The American Association for Cancer Research Task Force on the Treatment and Prevention of Intraepithelial Neoplasia has recommended prioritizing precancer research (307).

23.5 Second Priority: Cancer Prevention

One of the axioms of cancer research is that if we knew what caused human cancers we could stop them from occurring.

Let us look at one example for which prevention seems to have been helpful: tobacco. In the 1800s, there were no chain smokers, and there was very little lung cancer. People smoked in the 1800s, but the prevalence of smokers and the intensity of the

List 23.4.2 Precancer lesions that are good trial candidates.

Multifocal squamous dysplasia in oral cavity
Barrett esophagus with dysplasia
Ductal carcinoma in situ of breast
Prostatic intraepithelial neoplasia
Adenomas of colon
Dysplasia of ulcerative colitis
Small pulmonary lesions found on spiral CT
Endometrial intraepithelial neoplasia

activity were trivial in comparison to the cultural shift that occurred in mid 20th century, when mass-produced, inexpensive cigarettes were marketed with demonic zeal. Nineteenth century medical literature bears testimony that lung cancer was once a rare disease. A 19th century physician could spend a career tending to the sick and never once encounter a lung cancer.

Physicians of the 20th century watched as a rare cancer climbed in incidence so that shortly after mid century, bronchogenic cancer became the number one killer among all other cancers. As the incidence of lung cancer rose, a growing number of scientists, physicians, and lay people strongly suspected that tobacco was to blame. When I worked as a service pathologist at the Baltimore VA Hospital, Friday was our day for receiving the pneumonectomies for lung cancer. A pneumonectomy was the most typical procedure for lung cancer and involved resecting the lobe of lung containing the cancer. I would always check the patient's history to see whether he or she was a smoker. I cannot recollect a single instance when cancer occurred in the absence of smoking.

Most studies conclude that cigarette smoking accounts for about 90% of the incidence of bronchogenic carcinoma. These studies cannot fully account for second-hand exposure to cigarette smoke. Many nonsmoking baby boomers were passive smokers as children growing up in households in which one or both parents smoked. When they entered the workforce in the late 1960s and early 1970s, many of the boomers were exposed to heavy concentrations of cigarette effluvia from their coworkers. It seems likely that cigarettes contribute to all but a small percentage of the cases of bronchogenic lung cancer.

In the year 2000, the estimated number of deaths worldwide from lung cancer was 1,211,500, accounting for 17.5% of all cancer deaths (38). Lung cancer is one of the most lethal forms of cancer. About 85% of people who contract lung cancer will die of the disease, a statistic that has not improved to any extent over the past several decades.

Lung cancer is a unique and perfect case in which a carcinogen was introduced at a moment in time and resulted in the specific increase in the incidence of a tumor that occurred only rarely before the advent of the carcinogen; lung cancer is still the most frequent or the second most frequent cause of cancer death in every industrial nation. Worldwide, smoking is more popular than ever, and we can expect an increase in the incidence of lung cancer throughout Asia and in most developing nations.

Cancer is the second leading cause of death in the United States, and the lifetime risk of developing cancer is approximately 41% (301, 308). Despite the high risk of cancer occurrence, cancer prevention may not have the life-extending effect that people expect. In 1978, Tsai and coworkers calculated the increase in life expectancy that would occur if cancer was eliminated as a human disease. They predicted that the elimination of cancer would extend the human life by no more than 2.5 years (309). A more modest reduction in cancer deaths (30% reduction) would increase life expectancy by 0.71 years. A similar 30% reduction in motor vehicle crash deaths would extend life expectancy by 0.21 years (309). Given the difficulty we have had in reducing the cancer death rate by even a few percentage points since 1969 (3), a cynic might think that our efforts might have been better spent teaching U.S. citizens how to drive safely.

Why is the benefit so minor? Cancer is a disease that occurs predominantly in the geriatric population. Older people who do not die of cancer often die from some other disease. Eliminating one disease of the older population allows other diseases of the elderly to take its place. The net effect of eliminating any single disease from the geriatric population is minor.

The role of simple public health measures in the reduction of deaths from cancer and other diseases has been proven. In Finland, in about 1970, the population had reached its highest incidence of cardiovascular diseases. As a result, public health measures were introduced to improve diet, exercise, and other health habits (particularly smoking) in the population. The multidecade effort, called the North Karelia project, stands as one of the greatest successes in public health intervention (List 23.5.1).

These findings emphasize that approaches aimed to improve the general health of populations, combined with measures that reduce the incidence of cancer, may yield results that vastly exceed the benefit that would follow from advances in cancer treatment.

Sometimes, reductions in the cancer death rate may occur unintentionally. Statins are widely used drugs that reduce the blood levels of cholesterol and various other blood lipids. To the best of my knowledge, nobody expected that the use of statins would have any effect on the incidence of cancer.

List 23.5.1 Results of the North Karelia project (310).

An increased life expectancy (7 years longer for men and 6 years longer for women).

Lung cancer death rate reduced 70% in North Karelia.

Heart disease mortality reduced 65% in men.

In a recent study involving nearly a half-million male patients conducted between 1998 and 2004, statin use exceeding 6 months was linked to a significant lung cancer risk reduction of 55%. Participants who took a statin longer than 4 years had a 77% reduction in lung cancer risk (311). Let us imagine that the report is accurate and that we can eliminate deaths from lung cancer by 77% just by prescribing statins to everyone at risk for lung cancer. In the United States, about 90,000 men and 70,000 women will die of lung cancer this year (312). If this were reduced by 77%, we would prevent the cancer deaths of about 123,000 people. This number is equivalent to the total number of cancer deaths attributed each year to prostate cancer, breast cancer, and colon cancer combined (312). This shows us that small reductions in the death rates of the commonly occurring cancers may yield enormous reductions in the cancer death rate (311).

Interventions that delay the onset of cancer will also be very important. Cancer is predominantly a disease of the older population. If the average age of cancer occurrence is pushed back another 10 or 20 years, many people in the population at greatest risk for developing cancer will have died of other causes. Cancer prevention should have a high priority in cancer research, and researchers should search for preventive measures that reduce the incidence, severity, and age of onset of the major causes of cancer deaths. The North Karelia project proved that general public health measures that enhance the health of a population will reduce the deaths from cancer and other diseases and will greatly extend life expectancy.

23.6 Third Priority: Treatments for Tumors of Primitive Cells (Almost All of Which Are Tumors of Childhood)

Some of the most effective cancer cures have been against the primitive childhood cancers. Between 1975 and 1995, the overall mortality of childhood cancer has declined 40% (184). Although the death rates declined, the overall incidence of childhood cancers increased 0.8% per year.

Primitive tumors occur in a young population because adults have lost populations of primitive cells. Nonetheless, all of the populations of fully differentiated cells in adults derived from primitive populations of cells and carry pathways inherited from primitive cells. Lessons learned from treating stem cell tumors will help us understand the pathways inherited through the stem cell lineage. These lessons will help us design drugs that may be effective against many different tumors within a developmental class.

23.7 Fourth Priority: Vaccines Against Oncogenic Viruses

Worldwide hepatocellular carcinoma is the third greatest cause of cancer deaths, accounting for 611,000 deaths annually, representing 8.7% of the total number of humans killed by cancer. In the United States, the incidence of hepatocellular cancer is about five new cases per 100,000 population per year and accounts for less than 2% of diagnosed cancers. In the Far East and sub-Saharan Africa, hepatocellular carcinoma has an incidence of 150 new cases per 100,000 population per year and comprises about half of diagnosed cancers.

Countries with high rates of endemic hepatocellular carcinoma have high rates of chronic viral hepatitis (usually hepatitis B and C viruses). With few exceptions, wherever the prevalence of chronic hepatitis is low, the incidence of hepatocellular carcinomas is low. Effective vaccinations against hepatitis B are the most effective way of reducing the number of deaths worldwide from hepatocellular carcinoma.

In countries that have not deployed Pap smear screening, cervical cancer is the leading cause of cancer deaths among women (72, 73). A vaccine is now available that is proven active against the types of human papillomaviruses most likely to cause cervical cancer. This vaccine is likely to greatly reduce the number of deaths due to cervical cancer in populations that use the vaccine. Vaccines are the simplest and most effective way of reducing the number of cancer deaths from cancers that are causally linked to viruses.

23.8 Fifth Priority: Epigenomic Drugs for Germ Cell Tumors

Germ cell tumors are very peculiar. They arise from germ cells that undergo epigenomic erasure and emerge as totipotent cells capable of producing benign or malignant teratomas, embryonal tumors, nongestational trophoblastic tumors, and yolk sac tumors. Germ cell tumors that have not transitioned into totipotent neoplasms (i.e., dysgerminomas and seminomas) are among the most curable cancers of woman and man, respectively. If the transition from curable tumor to an aggressive form involves epigenomic transformation of germ cells, it would seem reasonable that agents that impede this transformation would be therapeutically useful.

Although germ cell cancers represent a small percentage of all cancers of humans, they provide a unique opportunity to observe epigenomic events in carcinogenesis. The lessons learned from germ cell tumors may apply to all cancers. All genes operate in an epigenetic milieu that can enhance or inhibit cellular pathways. Would it not make sense to learn to control the epigenome through a tumor whose malignant phenotype is expressed through an epigenomic phenomenon?

The mechanism whereby all-trans retinoic acid reverses the neoplastic phenotype of promyelocytic leukemia seems to be mediated through the epigenome (27, 28).

The histone deacetylase inhibitors have shown great promise in the (epigenetic) treatment of a variety of hematologic diseases, including cutaneous T cell lymphoma and multiple myeloma (313). Epigenomic modification should be an active area of cancer research.

23.9 Sixth Priority: Treatments for Tumors of Mesoderm, Neuroectoderm, and Neural Crest

Tumors of classes mesoderm, neuroectoderm and neural crest share several common properties (List 23.9.1).

List 23.9.1 Some common biological properties shared among some tumors of mesoderm, neuroectoderm, and neural crest origin.

They tend to be characterized by simple translocations or fusion genes that are the principal drivers of their neoplastic phenotypes.
They contain, among their class members, tumors that can be cured.
They contain, among their class members, tumors that occasionally regress spontaneously.
They occur at all ages, in both genders, and most have no known cause.
They are uncommon or rare. The most striking exceptions are melanoma (neural crest), and glioblastoma (neuroectoderm). The well-known tumors of class mesoderm (leukemia, lymphoma, and osteosarcoma) are all uncommon tumors despite their notoriety.

Would it not make sense to build on successes by trying to understand the fundamental properties of these tumors that render them vulnerable to treatment?

23.10 Seventh Priority: Nonsurgical Treatment for Benign Neoplasms

Because benign tumors do not metastasize, when a benign tumor is totally excised, the patient is cured. With few exceptions, the treatment of choice for benign tumors is surgery.

The drawback of treating benign tumors with surgery is that we miss an opportunity to discover the pathways that control neoplastic growth when we cure benign tumors by excision.

We are in the midst of a revolution in cancer chemotherapy. Drugs are being developed that target the key pathways that fuel the malignant phenotype. Unfortunately, the tumors against which these new drugs are tested are, for the most part, highly advanced epithelial tumors that harbor a multitude of genetic and epigenetic changes. Common sense would dictate that a drug that targets a single pathway will prove ineffective against a heterogeneous tumor population that is constantly changing and that has aggregate alterations too numerous to count!

Some problems must be approached stepwise, building experience with simple systems and working up to a complex structures. Benign tumors are the training wheels of modern cancer chemotherapy. They are characterized by simple genetic alterations that seem to be the only drivers of their proliferative phenotype. They are the perfect model for studying effects of a drug that targets a genetically altered pathway. When you treat a benign tumor, you can sample the tumor over time (perhaps by fine needle aspiration) to determine the effects of a drug on its target pathway and on ancillary

List 23.10.1 Benign tumors for which patients may be willing to forego surgery.

1. Uterine fibroids
2. Papillary adenomas of kidney
3. Gastric adenomas
4. Small GISTs
5. Noninvasive transitional cell tumors of bladder
6. Meningiomas
7. Choroid plexus adenomas
8. Any benign tumor occurring in patients who are poor surgical risks

pathways. You can measure the effect of several different drugs, targeting different pathways or different proteins in a single pathway, to determine their cumulative effects. You can use the simplicity of the system and the low-risk nature of the study to gain crucial information that will prepare the oncologist for malignant neoplasms.

Can we ask patients with benign tumors to submit themselves to experimental protocols when time-tested surgical procedures are available? Actually, there are many important benign conditions for which patients may defer surgery (List 23.10.1).

Benign tumors may occur in dangerous anatomic locations. Removal of such tumors often raises significant operative risks. In addition, large benign tumors may occur in organs that patients would prefer not to lose. Uterine fibroids are very common in women of child-bearing age. They can produce discomfort and dangerous uterine bleeding, and they sometimes become malignant. Most women with uterine fibroids do not choose to have surgery, but all women with uterine fibroids would like to have a nonsurgical treatment option.

Oncologists traditionally focus their practices on treating advanced cancers. Metastatic cancers have received the greatest attention by oncologists and clinical trialists. As a new generation of low-toxicity drugs that target molecular pathways becomes available, the cancer community should develop treatment protocols and trials for the benign tumors.

23.11 Eighth Priority: New Diagnostic and Predictive Markers for Cancers

In a recent article entitled "International Web-based consultation on priorities for translational breast cancer research," the authors asked scientists to choose the most

List 23.11.1 Reasons why some cancer patients may not benefit from chemotherapy.

The original tumor, though malignant by histopathological criteria, was indolent and would not have produced any clinical problems for the patient, regardless of mode of treatment.

The original excisional biopsy of the patient's primary tumor was curative. The tumor had not metastasized prior to the removal of the tumor.

The original excisional biopsy of the patient's primary tumor was curative. The tumor had metastasized prior to the removal of the tumor, but the metastatic cells lack the phenotype required for progressive growth (i.e., the metastatic cells are harmless to the patient).

The original tumor had metastasized widely prior to excision of the primary tumor. The metastatic lesions will grow and will be refractory to any available treatment.

important priority for breast cancer research. There were 420 voters from 48 countries. Half of the voters were clinicians, and the remainder were academics, research scientists, or pathologists. The authors found that "the highest priority was to identify **molecular signatures** to select patients that could be spared chemotherapy."

It may seem odd to think that the highest priority in breast cancer research is to find a subset of patients who will not benefit from treatment, but there it is. When oncologists subject cancer patients to chemotherapy, they know that many of those patients do not need therapy (List 23.11.1).

Oncologists know that chemotherapy after excision of the primary tumor will only benefit a narrow subset of cancer patients: those patients whose tumor had metastasized prior to the excision of the primary tumor, for which the metastatic cells are capable of growing into invasive cancer and for which the metastatic cancers are responsive to the intended therapy. The dilemma to oncologists is that they do not know who will benefit from toxic chemotherapy.

In Chapter 7, we discussed gene expression arrays as a technique that prepares a profile, or signature, for each tested cancer. A cancer signature consists of the relative levels of thousands of different genes expressed in a cancer. The signature for each cancer is unique. Mathematical methods called clustering algorithms can group different tumors based on the similarities of their gene expression signatures. In the case of breast cancer, studies have found clusters of breast cancers (i.e., breast cancers that share similar gene expression profiles) that do not benefit from cancer chemotherapy.

From the thousands of genes that compose a cancer signature, a small number of genes may correlate highly with those features that distinguish one cluster from another. Such genes can be used as a panel for a predictive test (i.e., a test that predicts how a cancer will respond to treatment). Using this approach, it may be possible to develop panels of genes that will tell us which tumors will not benefit from chemotherapy.

The need to select a population of cancer patients who will not benefit from treatment applies to many types of cancer. Prostate cancer is extremely common, and small prostate cancers are found incidentally at **autopsy**. By the age 80 to 90 years, 70% to 90% of men have prostate cancer confirmed at autopsy (314, 315). The vast majority of men who have prostate cancer have indolent lesions and will not die of prostate cancer. How do we choose those men who will not benefit from treatment?

It is sobering to think that in 2008 one of our highest priorities in cancer research is to identify those patients that we cannot help, but the problem of overtreatment cannot be avoided.

New methods, applying mathematical techniques to complex clinical and biological datasets, can subdivide cancers into groups that have common biological behaviors. We should support studies that search for diagnostic tests and **predictive markers** for biological subtypes of neoplasms.

23.12 Ninth Priority: Use Experience Obtained from Items 1, 3, 6, and 7 to Develop Targeted Drug Treatments for Endoderm/Ectoderm Tumors

A common theme developed throughout this book is that tumors of Class Endoderm/Ectoderm are the most common and the most complex tumors of humans. We cannot cure these tumors after they have metastasized because these tumors simply have too many options to resist chemotherapy: large number of cells, heterogeneous population, and a constantly changing genome and epigenome.

Advances in the treatments of precancers, stem cell tumors, tumors of the other developmental lineages, and benign tumors are likely to precede major advances in tumors of Class Endoderm/Ectoderm. Failing miracle breakthroughs, if we make some progress in our prioritized efforts, we can apply those achievements to develop new approaches to the treatment of tumors of Class Endoderm/Ectoderm (List 23.12.1).

List 23.12.1 Areas that will benefit the treatment of Class Endoderm/Ectoderm cancers.

Continue to develop effective treatments for stem cell tumors, almost all of which are tumors of childhood and almost all of which operate through simple pathways that can be targeted by relatively nontoxic drugs. Agents that are effective against stem cells of Class Endoderm/Ectoderm may be effective against all tumors of Class Endoderm/Ectoderm.

Develop effective treatments for tumors of mesoderm, neuroectoderm and neural crest tumors. Pathways that are important to more than one class of tumor may also be effective against tumors of Class Endoderm/Ectoderm.

Develop effective nonsurgical treatment for benign neoplasms. Malignant tumors of Class Endoderm/Ectoderm often have a benign counterpart. Drugs that cure benign tumors may be effective in some cancers.

Develop new diagnostic and predictive markers for cancers. The best therapy for a tumor may depend on the predictive subclass of a tumor.

Chapter 23 Summary

Despite decades of institutional experience in developing research funding goals, the general approach to funding cancer research is the same now as it was in the early 1970s: attack cancer tumor by tumor, dividing available funds in rough proportion to the number of people who die from each tumor. The cancers that kill the most people—lung cancer, colon cancer, breast cancer, and prostate cancer—receive the bulk of cancer funding. Rare tumors receive very little funding.

The only problem with this simplistic approach is that it has failed miserably. We still do not know how to cure the most common cancers, despite decades of funding. In the past decade, hard-won knowledge of the genetics of cancer has served to highlight the incredible genetic complexity of the common epithelial cancers. Increases in survival of the common tumors of humans have resulted primarily from improvements in prevention and early detection, not from advances in treatment. If we can diagnose cancers when they are small and before they have had the chance to metastasize, our chance of curing patients is improved. This is something that we have known for well over a century.

When you think about the real successes in treating advanced, metastatic cancers, virtually all of the success has occurred in the realm of rare tumors: pediatric neoplasms, certain lymphomas, germ cell tumors, some leukemias, and gastrointestinal stromal tumors.

In this chapter, we advanced nine priorities for cancer research: treating precancers, preventing the occurrence of metastatic cancers with drugs that interfere with one or more of the steps of carcinogenesis and tumor progression, developing drugs that target growth-related pathways in stem cell tumors, developing vaccines against viral and microbiological agents that produce cancers, developing drugs that target the epigenome, developing treatments that target simple pathways used by tumors of non-endoderm/ectoderm origin, developing nonsurgical treatments for benign neoplasms, developing new diagnostic and predictive markers for cancers, and using the knowledge from these efforts to develop new treatments for the common cancers of humans.

Each of these priorities benefits from our knowledge of the different biological classes of tumors or of the different tumor properties that characterize stages of tumor development that apply to classes of tumors.

CHAPTER 24 Finding Neoplasms: Suggestions for Cancer Researchers

Labs, rads, and meds drive patient care; everything else is supporting cast.

—Anonymous

24.1 Background

If you are a cancer researcher or if you aspire to be a cancer researcher, you will eventually need access to samples of human tumors. I have encountered cancer researchers who believe that the best way to procure tissues for their research is to write down a list of disease names, along with the quantity of tissues needed, and to drop the list off with whoever is sitting at the pathology department reception desk. Although stunning in its simplicity, this bold gambit never produces the desired results.

In all hospitals, the pathologist is the person who is charged with receiving any and all tissue removed from patients. The reason that the pathologist is given this responsibility is that the pathologist must render a diagnosis on all excised tissues. When tissues are received in the pathology department, each tissue is assigned a unique identifier number, the accession number. The pathologist describes all of the accessioned samples and prepares a document that includes the gross (seen with the unaided eye) description. Then the pathologist cuts samples of tissue that are most likely to yield diagnostic material. The tissue is prepared for microscopic examination by histotechnicians. The pathologist views stained tissue samples under a microscope. If the first tissue samples do not yield a diagnosis, the pathologist must go back to the specimen and select additional samples. After a diagnosis is rendered, the pathologist is responsible for curating the tissue blocks, glass slides, and pathology reports on the specimen and on all of the other specimens obtained by the hospital. In the United States, approximately 25 million human tissue specimens are excised and archived each year.

The acquisition of human tissues and data for research is one of the most important skills required of every cancer researcher. The purpose of this final chapter is to describe the connections between patients, cancer researchers, pathologists, tumor tissues, and tumor data.

24.2 What Do Pathologists Do?

Anatomic pathologists are the people who examine tissue samples and determine the names of any lesions present in the samples. A typical pathologist renders between 100,000 and 200,000 tissue diagnoses in his or her career. In addition, he or she may be asked to consult on cases encountered by other pathologists or clinicians who need to get an additional opinion on a difficult case before rendering a final diagnosis.

Pathologists also participate in quality-assurance exercises in which they review the diagnoses rendered by their colleagues. When a cancer patient dies, and the family provides permission for an autopsy, the pathologist determines the extent of disease at the time of death and evaluates any other pathologic conditions that may have resulted from the cancer or that may have contributed to the patient's morbidity (state of ill health).

The pathologist is expected to stay current in her field by reading case reports and other scientific articles and by reviewing collections of cases of special interest. Every cancer case receives a detailed examination by a pathologist. If the case is challenging or biologically surprising, it will be reviewed by all of the pathologists in the department. All of the received tissues are archived within the pathology department. Some venerable hospitals maintain tissue archives that date back to the times of the American Civil War.

The anatomic pathologist embodies all the memories of a medical center's experience with cancer tissues. The sheer volume and range of cancer cases encountered by anatomic pathologists endow them with a unique set of skills (List 24.2.1).

List 24.2.1 Anatomic pathologists acquire a unique understanding of these items:

1. The typical presentation of every neoplasm
2. The full range of biological potential of every class of neoplasm
3. Expertise in uncommon variants of common tumors
4. Expertise in uncommon presentations of common variants of common tumors
5. Familiarity with rare tumors
6. Familiarity with lesions that can be mistaken for cancer
7. A deep sense of the limits of knowledge in the field of tumor biology

List 24.2.2 Categories of cancer-related articles in anatomic pathology journals.

1. Research on the biological properties of specific neoplasms. Often, this involves validating diagnostic markers that can distinguish a neoplasm from other neoplasms or from distinct biological variants of the same type of neoplasm.
2. New findings based on a review of a collection of a specific type of neoplasm or of a specific variant of a neoplasm.
3. Case studies of biologically interesting or rare neoplasms or diagnostic cases that are difficult or that could be mistakenly diagnosed.

We are told that pathologists render their diagnoses based on a histological examination of tissue (i.e., by viewing a stained thin section of the tumor sample under the microscope). In most cases, the pathologist is fairly sure of the diagnosis well before the microscopic examination commences. Tumors often present with a characteristic demographic profile (age, gender, and ethnicity), clinical details (family history, associated diseases, and clinical presentation of the tumor), anatomic specificity (exact location of the tumor), and gross pathology (size, shape, consistency, color, presence or absence of necrosis and blood, and border of the lesion). With all of this information in hand, the diagnosis is virtually certain even before histologic examination.

Pathologists seem to care more about cancer than any other disease. Open any general anatomic pathology journal (a journal devoted to all the areas of anatomic pathology) and read the titles of the articles. You will find that cancer articles dominate. The disease that absorbs most of a pathologist's time is cancer. Among the cancer articles, most will fall into one of several narrow categories (List 24.2.2).

Because of the special relationship between pathologists and tumors, pathologists have chosen most of the names by which the tumors are known. When a new tumor is "invented," pathologists are, in most cases, the first-round epidemiologists and biologists who describe the new entity.

24.3 Times Are Changing

Today, oncologists want predictive information about tumors (which tumors will respond to which specific treatments), not just diagnostic information (the name of their patient's tumor, rendered by a pathologist). Predictive markers must be tested in clinical trials, and clinical trials are conducted by oncologists, not pathologists. This trend has somewhat marginalized the role of pathologists in tumor diagnosis.

Forward-thinking pathologists have been searching for new roles, some of which require new sets of expertise (List 24.3.1). In the immediate future, pathologists will participate in cancer research by developing, validating, implementing, interpreting, and finding new uses for predictive tests.

List 24.3.1 Present and near-future roles of pathologists.

1. Review. The gold standard for any new diagnostic test is the pathologist's diagnosis. Pathologists review the cases that accrue into clinical trials to ensure that the patients actually have the neoplasm that the protocol is designed to treat. All new diagnostic methods are built on a foundation of old diagnoses rendered by anatomic pathologists. As more and more clinical trials and newer diagnostic and predictive tests are developed, the anatomic pathologist will play a crucial role in every effort.
2. Tissue archivists. The tissue samples prepared by pathologists are archived in vast paraffin repositories. A large hospital can accumulate millions of paraffin tissue blocks. Archived tissues are an invaluable resource for researchers who are studying the neoplasms that occur in humans. Tissue samples can be used to develop and test new diagnostic methods. They can also be used to determine the effects of new chemotherapeutic agents on human cancers.
3. Informaticians. Pathologists collect a large portion of the textual data produced by medical centers. Pathologists often quote the 70% rule: "70% of the data collected in a hospital are pathology data, and 70% of the clinical decisions rendered in a hospital are rendered based on pathology data, and 70% of patients seen in a hospital receive some service that generates new pathology data." Pathologists create and understand the most important data collected in hospitals. They are in the best position to use this information for research and treatment.
4. Validation and outcomes experts. Clinical trials conducted by industry and government are just another form of experiment. The findings obtained by clinical trials must be confirmed again and again by studying the outcomes that follow clinical implementation. Pathologists render the diagnoses and monitor the effects of new diagnostic and predictive tests and of new cancer treatments.
5. Molecular diagnosticians who understand morphologic diagnostics. Molecular biologists know a lot, but pathologists know what the molecular biologists do not know. Expertise in anatomic human pathology is required to find the clinical relevance of molecular discoveries.
6. Ethicists. The history of the development of new tests and therapies has been marred by deceitful and illegal incidents (281). Pathologists, who are entrusted with tissue repositories, have the best opportunity to observe ethical lapses and to enforce ethical values in human subject research activities. Pathologists who have the responsibility of operating tissue repositories should be active members of institutional review boards.

24.4 Who Owns the Tissues and Reports that Are Archived Within Pathology Departments?

Pathologists archive all the tissues blocks and associated pathology reports generated within hospitals. Tissue archivists must have administrative skills and must work with a wide variety of people who are not members of the pathology department. Tissue archivists must be familiar with rules of confidentiality, ensuring that research with archived tissues is conducted in a manner that does not betray the confidentiality of patients.

Recently, the role of the tissue archivist has been complicated by issues of rights and ownership.

Three entities lay claim to the tissues slavishly archived within pathology departments: pathologists, hospital administration (or university administration in the case of university-owned medical centers), and patients. All three entities use archived tissues and associated data for a variety of necessary purposes (patient care, research, preparing statistical reports for health care agencies, etc.).

Who actually owns the tissues collected by pathology departments? Ownership is a mercantile concept conferring the right to sell. If someone owns a cow, that means that they have the right to sell the cow. I own my house, even if I have a huge mortgage that I must pay, and so I have the right to sell the house. In law, there does not seem to be anyone who is legally assigned the right to sell human tissue specimens and human medical records. Even the patient has no established right to sell his or her own tissues. Most countries have deferred assigning legal rights of sale on archived tissues for fear of establishing precedents for trade in flesh.

There are many examples where items are used, shared and regulated, without assigning legal ownership. We breathe air, but nobody owns the atmosphere. We also can be required not to pollute the air, even though we do not own the air. Basically, we can have a system of rights and obligations without assigning ownership. In the case of tissues and medical records, patients have the right to ask hospitals to send their medical records to other medical centers or to themselves. Hospitals are expected to archive tissues, medical reports, and patient charts to serve the patient and society. State health departments, the Centers for Disease Control and Prevention, and cancer registries all expect medical centers to deliver medical reports on request. All of these customary activities proceed without an assignment of ownership.

Problems may arise, however, when anyone seeks to sell or transfer records and tissues. There are not many legal precedents in the biorepository field, and when a case comes under review by an appellate court, it can have great impact on future activities within the field.

In 2006, the U.S. Court of Appeals for the eighth circuit heard arguments for an interesting case litigated by Washington University (appellee) and William J. Catalona

List 24.4.1 Tissues that are routinely destroyed by pathology departments.

Institutions regularly dispose of tissues removed during surgical procedures. When a large specimen, such as a colon, is received in a pathology department, samples are routinely embedded in paraffin and saved for at least 5 years. The unsampled colon (the bulk of the specimen) is saved for several weeks, sufficient time to ensure that the pathologist has rendered a final diagnosis on the specimen, and then the specimen is discarded.

Institutions regularly dispose of archived paraffin-embedded tissues. Most institutions archive paraffin-embedded tissues for at least 5 years. At that time, some medical centers conclude that the tissues are no longer of any importance to the patient. To avoid the expense of continued storage, some institutions simply dispose of archived material after 5 years.

(appellant) (316). Dr. Catalona, a cancer researcher at Washington University, saved tissues in a biorepository over several decades. When he took a job at another institution, he sent letters to many of the people whose tissues were stored in the biorepository. The letter asked for permission to transfer their tissues to his new employer. He received a number of letters granting the permission. Washington University stepped in and informed Dr. Catalona that the tissues would need to remain in the Washington University biorepository. Washington University asked the court for a declaratory ruling that the tissues were owned by Washington University and could not be taken by Dr. Catalona. The appellate court upheld that the tissues were owned by Washington University (316).

There was one point in their ruling that is particularly troubling. The court noted that the tissues were donated to Washington University by patients (which assumes that patients were the original owners of the tissues). Once donated (according to the court), the tissues were owned by Washington University (not Dr. Catalona); in other words, the court ruled that the tissues were property owned by Washington University. The Washington University v. Catalona decision seems to break a long-held legal tradition of not assigning ownership to human tissues.

What happens next? Have archived tissues become property that can be bought and sold? In many instances, stored surgical tissues have no value to patients and are typically destroyed (List 24.4.1). If hospitals routinely dispose of tissue, why cannot hospitals transfer the tissues (that would otherwise be destroyed) to a third entity (i.e., a party that is neither the patient nor the medical care facility that excised the tissue) (List 24.4.2)?

List 24.4.2 Questions that institutions should ask before transferring tissues and medical records to an external tissue repository.

Would the transfer to a third party constitute a sale of human tissue?
Would the transfer to a third party harm any of the patients from whom the tissue was excised?
Would the transfer to a third party benefit society?
Do any of the institutional staff encouraging the transfer of tissues and data have relevant conflicts of interest?

When tissue samples are transferred from a medical center to a private or public tissue bank, there is a considerable amount of work that goes into retrieving the tissue specimens, annotating the specimens with deidentified clinical information, and ensuring that the quantity and quality of the archived tissues is sufficient for the intended uses of the tissue. At what point does fair compensation (for preparing research tissues) become profiteering through the sale of human flesh? When tissues and records are provided to a third party, the privacy risks are transferred, along with the tissues. Medical institutions that transfer tissues and records to a third party are obligated to protect their patients, and this is typically done by anonymizing the transferred materials (removing the links that connect a tissue or record to a specific patient). An institutional privacy board or an institutional review board should verify that the transfer of material (tissues and records) is planned in a manner that protects patients from harm and provides a societal service. Institutional review boards may also determine whether any of the medical center staff have conflicting interests in the transaction (e.g., are they employed by the party receiving the tissues, and can they profit from the arrangement?). The purpose of most tissue repositories is to facilitate **translational research** that will benefit society. There is ample reason to support efforts to build a translational research infrastructure that does not harm patients.

24.5 What Role Will Computers Play in Cancer Treatment?

Consider these two stories.

Story 1

A number of years ago, while working in Baltimore, I hired a young woman as a laboratory technician and introduced her, in turn, to the staff members in Pathology. When the senior cytology supervisor met her, the two exchanged a few words, and then the supervisor asked the young woman whether she might be the daughter of someone the supervisor had known in high school, a woman with a different last name from the new laboratory technician. She was right. Somehow, our cytologist had picked up enough clues, from the visitor's looks, the way she spoke, her age, her body language, to identify her as the daughter of an old friend. It will be a long time before a computer can scan the image of a person and correctly identify the person's mother!

Story 2

In 1994, Betsy Lehman, an award-winning health columnist for *The Boston Globe*, was treated for breast cancer at the world-renowned Dana Farber Cancer Institute. She was given four times the intended dose of the cancer drug Cytoxan for 4 consecutive days and died. The treatment protocol that contained the dosage error had been reviewed by pharmacists, physicians, and nurses before Mrs. Lehman's chemotherapy. None of the people who approved her medication dosages found the error.

Story 1 tells us that humans are very intelligent and can perform tasks that are well beyond the abilities of modern computers. Story 2 tells us that intelligent humans make very stupid mistakes.

It would be nice if we were capable of designing computers with sufficient intelligence to see the mother hidden in the face of a daughter. We are not. We cannot use computers to render histological diagnoses. We cannot create a comprehensive and meaningful classification of neoplasms by analyses of gene expression data from tumor samples. We cannot determine which chemicals are important carcinogens by feeding structural information into a computer. We cannot make important scientific discoveries just be computationally parsing medical databases.

We are capable of designing software that can recognize an overdose on a doctor's order or an inappropriate treatment schedule or a discrepancy between the name on a bar-coded wristband and the name on a bar-coded intravenous bag.

Do you think of computers as agents that will help us become smarter and smarter, or do you think of computers as agents that will keep us from acting stupidly? In my opinion, the greatest role of the hospital computer will be to save us from ourselves by simplifying information (through the storage and retrieval of patient data) and by reducing hospital errors.

24.6 Pathologists and Nonpathologists

Laymen, biologists, epidemiologists, statisticians, and clinicians often treat pathology as a research support service. When a pathologist is provided a list of tumors to fetch for a research protocol, it is unlikely that the process will end well (List 24.6.1).

List 24.6.1 Reasons why the service-oriented approach does not work with pathologists.

1. Pathologists are under the impression that they are professionals and that the diagnoses they render are as important to the success of a research venture as any measurements conducted by a biologist. Marginalizing the work of the pathologist will engender deep resentment. Pathologists are known for their long memories and will carry grudges for decades.
2. Pathologists, like other physicians, have fiduciary obligations to patients. Data and tissues requested from a pathology department usually fall under "human subject research." In the United States, human subject research falls under the so-called Common Rule (317) and the Health Insurance Portability and Accountability Act (318). The pathologist has an obligation to protect patients from harm resulting from the use of tissues and tissue-related data. All good pathologists take these protective responsibilities very seriously. Sometimes a researcher will attempt to access data and remove tissues from a pathology department without first demonstrating to the pathologist that patient confidentiality has been protected and that institutional safeguards have been followed. In such cases, the consequences can be catastrophic for the research endeavor and for the researcher.
3. Cancer research using human tissues is often a waste of time and money if done without the active involvement of a pathologist. It is easy to spot cancer research conducted without pathologists. Pathologists never render final diagnoses with nonspecific terms, such as "brain tumor" or "lung tumor." Because there are hundreds of different types of tumors that can occur in the brain or the lungs, research papers that draw conclusions on "brain tumors" or "lung tumors" have very little value. Oncologists would never dream of conducting and evaluating a large clinical trial without the involvement of one or more statisticians. It always amazes me that researchers will publish the results of studies on human cancer tissues without the close involvement of pathologists.

Chapter 24 Summary

The field of pathology in general, and tumor biology in particular, has shrunken over the past few decades as disease research (formerly the domain of pathology departments) has migrated to research institutions (led by molecular biologists) and as new disease tests use complex technologies best performed in reference laboratories or in specialized laboratories associated with research institutions.

Today, cancer research is performed by clinical scientists from many different scientific disciplines. These scientists must interact with anatomic pathologists, who are tasked with the acquisition, curation, and disposition of all human tissues obtained in hospitals. Consequently, modern pathologists have become tissue bankers and information managers. A good pathologist is adept at rendering histology-based cancer diagnoses and merging his or her own results with those of other laboratories. The pathologist must somehow ensure that composite data are interpreted correctly and used in an intelligent and timely manner by the clinical staff. The pathologist must also use the residual data and tissues to support a wide variety of research efforts conducted by nonpathologists. The types of activities supported by curated tissue and data archives may involve developing new tests and new therapeutic protocols, validating new and existing tests, and accelerating the evaluation of cancer therapies. The best results for researchers, pathologists, and patients will involve the active professional involvement of pathologists in the design and implementation these efforts.

CHAPTER 25 Principles

Comfort the afflicted and afflict the comfortable.

—Finley Peter Dunne (among others to whom the quote is attributed)

25.1 Background

Despite all we now know, carcinogenesis is still very much a mystery. Prior attempts to explain carcinogenesis have been relegated to a host of theories (List 25.1.1).

It very difficult to encapsulate all of the properties of every cancer within any single theory (List 25.1.2). Nonetheless, neoplasms conform to a basic set of principles. The purpose of this short chapter is to list the principles developed throughout this book.

List 25.1.1 Some theories of carcinogenesis.

The somatic mutation theory. Cancer arises from single somatic cells that have accumulated mutations in oncogenes and tumor suppressor genes (319).

The field theory. Developmental or acquired changes in the organization of a tissue lead to neoplastic growth (280).

The stem cell theory (320). Mutations in stem cells lead to the growth of a mixed-cell neoplasms consisting of true neoplastic stem cells (the minority population within the tumor) and the differentiated cells that derived from the stem cells. Curing cancer would require eliminating the minority stem cell population.

List 25.1.2 Reasons why unifying theories of cancer have not been validated.

Cancer alters every aspect of cells. When everything is changed, it is very difficult to find a single alteration that accounts for the neoplastic phenotype.

Carcinogenesis may be different for different cancers. There is no reason to assume that carcinogenesis in the 129 mouse (which develops tumors simply by transplanting primitive gonads) follows the same process as squamous cell carcinoma produced by prolonged exposure to ultraviolet light.

Cancers are diverse. There is no reason to think that a nevus (hamartoma), a seminoma (germ cell tumor), a rhabdomyosarcoma (tumor of Class Mesoderm), and a colon cancer (tumor of Class Endoderm/Ectoderm) all share the same method of carcinogenesis.

25.2 Principles of Neoplasms

Ten simple principles explain the development and diversity of neoplasms (List 25.2.1).

List 25.2.1 The 10 principles of neoplasms.

1. The properties of neoplasms are the properties of normal cells.
   Corollary: The neoplastic phenotype is simply an alteration of the normal phenotype.
2. Initiation is common and cancerization is rare (i.e., most initiated cells will never complete all of the steps required to produce a clinically detectable invasive cancer).
   Corollary: There are opportunities to halt the development of cancer by interfering with the steps in carcinogenesis.
3. The phenotype of a neoplasm is determined by its cell lineage.
   Corollary: The developmentally restricted cell types of the body account for the diversity of neoplasms.
   Corollary: Oncogenes cause tumors in a narrow range of cells. Oncogenes can only activate pathways that produce a neoplastic phenotype in those cells in which the key pathway is available. Pathways are inherited through cell lineage.
4. Every neoplasm is unique, but every neoplasm belongs to one of several thousand species of neoplasms, and every species of neoplasm belongs to one of several dozen classes of neoplasms.
   Corollary: Neoplasm speciation follows principles that are similar to animal speciation.

(continues)

List 25.2.1 (continued)

5. Progression is the rule for cancers and is the exception for benign tumors.
   Corollary: The most important difference between the development of benign tumors and the development of cancers is the likelihood of progression.
6. It is easy to cure a precancer and hard to cure a cancer.
   Corollary: Eradicate the precancer and avoid the cancer. People will gladly accept treatment for precancers, even though they are not certain that they have a precancer, if the treatment is relatively safe.
7. As a general rule, rare tumors have a single cause and common tumors have many causes.
   Corollary: Among the rare tumors, we find the cancers that have a single characteristic genetic alteration or a single environmental carcinogen.
   Corollary: Among the rare tumors, we find the cancers that can be treated by targeting a single pathway.
8. We will not succeed in curing the most common cancers until we learn how to cure the rare cancers.
   Corollary: Rare tumors teach us how to treat common tumors.
9. The malignant phenotype is expressed through the aberrant control of one or several normal pathways.
   Corollary: Neoplastic growth is arrested when normal control of aberrant pathways is restored.
   Corollary: The pathways in a tumor cell are inherited through the developmental lineage of the cell.
   Corollary: Cancer cells that have a common lineage will tend to express their malignant phenotype through common pathways.
   Corollary: A drug that arrests the growth of a neoplasm is likely to arrest the growth of other neoplasms that belong to the same class.
10. A neoplasm classification provides a conceptual framework for all neoplasms.
    Corollary: A neoplasm classification allows us to grasp the fundamental properties and relationships among all neoplasms.
    Corollary: A neoplasm classification allows us to ask questions about general classes of neoplasms.
    Corollary: A neoplasm classification provides relationships among all the diverse types of neoplasms.

Chapter 25 Summary

The two major phases in the life of a cancer are carcinogenesis and progression. These two multistep processes precede (in the case of carcinogenesis) and follow (in the case of progression) the emergence of an invasive cancer. Because these processes are multistep, there are many opportunities to interfere with the cellular pathways that control these processes. Because cellular pathways are inherited through the developmental lineage, a neoplasm classification based on cell lineage will contribute enormously to the development of new chemotherapeutic agents that will be effective against classes of neoplasms. The current version of the Developmental Classification of Neoplasms, containing over 5,800 neoplasm concepts and over 130,000 neoplasm terms, is available as a free, open-access document from this book's listing at the Jones and Bartlett web site (http://www.jbpub.com) and at the author's own web page (http://www.julesberman.info/).

Glossary

Ab initio—Latin term meaning *from the beginning.* In this book, we use the term *ab initio* to indicate that a neoplastic property or condition was present from the very start of carcinogenesis.

Activated oncogene—Oncogenes, the genes that are responsible for producing the neoplastic phenotype, are present in animal cells as normal genes that contribute to the normal phenotype of cells. They need to be activated before they become a problem. Activation can occur through mutation, through amplification (increase in gene copy), through any mechanism that causes overexpression, including translocation to a highly expressed region of the genome or heightened local transcription, by fusion with another gene that enhances the activity of the oncogene, or by additional mechanisms that we have not yet discovered. Oncogenes that have not been activated are called proto-oncogenes.

Adrenocortical adenoma—A benign tumor arising from the steroid-producing cells of the cortex of either of the paired adrenal glands.

AIDS—A syndrome of diseases resulting from infection with the HIV virus, which attacks lymphocytes. The diseases that occur are typically opportunistic infections caused by endogenous microorganisms that proliferate in immunodeficient patients. Opportunistic viruses produce several neoplasms in AIDS patients, including Kaposi sarcoma (HHV-8), cervical carcinoma (human papilloma virus), and lymphoma (Epstein-Barr virus).

Allele—one of two or more forms of a gene within the same genome.

Alternative RNA splicing—In eukaryotes, the genome is composed of exons and introns. The transcribed exons contain sequences that are found in mRNA. The introns are sequences that intervene between exons. The transcribed intron sequences are not found in mRNA. The complex mechanism involved in cutting specific locations in transcribed RNA, and assembling spliced sections into mRNA molecules, results in a number of RNA species that is much larger than the number of available DNA genes. It is estimated that about 60%

of normal human genes have more than one form of transcribed RNA. The posttranslational modification of peptide sequences further amplifies the number of available protein species that may be found in a eukaryotic cell. Mutations in the genes that code for molecules involved in normal RNA splicing will result in an increase in the number of different RNA molecules. A change in a single RNA splicing regulator may potentially result in the production of thousands of alternate protein molecules. Genetic instability characterizes tumor progression, and it is not surprising that some of the mutations found in cancers may alter RNA splicing pathways. Cancer cells have many more different transcribed RNA sequences than normal cells. It is likely that some alternately transcribed protein products may provide a growth advantage to cells and contribute to the cancer phenotype. It is also feasible that these growth advantages may be tumor type specific or tumor class specific and would thus serve as diagnostic or prognostic markers.

Ames test—A screening test for possible carcinogens based on the observation that most carcinogens are also mutagens. A chemical is added to a culture of *Salmonella* bacteria; if the number of new mutations in the bacteria is elevated (above the level found in control cultures), the screen is "positive," and the chemical may be a carcinogen (321).

Anatomic pathologist—These are physicians who have been trained to examine and evaluate lesions by their gross and microscopic appearances. Anatomic pathologists perform autopsies and render diagnoses on tissue biopsies. They provide intraoperative consultations to surgeons by performing rapid analyses of quick-frozen sections of sampled tissues removed during surgery. All cytopathologists have received training in anatomic pathology. All forensic pathologists (the physicians who perform autopsies for the Medical Examiner's Office) received training as anatomic pathologists. See glass slide.

Aneuploidy—The presence of an abnormal number of chromosomes (for the species) in a cell. Poorly differentiated cancer cells are virtually always aneuploid. The individual cells tend to have more chromosomes than normal cells, and the individual chromosomes tend to have many different abnormalities. The number of chromosomes and the aberrations within chromosomes can differ from cell to cell. The presence of aneuploidy in a tumor cell population is often an indicator of a poor prognosis. Occasionally, aneuploidy is found in benign lesions or lesions that have a tendency to regress, such as keratoacanthoma.

Aneurysmal bone cyst—A cystic lesion of bone once thought to be a reactive (i.e., nonneoplastic) process. Because many of these lesions have a characteristic chromosomal translocation t(16;17)(q22;p13) that creates a CDH11-USP6 fusion gene, aneurysmal bone cyst is now considered a type of neoplasm.

Angiogenesis—Tumor cells require a source of oxygen for survival. All solid tumors are vascularized by new vessels that arise in concert with the growth of the tumor. Tumors can induce vascular growth in nonneoplastic connective tissue by secreting angiogenesis factors. Because all solid cancers require vascular support, drugs that target nonneoplastic vascular cells within tumors may prove to be effective anticancer agents against all types of solid cancers. Currently, there are no tumors that can be cured with angiogenesis inhibitors. It would be useful to test angiogenesis inhibitors on precancers. When a precancer progresses into an invasive cancer, it induces new vessel growth in the adjacent stroma (connective tissue). By inhibiting the very earliest vascular growth in precancers, we might arrest the transition of precancers into invasive cancers.

Anlagen—Embryologic precursor structures.

Annotation—In the field of biomedical informatics, annotation is the addition of descriptive data to a data record or to a data object. An example of annotation would be the insertion of clinical and pathological descriptive data into the header section of an electronic image of a pathological lesion. Another example might be the addition of a relevant reference citation to a database record of a particular protein. Annotations, added over time, enrich the information value of data elements.

Anucleate cells—Cells that have no nuclei. Examples are red blood cells and keratotic squamous cells. Anucleate cells perform a function using the cellular machinery in their cytoplasm (keratin formation in the squamous cell and oxygen transfer in red blood cells), and then they die. It is possible to have proliferating populations composed largely of anucleate cells. Epidermoid cysts are nodules composed of the debris of anucleate squamous cells surrounded by a thin rim of proliferating nucleated squamous cells. Epidermoid cysts can grow to be large, approaching the size of an egg, thus demonstrating that a small population of proliferating cells can give rise to a mass composed almost entirely of anucleate cells. Polycythemia vera is a blood condition characterized by an increase in the number of circulating mature red blood cells. The red blood cells are anucleate. The population of anucleate cells is created by an erythroid stem cell population in the marrow. Malignant tumors may also be composed largely of anucleate cells. Examples are well-differentiated squamous cell carcinomas and erythroleukemia. The occurrence of neoplasms populated by anucleate cells is proof that the stem cell theory of neoplasia, which holds that stem cells are the cells of origin of neoplasms and that the differentiated cells in tumors do not have all the properties of malignant cells, is true in some instances.

Aplastic anemia—A profound reduction in all types of circulating blood cells and in bone marrow stem cells. The most severe cases of aplastic anemia have a high mortality rate. A reduction of all blood cell lineages (whether profound or mild) is called pancytopenia. When an isolated lineage is reduced, the anemia is named after the cell involved: reticulocytopenia, red cells; neutropenia, white blood cells; thrombocytopenia, platelets; and lymphopenia, lymphocytes.

Apoptosis—Programmed cell death. Multicellular organisms have a variety of methods whereby they determine which cells should die and how they should die. The most common method of cell death and cell removal is through terminal differentiation. Most fully differentiated cells are postmitotic, meaning that they can never again divide. The keratinocytes in the upper levels of the skin are postmitotic. Over time, these cells increase their intracellular keratin, shrink their nucleus until it no longer exists, and are eventually sloughed off of the skin. Sloughed keratinocytes account for the bulk of house dust.

In the gut, fully differentiated enterocytes eventually slough into the gut and contribute to the bulk of fecal material that is eventually expelled from our bodies. In adenocarcinomas, malignant glands may slough dead cells into their lumen (the gland spaces), where the necrotic debris accumulates (Figure G.1). Apoptosis generally involves a triggering event (such as a toxic change in cellular homeostasis or a developmental signal to reduce the number of cells in a population) followed by a series of biochemical events involving the shrinkage of cellular compartments and the excision of DNA into small pieces, all accompanied by a cascade of lytic enzymes that reduce the cell to detritus that can be engulfed by macrophages. Several tumor suppressor genes modulate cellular apoptosis.

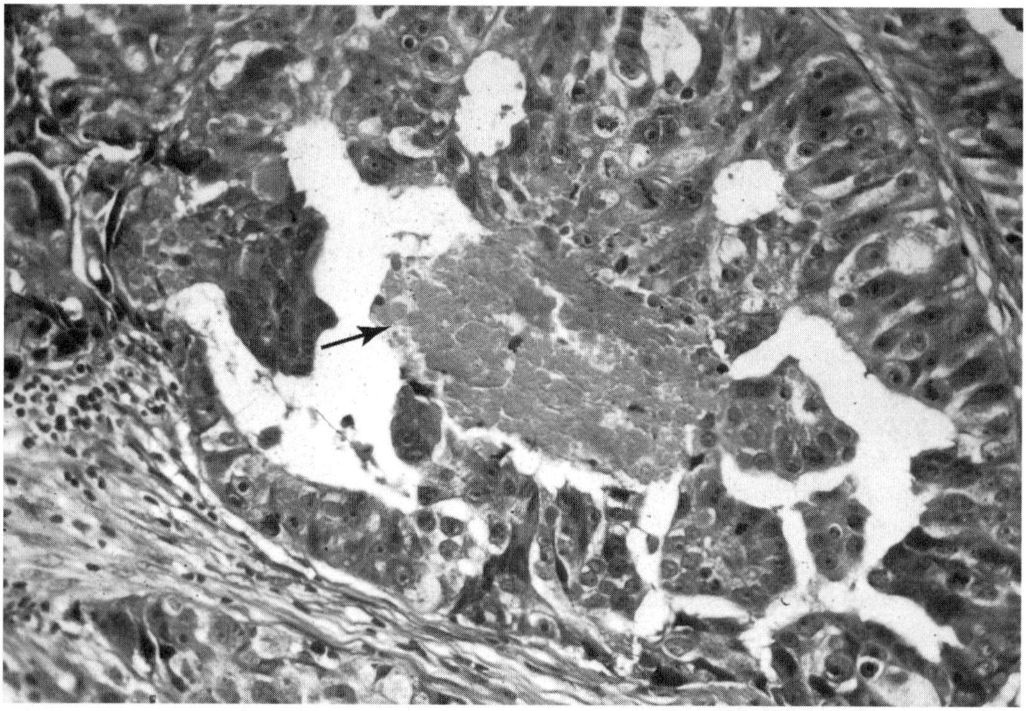

FIGURE G.1 Necrotic debris in the center of a malignant gland from a breast adenocarcinoma.

APUDoma—Tumors derived from cells with the APUD property (amine precursor uptake and decarboxylation). APUD cells are specialized endocrine cells that metabolize amines to short amine-based molecules or short polypeptides (e.g., serotonin and epinephrine). They include the precursor cells for carcinoid tumors, small cell carcinoma, pheochromocytoma, gastrinoma, islet cell tumors, and many others. All APUDomas were once thought to be of neural crest origin. This hypothesis has been discredited, and the term APUDoma has lost much of its popularity. See Chapter 5.

Archived tissue block—All tissues removed from patients (e.g., by surgeons during operations, by dermatologists who sample small skin lesions, by phlebotomists who draw blood, by patients themselves when they collect urine specimens) go to the pathology department where they are examined. In many cases, samples of the tissues are fixed in formalin and then processed to produce a paraffin-infiltrated tissue encased in a block of paraffin. These blocks, often referred to as cassettes (because plastic cassettes hold the paraffin block), are used as the source of thin tissue sections that can be mounted and stained on glass slides. Pathologists look at these glass slides under the microscope and reach a diagnosis based on a correlation of clinical, gross, and microscopic features of the lesions. Unlike radiologists, who look at visual representations of physical lesions, pathologists look at the actual cells taken from the patient. In most cases, after some of the block has been used to produce tissue sections mounted on glass slides, the majority of the embedded tissue remains in the block and serves as a permanent sample of the biopsied tissue. Some laboratories have paraffin-embedded blocks that have been saved for over a century. Provided that the tissues were properly fixed after they were removed from the patient, these ancient paraffin-embedded tissues are perfectly suitable material for many types of modern research studies. There are about 25 million surgical pathology specimens collected each year in the United States. Over the past century, a distributed archive of human tissue blocks has accumulated that has enormous value for biomedical researchers.

Astrocyte—These, along with oligodendrocytes, are cells of neuroectodermal origin that serve as supportive cells for neurons. Special stains, employing metal impregnation, show us that astrocytes resemble twinkling stars (asters) and have fine cytoplasmic projections that extend between neurons and vessels. It is thought that astrocytes protect neurons, contribute to the blood–brain barrier, help regulate blood flow within the brain, and have a role in cell-to-cell signaling. Astrocytomas, the most common brain tumor of humans, have an astrocytic lineage within Class Neuroectoderm. See oligodendrocyte.

Ataxia telangiectasia—Also known as Louis-Bar syndrome or as Boder-Sedgwick syndrome, and often misspelled as ataxia telangectasia. Ataxia telangiectasia is an inherited syndrome featuring cerebellar ataxia, telangiectases (small focal vascular abnormalities marked by dilatation of vessels), immune defects, and a predisposition to malignancy (lung, gastric, lymphoid, and breast cancers). The cells of persons with ataxia telangiectasia are highly susceptible to radiation-induced clastogenesis. The gene that is mutated in ataxia telangiectasia is ATM. The normal ATM gene product phosphorylates DNA repair enzymes following DNA breakage.

Atrial myxoma—See cardiac myxoma.

Atypia—Morphologic cellular aberrations. Cancer cells have atypia. The changes that are most indicative of the cancerous state are found in the morphology of the nucleus (not the cytoplasm) of cancer cells. See nuclear atypia.

Autonomous growth—When a population of cells grows without external stimuli and despite external inhibitory factors, it is said to have autonomous growth, determined only by the intrinsic properties of the growing population. Of course, no growth of cells is autonomous. Populations of cells growing in the body always depend on the blood to provide oxygen and nutrients. Even cell lines growing in tissue culture must have a source of nutrients and fresh culture medium containing growth factors. The criteria for autonomous growth are liberally interpreted. A hormone-responsive breast tumor (one that increases its rate of growth in the presence of hormone) might be considered autonomous if it grows when the hormone is removed or blocked. A gastric MALToma that regresses after the patient is treated for *Helicobacter pylori* infection (the presumed cause of the MALToma) would be considered nonautonomous (246). The term "semiautonomous" is sometimes used to express the circumstantial ambiguity of the concept.

Autopsy—Also known as postmortem examination. The closely related term, necropsy, is usually reserved for veterinary autopsies. The word autopsy comes from Greek root terms auto (self) and opsis (sight), indicating that the autopsy is a method for mankind to inspect itself. A typical hospital autopsy consists of a systematic review of organs, cataloging the pathologic changes present in each organ, and synthesizing a cause of death based on pathologic findings correlated with clinical history. Autopsies generally take a few hours to conduct, during which time small samples of tissues are fixed in formalin and embedded in paraffin (wax). Thin slices of the paraffin-embedded tissues are mounted on glass slides and stained so cells can be visualized under a microscope. Although the preliminary results of an autopsy are usually ready within a day, the final autopsy report, based on the examination of the microscopic slides, may take weeks or even months. The final autopsy report may be many pages in length, with detailed descriptions of clinical observations, laboratory studies, and postmortem findings. Usually 20 to 100 paraffin-embedded tissues are collected in an autopsy, and these collected tissues are usually referred to as "tissue blocks" or "archived blocks." Most hospitals save these blocks indefinitely, and several academic institutions have archived blocks that date back to the nineteenth century. The millions and millions of archived autopsy blocks distributed in hospitals throughout the world have enormous research value because they contain preserved cells and because they are annotated with clinical data (the autopsy reports). In the 1960s, it was common for hospitals to perform autopsies on 30% to 50% of hospital deaths. Worldwide, the rates of autopsies have plummeted. Today, in may hospitals, less than 1% of hospital deaths are followed by an autopsy. "*Hic locus est ubi mors gaudet succurrere vitae*" ("this is the place where death delights to help the living") is a motto appearing in pathology morgues.

Basal cell carcinoma—This is the most common skin cancer. About 600,000 new cases occur each year in the United States. They most frequently occur in the areas of the skin that are exposed to the greatest amount of sunlight. Basal cell carcinomas often occur on the face, and the tip of the nose (which receives more direct sunlight than other parts of the face). These tumors are locally invasive, but they rarely, if ever, metastasize. Unlike squamous cell carcinoma, which derives from the surface epithelium, basal cell carcinomas develop from hair follicles, located in the dermis (the tissue underlying the surface epithelium).

Benign tumor—With a few exceptions, benign tumors are tumors that grow and grow, without invading or metastasizing. As a general rule, benign tumors grow slowly and have limited growth potential. Benign tumors are often diploid or near diploid and do not accumulate the large number of genetic alterations that characterize cancer progression. Benign tumors that grow relatively fast, and benign tumors that are aneuploid and demonstrate genetic instability, are more likely to become cancerous than benign tumors that lack these features.

Biocurator—The curator of a biological or medical nomenclature. See curator.

Biomedical informatics—This is the scientific field that uses the data produced by research laboratories (sometimes called discovery data) and the data obtained from clinical repositories to obtain clinically useful results (e.g., new discoveries, tests, therapies, services, or procedures). Because biomedical informatics translates basic science into clinical reality, it is typically regarded as a translational or applied science.

Blast crisis (blast phase)—A rapid transformation from chronic myelocytic leukemia (in which the vast majority of circulating neutrophils are mature cells) to a phase where the blood and the bone marrow are dominated by blast cells, creating a clinical and pathologic picture resembling acute myelocytic leukemia.

Blastema—The collection of stem cells from which all of the tissues of an organ or a body part will develop.

Blast (blast cell)—The early stem cells in a well-defined lineage that are the progenitor for an identifiable, differentiated cell type. In the bone marrow, most blast cells are myeloid. Marrow blasts account for about 1% of the nucleated cells in a normal bone marrow. In acute myeloid leukemia, blasts account for the majority of nucleated cells. In lymph nodes, lymphoblasts develop into mature lymphocytes. The number of blast cells in lymph nodes increases in lymph node hyperplasia and in immunoblastic lymphomas.

Blastoma—A tumor in which the cells have a morphology that is similar to the morphology of the normally encountered blast cells in the developing organ in which the tumor arose. Some people restrict the term blastoma to tumors that occur in infants and children and that are composed of cells with embryonic morphology. These tumors would include ependymoblastoma, esthesioneuroblastoma, hepatoblastoma, medulloblastoma, nephroblastoma, neuroblastoma, pancreatoblastoma, and retinoblastoma. Others use the "blast" prefix to include tumors occurring in adults that are composed almost exclusively of tumor stem cells. These blastic tumors would include lymphoblastic lymphoma, immunoblastic lymphoma, and blast phase leukemias.

Brain tumor—A tumor arising from neuroectodermal cells and from within the skull. In common usage, this would include tumors deriving from the brain covering layers (the meninges) though the meninges do not derive from neurectoderm. The spinal cord is an extension of the brain down into the vertebral canal, so the same types of tumors that occur in the brain will also occur in the spinal cord.

BRCA—BRCA1 and BRCA2 are breast cancer tumor suppressor genes that are involved in DNA repair. Mutations in these genes are inherited in a small percentage of women with breast cancer. In addition, BRCA1 and BRCA2 mutations are found in other tumors, particularly ovarian and prostate cancer. People born with inactivated BRCA2 have a variant type of Fanconi anemia.

Bronchogenic carcinoma—A tumor arising from the cells lining the bronchus. The term bronchogenic carcinoma usually refers only to squamous cell carcinomas arising in the bronchus. Technically, it applies to the adenocarcinomas that arise from the bronchus, to virtually all small cell carcinomas, and to various rare tumors of bronchial origin (Figure G.2).

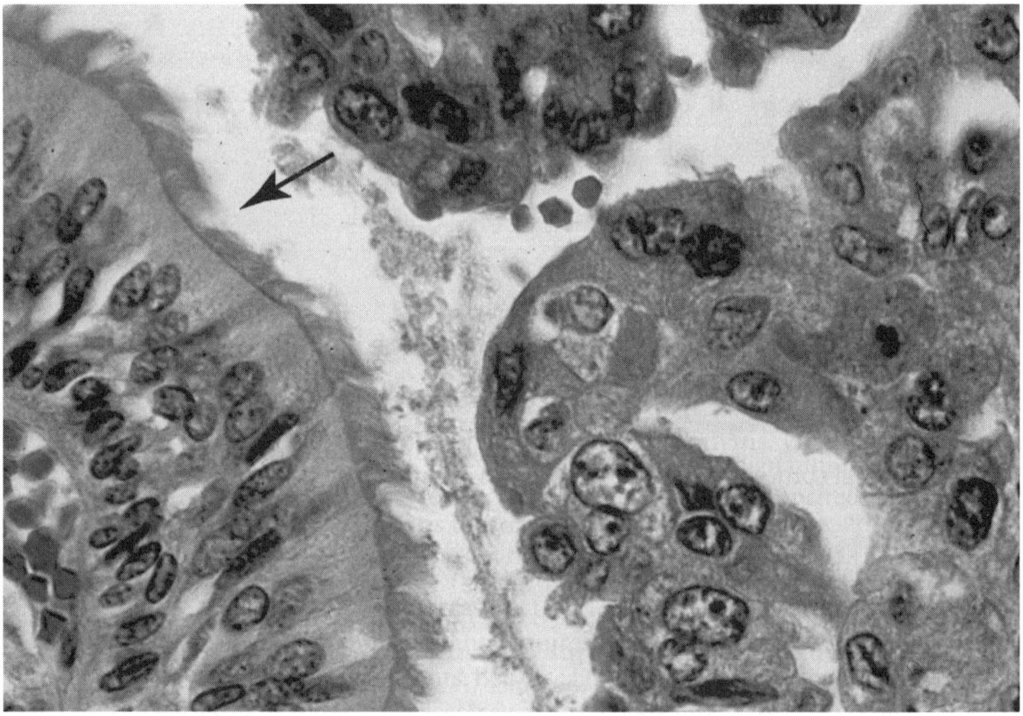

FIGURE G.2 Bronchogenic carcinoma. The mucosa on the left is normal bronchial lining, with numerous cilia protruding from the apex of the lining cells. On the right is carcinoma.

Carcinoid—A tumor of neuroendocrine cells of the gut. The gut neuroendocrine cells derive from endoderm. Carcinoid tumors are common tumors that occur anywhere along the embryologic gut, including the foregut (lungs, pharynx, and first third of adult gut), midgut (small intestines), and hindgut (colon and rectum). Carcinoids may secrete serotonin-like hormones that are normally deactivated in the liver. When carcinoids metastasize to the liver, the normal deactivation of serotonin-like molecules, received through the portal vein, is bypassed. Subsequently, patients with carcinoid metastatic to liver may develop a paraneoplastic syndrome caused by secreted tumor hormones.

Carcinogen—The term carcinogen is used differently by different people. The confusion arises because carcinogenesis is a multistep process that can be modified at many different biological stages. Some people use the term carcinogen to mean a chemical, biological, or physical agent that can cause normal cells to transform, eventually, into cancer cells. Sometimes, the term *complete carcinogen* is used to emphasize the self-sufficiency of the substance as an agent of carcinogenesis. Others in the field use the term carcinogen to mean anything that will increase the likelihood of tumor development under any set of circumstances. An agent that causes tumor growth in an initiated cell population (i.e., a promoter), an agent that reduces the latency period, or an agent that facilitates any step in carcinogenesis are all carcinogens under this alternate definition.

Carcinogenesis—The process by which cells acquire attributes that confer a malignant phenotype. This is a synonym for the deprecated term cancerogenesis.

Cardiac myxoma—Same as atrial myxoma. Cardiac myxoma is a mass found attached to the walls of an atrium of the heart. Histologically, cardiac myxoma has many of the features of an organizing thrombus. It sometimes dislodges from the atrium and travels through the arterial system to cause an abrupt ischemic event in an extremity, a biological sequelae encountered with thrombi. It can occur sporadically or in families with Carney's syndrome. Mutations in PRKAR1A have been found in some cardiac myxomas (61), as well as aneuploidy (62), both suggesting a neoplastic origin for this lesion.

Cell type—The number of different kinds of cells in an organism varies based on how you choose to categorize and count the different cells of the body, but most would agree that there are about 200 different cell types in the body. Because we can classify thousands of different cancers in humans, we can safely say that one cell type may correspond to many different types of cancer. For example, both follicular carcinoma of thyroid and papillary carcinoma of thyroid seem to derive from the epithelial cells that give rise to normal thyroid glands. Follicular carcinoma of the thyroid and papillary carcinoma of thyroid behave differently from each other, and each has its own distinct genetic defect.

Cerebellar hemangioblastoma—A rare benign tumor of vascular origin that is sometimes seen in cases of Von Hippel Lindau disease. Morphologically, it can resemble metastatic clear cell carcinoma of kidney. Because either renal cell carcinoma or cerebellar hemangioblastoma can occur in Von Hippel Lindau disease, it is very important that pathologists not mistake one neoplasm for the other when evaluating a cerebellar biopsy.

Chemokine—A cytokine that stimulates white blood cells to move to a tissue target.

Chinchilla—A rodent native to the Andes mountains of South America.

Choristoma—Choristomas are closely related to hamartomas. Whereas hamartomas are malformed, overgrowths of tissues that are normally found in the location of the hamartoma, a choristoma is a malformed overgrowth of tissues that are not normally found in the location of the choristoma. Choristomas are different from ectopias (aberrantly occurring tissues) in that the aberrant tissues in a choristoma grow as a mass. Both hamartomas and choristomas are characterized by limited growth (i.e., the lesions grow to a certain size and then stop growing), and this feature helps to distinguish hamartomas and choristomas from benign tumors and cancers, both of which grow continuously.

Cirrhosis—A liver condition in which fibrous tissue proliferates in liver acini (the glandular portion of the liver) and around ducts and vessels. This results in the death of liver cells and a marked reactive overgrowth of hepatocytes with distorted and enlarged liver acini. These changes result in a marked decrease in liver function and a high mortality rate. In the United States, about 80% to 90% of hepatocellular carcinoma cases arise in cirrhotic livers.

Clade—A clade consists of a class of organisms that share the same direct ancestor and that includes every direct descendant of that ancestor. A classification that consists entirely of clades is monophyletic, meaning that organisms with different ancestors will never occupy the same clade. The Developmental Classification of Neoplasms is not strictly cladistic.

Cladistics—The assignment of classes and subclasses within a classification by strict order of lineage (i.e., through clades). The commonly used classification of living organisms is not strictly cladistic because assignment of organisms to classes depends on both lineage and similarity.

Class—A class is a group of items that share some common feature or features. A class can have a subclass (sometimes called child) and a superclass (sometimes called parent). In classifications, a class can have only one superclass. In ontologies, a class can have more than one superclass. In classifications and ontologies, a class can have more than one subclass. The classification of organisms has named rankings (Kingdom, Phylum, Class, Order, Family, Genus, and Species). In the Developmental Classification of neoplasm, these terms are not used. Only the terms class and subclass are used. The term superclass is inferred and not actually used as a designation.

Classification—There are many definitions for classification. Some people consider any grouping of objects to be a classification. Some people use the term classification to mean the method used to group objects (e.g., classification by clustering). Some people use the term classification to refer to the act of assigning an object to its correct group (e.g., I agree with your classification of this tumor as a squamous cell carcinoma.). In this book, the definition of classification is taken from the work of Ernst Mayr, one of the most influential evolutionary biologists of the twentieth century (119). A classification is a hierarchical and comprehensive grouping of all of the members of a well-defined knowledge domain. In a classification, a member can only be assigned one class. Each class can be the subclass of exactly one immediate ancestral class. The top class in a classification is the only class with no ancestor. This creates a simple lineage for each member that extends to the top class in the hierarchy. The oldest and most important classification in biology is the classification of all terrestrial organisms (sometimes referred to as the tree of life). See ontology and taxonomy.

Clastogen—Clastogens are agents (chemical or physical) that break chromosomes. Examples of clastogens are benzene and ionizing radiation. Clastogens are different from mutagens. Clastogens break up chromosomal material inducing lesions that are visible under a light microscope (chromosome aberrations, sister chromatid exchanges, and micronuclei). Mutagens produce small changes in the sequence of base pairs, sometimes involving single bases (point mutations).

Clinical pathologist—A clinical pathologist works in a hospital laboratory or in a so-called reference laboratory that does testing for hospitals and individual clinicians. Clinical pathologists supervise the personnel and the special automated testing equipment that performs simple or complex tests on biological samples (most commonly, blood). The kinds of tests that clinical laboratories perform usually yield a number (e.g., prostate specific antigen of 4, glucose of 85, and potassium of 6). The association of a test with a quantifiable value often separates the job of a clinical pathologist from the anatomic pathologist, whose job is to provide a textual diagnosis on a submitted piece of tissue.

Clinical trial—Today, the term clinical trial is virtually synonymous with prospective randomized clinical trial. Randomization refers to the random assignment of patients to the treated or to the untreated (or standard treatment) group. A prospective trial begins with the accrual of patients and ends after all of the accrued patients have been treated, evaluated through the treatment period and evaluated again after some agreed follow-up period. In addition to being randomized, many clinical trials are blinded so that neither the treating physician nor the patients know the assignment group (which patients receive which drugs) during the trial.

Cloacogenic carcinoma—A tumor that arises from the anal transition zone (between rectal mucosa and anal skin). The tumor consists of nests of small round cells simulating the appearance of a basal cell carcinoma of skin. It is very important to distinguish these two tumors on pathologic examination because the prognosis and treatment of these two tumors are quite different.

Cluster analysis—The different members of a collection can be grouped by their similarities. Cluster analysis is a mathematical process by which the properties of different objects (usually data measurements) can be compared. Objects are clusted into groups that share the most properties, with the closest measurements associated with each property. There are many different mathematical approaches to cluster analysis, and they can all produce different results (i.e., different groupings from the same set of objects). Biologists should not think that the groups produced by clustering methods constitute a valid biological classification. Biological entities (genes, proteins, cells, organs, and organisms) may share many properties and still be fundamentally different. For example, two genes may have the same length and share some subsequences, but both genes may have no homology (i.e., no shared ancestry). Another set of genes may be structurally dissimilar but may belong to the same functional group. At best, cluster analysis produces groups that can be used to start piecing together a biological classification.

Complete carcinogen—In animal model systems, some chemical carcinogens yield a large number of tumors after a single exposure. Other carcinogens (incomplete carcinogens) cannot yield tumors unless additional measures are taken. This may involve multiple exposures with the carcinogen or posttreatment promotion with additional agents. The idea of a "complete" carcinogen has implied, to many people, that the initial treatment was sufficient to create all of the mutational events necessary for the expression of the malignant phenotype. Complete carcinogens may not actually exist. Even after treatment with a strong carcinogenic agent, a series of additional steps must take place over time.

Complete mole—There are two types of hydatidiform mole: complete mole and partial mole. A complete mole is characterized by edematous placental villi and little or no embryonic tissue. Complete moles can develop into either invasive moles or choriocarcinoma. Complete moles occur after an anucleate ovum is fertilized by two sperm. The cells of a complete mole are diploid with both sets of chromosomes having a paternal origin. See hydatidiform mole, invasive mole, and partial mole.

Connective tissue—Tissues whose primary role is connecting one tissue with another, keeping tissues from falling apart, and providing a physical structure to the body. Connective tissues include fibrous tissue, bone, cartilage, muscle, and vessels.

Cost of cancer medication—The cancer death rate will rise if the cost of cancer treatments is not affordable. Paul Goldberg reported in 2005 how one drug company raised the cost of procarbazine, a chemical that had been approved in 1969 as a treatment for Hodgkin disease (322). For many years, a bottle of 100 pills of procarbazine cost about $70. In 2005, the price went up to $650, and this was soon followed by an increase in price to $5,568 per bottle. Alex Berenson, a reporter for the *New York Times*, reported that the cost of nitrogen mustard, a chemical that had been used for decades to treat mycosis fungoides, had jumped from $78 to $548 in February 2006 (323). In each case, the price of these effective, generic cancer drugs was increased because there was little or no competition from other companies. When a tumor is rare, that means that the market for effective drugs is small. Assuming that a company will provide the drug to a small market, the lack of any competition within the industry provides an opportunity for the manufacturer to charge any price that the small market will bear. The Food and Drug Administration can approve or reject drugs, but it does not regulate prices. Medicare is not permitted to consider cost when it decides whether a treatment can be covered. In effect, the cost of curing the few cancers that we can actually cure may be beyond the means of the majority of the people who actually suffer from the cancers (324).

Cowden syndrome—An inherited disease syndrome characterized by hamartomatous neoplasms of the face, breast, colon and other organs. The skin lesions of the face are primarily trichilemmomas and fibromas (79). Palmar and plantar hyperkeratotic pits are also observed. Papillomas of the oropharynx are additional common findings. Cowden disease is caused by loss of function in the tumor suppressor gene PTEN, the phosphatase, and tensin homolog gene. PTEN mutations are commonly found in sporadically occurring brain, breast, and prostate cancers. See Lhermitte-Duclos disease.

CpG island—CpG islands are concentrations of CpG dinucleotides that have a GC content over 50% and that range from 200 base pairs to several thousand base pairs in length. There are about 29,000 to 50,000 CpG islands, and most of these are associated with a promoter (208). See DNA methylation.

Craniata—Craniata is the class of animals that have a skull encasing a brain. The significance of class Craniata in the Developmental Classification is that skulls are the product of the neural crest. Before the appearance of organisms of class Craniata, there was no neural crest or the neural crest was primitive and incapable of producing all of the cell types and derivative tissues found in organisms of class Craniata. With the appearance of the neural crest came all of the tumors that derive from neural crest, including many of the tumors of the connective tissue of the head, tumors of the peripheral nervous system, several endocrine tumors, and tumors of melanocytic origin. Tumors of neural crest are present in Craniates and absent in non-Craniates.

Curator—The word curator derives from the Latin *curatus*, the same root for curative, and conveys that curators take care of things. The curator must ensure that nomenclatures are comprehensive for their knowledge domain, and this can be quite difficult when the knowledge domain is growing rapidly. The modern curator must also ensure that the nomenclature is formatted and annotated in a manner that permits interoperability with other databases.

Cytogeneticist—A scientist who studies the chromosomes of cells, trying to identify markers for genetic diseases or trying to find new markers.

Cytologist—A cytologist is someone who has been specially trained to screen cytology specimens (particularly Pap smears). Their training is focused exclusively on the morphologic evaluation of cells. The diagnostic opinion of an individual cytologist is reviewed through a quality assurance process that often involves review by other cytologists and by a cytopathologist. The final diagnosis is typically countersigned by a laboratory director, verifying that the case was handled through a process that guaranteed a highly reliable diagnosis. See Pap smear.

Cytopathologist—Cytopathology is a subspecialty of anatomic pathology. Cytopathologists are trained to diagnose a wide variety of pathological processes, including cancer, by examining individual cells (rather than whole tissues). The scientific basis for cytopathology is that diseases are the clinical expression of abnormalities in cells of the body. If you have access to the diseased cells, you can often determine the disease just by close observation of cellular morphology. See cytologist.

Dedifferention—Most tumor cells lose some of the specialized functionality of their normal cellular counterparts. The developmental process through which cells gain specialized functionality, and through which they exclude the functionality expressed in other types of cells, is called differentiation. In general, as a tumor becomes more advanced (as measured by growth rate, degree of invasion, propensity for metastasis), it loses some of its differentiated properties, and it expresses pathways found in the normal stem cells or precursor cells for the putative histogenetic cell type of the tumor. For example, hepatocellular carcinomas often express alpha-fetoprotein, a protein produced normally by fetal hepatocytes. Some researchers believe that dedifferentiation supports the stem cell theory of cancer, which holds that during carcinogenesis, tumor cells revert to a stem cell phenotype. The Developmental Classification of Neoplasms essentially side steps the concept of dedifferentiation. In the Developmental Classification, tumor cells express pathways inherited through class lineage. Some of these pathways may be shared with known features of less differentiated cells of the same lineage (i.e., stem cells). The physiologic processes of dedifferentiation of tumor cells, and the acquisition of a new phenotype characteristic of a specific stem cell, are not conceptual components of the Developmental Classification.

Dermal bone—This is bone that arises by calcification of a preshaped mesenchymal matrix (e.g., the cranial bones). All dermal bones in humans are of neural crest origin. Dermal bone formation is synonymous with membranous bone formation. See enchondral bone.

Diagnosis—In the field of tumor biology, diagnosis is the process by which a tumor is assigned a precise term or name selected from the taxonomy of all neoplasms. Tumor diagnosis is performed by pathologists and always involves examining a histologic sample of the tumor. Sometimes the word classify is used when the word diagnosis or identify should be used. For instance, a pathologist might say, "I classified the tumor as a basal cell carcinoma," when they have simply identified the tumor as a basal cell carcinoma. The term classification should be reserved for the process of assigning taxonomic species of tumors to classes and subclasses.

Differentiation—The process by which a population of cells comes to be different (defined and separable) from other cells. The primary mechanism of differentiation in normal cells involves inherited epigenetic modifications of chromosomes.

Diploid—Humans are diploid organisms. In females, each chromosome is paired with a matching chromosome. In males, every autosomal chromosome (nonsex chromosome) is paired to a matching chromosome. The single sex chromosome (y) is paired to a larger x chromosome. In humans (male and female), the paired (diploid) number of chromosomes is 46. The un-paired (haploid) number of chromosomes is 23.

Disomy—See uniparental disomy.

DNA methylation—The most studied epigenomic change is DNA methylation. DNA methylation occurs primarily on cytosine, usually at locations for which cytosine is followed by guanine. This type of methylation, which accounts for most of the DNA methylation in developed human cells, is designated "CpG." Approximately 1% of DNA is methylated in human somatic cells. In the cancers that have been studied to date, there is a genome-wide decrease in cytosine methylation, but there is an increase in methylation in so-called CpG islands. See CpG island.

DNA repair—For every type of DNA damage, there is a normal cellular repair mechanism. Several tumor suppressor genes control normal DNA repair. The inactivation of DNA repair increases the likelihood that an oncogene will be activated and increases the number of genetic aberrations accumulated during tumor latency, precancer, and tumor progression.

Dormancy—This is the period after a metastatic cell has seeded a distant site and before the seeded cell proliferates into a clinically detectable size. There is abundant evidence that most seeded metastatic cells will fail to produce a clinically detectable mass. The cells that have a very short dormancy presumably arrive at their seeded site with the properties necessary for growth in the target microenvironment. The cells that have a long dormancy must acquire properties, possibly through tumor progression, that will enable them to grow in the target microenvironment. Most of the people who die from their cancers succumb from their metastases, not their primary tumor. If we knew how to arrest or kill dormant cells, we would be able to save many lives.

Dysplasia—The term dysplasia means abnormal growth. It can be applied to many different conditions, including disorders of skeletal growth and a variety of malformative conditions in infants. Used in the context of this book, dysplasia refers to the nuclear atypia characteristic of neoplastic cells. Dysplasia is found in precancers, benign tumors, and invasive cancers. There are very few neoplasms that have morphologically normal nuclei. Every invasive cancer has dysplastic nuclei discernible to cytopathologists.

Dyserythropoiesis—A disorder in the production of red blood cells. It is caused by mutations in erythroid stem cells leading to red cells and red cell precursors that have an abnormally short life span. Dyserythropoiesis is characterized by hypercellularity, increased rates of cell death, and shortened cell life spans (often referred to as high cell turnover). Dyserythropoiesis, as its name implies, always involves the red cell lineage, and can involve all the myeloid lineages to some extent. Dyserythropoiesis is the bone marrow equivalent of dysplasia found in precancers of all other tissues, and it is characteristic of the myelodysplasias (preleukemias). The circulating blood cells found in myelodysplasias are the progeny of dysplastic cells and provide us with an opportunity to see the results of dysplastic growth in a differentiated, nondividing cell population.

Ectoderm—One of the primary embryonic layers. Derivatives of the ectoderm in the developed human include the skin epidermis and the skin appendages.

Ectomesenchyme—Mesenchyme derived from neural crest. This would include the skull bones, most of the connective tissue of the head, and some of the connective tissue in the neck. Because connective tissue in the head has a neural crest lineage, tumors of the connective tissue in the head, including osteosarcomas of the skull bones, are technically neural crest-derived tumors, not mesodermally derived tumors. See paraphyletic classification.

Embryonic stem cell—See ES cell.

Enchondral bone—Bone that forms through ossification of growing cartilage. Long bones (e.g., the femur) are all enchondral bones. With very few exceptions (confined to parts of bones of the shoulder-girdle), enchondral bone formation indicates a mesodermal origin of bone, and dermal bone formation indicates a neural crest origin of bone. See dermal bone.

Endoderm—One of the primary embryonic layers. Derivatives of the endoderm in the developed human include the lungs, the gastrointestinal mucosa, the liver, and the pancreas.

Endodermal sinus tumor—A tumor of germ cell origin in which the tumor has differentiated to resemble cells of the endoderm sinus, a primitive structure (that does not persist in the adult organism) that functions in early embryonic life to produce fetal blood proteins. The endodermal sinus is considered by embryologists to be analogous to the yolk sac found in egg-laying animals. The endodermal sinus tumor is also known as yolk sac tumor and as Teilum tumor (after the pathologist who provided an early description of this tumor).

Enhancer—DNA sequence that can enhance the activity of a promoter. Enhancers do not need to be near the promoter site, but the enhancer and the promoter must be on the same chromosome. See promoter sites.

Enterocyte—An epithelial cell lining the gastrointestinal tract. All enterocytes are of endodermal origin.

Eosinophilia—A term with two medical meanings. The word eosinophilia, meaning love of Eosin, may indicate that a tissue component has absorbed Eosin dye and appears pink under microscopic examination. The word "eosinophilia" may also refer to an increase in the number of circulating blood eosinophils, a type of granulated leukocyte (white blood cell) for which uptake of Eosin dye into the granules is characteristic.

Epigenomic instability—Epigenomic changes are modifications to DNA and to chromosomal material that do not change the sequence of nucleotides. These include DNA methylation changes and changes in the proteins that inhabit chromosomal material. In many of the cancers that have been studied to date, there is a genome-wide decrease in cytosine methylation, but there is an increase in methylation in so-called CpG islands. See CpG island.

Epithelial cell—Epithelial cells fit together tightly, as polyhedral units, connected by specialized junctions. Epithelial cells form the mucosal lining of ducts, glands, and most multilayered pavemented surfaces. Tumors of epithelial cells account for over 90% of the cancers that occur in humans.

Epithelial–mesenchymal transformation—In normal tissue development, there are times when cells with an epithelial phenotype transform into cells with a mesenchymal phenotype (and vice versa). An epithelium consists of polygonal cells that fit closely together forming a pavement, duct, or glandular structure. A mesenchyme is a collection of spindle-shaped cells fitting loosely together, usually extruding collagen or a calcified protein matrix, and sometimes filling with fat, forming either a thick or thin, hard or soft structure. Epithelial–mesenchymal transformation occurs normally in the development of the metanephros (kidney) as condensed mesodermal stromal type cells transform into the epithelial cells of the renal tubules. There are many examples when cells of epithelial tumors transform into a mesenchymal phenotype (e.g., metaplastic breast cancer and sarcomatoid squamous cell carcinomas), but these phenotypic transformations should not be construed as lineage transformations (e.g., ectodermal cells transforming into mesodermal cells). As far as we know, epithelial–mesenchymal transformations are simply phenotypic changes consistent with differentiation options conferred by a cell's developmental lineage. See transdifferentiation.

Epithelioid cell—Epithelioid cells are nonepithelial cells that appear as large, polygonal, cells that simulate an epithelial morphology. Epithelioid cells are found in granulomas (collections of histiocytic inflammatory cells), sarcomas (e.g., epithelioid sarcomas), lymphomas (e.g., Hodgkin disease), and maternal placenta (decidual cells). A synonym for "epithelioid" is "pseudoepithelial".

Eponymous neoplasms—A great many diseases carry the name of a person. In the past, it was considered proper to specify eponymous diseases using the possessive form (e.g., Hodgkin's lymphoma, Kaposi's sarcoma, Warthin's tumor). More recently, the fashion has been to use the nonpossessive surname. During a 1975 meeting to standardize the nomenclature of malformations, a recommendation was made to use "Down Syndrome" not "Down's Syndrome." The argument was made that "the possessive use of an eponym should be discontinued, since the author neither had nor owned the disorder" (325). The same applies to neoplasms, and the nonpossessive form of eponymous tumors is used consistently in this book.

Erasure—Germ cells arise from the endoderm and enter the genital ridges to form ovaries (in females) and testes (in males). Soon after entering the genital ridges, germ cells erase their epigenome. Erasure permits germ cells to contribute a nondifferentiated (totipotent) set of chromosomes to the preimplantation embryo. The mechanisms whereby erased cells regain epigenomic methylation patterns, including gamete-specific allelic imprinting, is an area of intense investigation (264).

ES cell—Cultured cells derived from the inner cell mass of an embryo. These cells are pluripotent and can differentiate into any type of cell, except for cells of the trophectoderm (which has separated from the inner cell mass before the extraction of the progenitor ES cell). ES cells are used in the creation of genetically engineered mice. Altered ES cells can be reimplanted into a normally developing embryo and will contribute to the resulting chimeric animal. If the ES cells have contributed to the germ cell population in the chimera, a new animal with ES DNA in its germline, can be bred.

Extraembryonic cell—Extraembryonic cells are derived from the blastocyst and give rise to tissues that do not persist in the developed human, such as the placenta, the umbilical cord, the yolk sac, and the amniotic sac.

Field theory—The field theory of carcinogenesis holds that cancer and all of the genetic changes observed in cancer cells are the result (not the cause) of disordered tissue growth. The theory suggests that carcinogens act by poisoning the stroma underlying epithelial cells, resulting in the loss of normal epithelial development (in a tissue field) and the growth of disorganized tissue (i.e., tumors). It is also known as the tissue organization field theory.

Funder—A cancer funder is someone who controls financial resources and who has some influence over which cancer projects are funded. Cancer funders include a great many diverse people. Presidents, congressmen, governors, mayors, and state legislators all have influence because they determine how taxpayer money is earmarked and spent. Board members of nonprofit research organizations are likewise important. The list of people who influence the cancer funders is long and includes lobbyists for professional organizations, appointed members of governmental advisory boards, administrators from academic centers, researchers who sit on National Institutes of Health study sections (that evaluate grant proposals), and representatives of patient advocacy groups.

Gamete—An egg in the female or a sperm cell in the male.

Ganglioneuroma—A tumor of the neural crest consisting of peripheral nervous system neurons admixed with Schwann cells. The analogous tumor of the neuroectoderm is the ganglioglioma, which consists of central nervous system neurons admixed with glial cells.

Gastrinoma—A gastrin-producing tumor arising from gut-derived endocrine cells found in the pancreas, stomach, or elsewhere in the alimentary tract. Gastrin stimulates stomach lining cells to secrete acid (hydrochloric acid) and can cause stomach ulcers. The occurrence of gastrinoma, hypersecretion of stomach acid, and ulcers is known as Zollinger-Ellison syndrome.

GEM—See genetically engineered mouse.

Gene transfer carcinogenesis—Gene transfer can correct x-linked severe combined immunodeficiency (SCID). SCID is caused by an inherited deficiency of adenosine deaminase and results in profound immunodeficiency. A retrovirus is used to carry the missing gene into the genome of autologous lymphocytes of children who have SCID. Tragically, some children who received gene transfer therapy developed a clonal proliferative process of T-cell lymphocytes. Examination of their proliferating T lymphocytes showed that the retrovirus happened to integrate in proximity to a promoter of the LMO2 oncogene, leading to overexpression of LMO2 (49). The overexpressed LMO2 resulted in the uncontrolled growth of cells, which eventually manifested as a T-cell lymphoma.

Genetically engineered mouse (GEM)—Mice with artificially inserted sequences of DNA that provide a new gene function or that produce the loss of function of a gene (by replacing a normally functioning gene with a nonfunctioning construct). New gene functions are achieved by inserting a gene into a fertilized mouse egg, along with a promoter that drives the expression of the gene. GEM scientists strive to find promoters that are organ specific. This limits the functionality of the new gene to a specific organ, permitting observation of the gene is selected cell types. Loss of function GEMs are usually created by culturing embryonic stem cells from a mouse, introducing a nonfunctional variant of a gene into the nucleus of a cultured embryonic stem cell using electroporation, and hoping that the foreign DNA will be substituted for its closely related native DNA sequence through homologous recombination. A marker gene is used to determine whether the recombination was successful. If so, the embryonic stem cell is reinserted into a mouse blastocyst. If the altered embryonic stem cell contributes to the germ cell lineage, and if an altered germ cell contributes to a next-generation conceptus, the offspring will have the altered genotype. Hybrids between two GEM mice will sometimes yield a mouse homozygous for the GEM knockout gene.

Genetic instability—Genetic instability is a condition in which the normal genome is altered and the accumulation of additional alterations is likely. Genetic instability is a phenotypic property of most cancers.

GERD—Gastroesophageal reflux disease. GERD is caused by the ascension of gastric acid into the esophagus. This condition is very common in the geriatric population because the tension in the lower esophageal sphincter tends to weaken as people age. Chronic acid irritation in the esophagus can lead to the replacement of the normal squamous mucosa of the esophagus with an abnormal glandular mucosa. This metaplastic glandular mucosa, occurring in the esophagus, is called Barrett esophagus. In rare cases, the sequential development of dysplasia and esophageal cancer may arise from a Barrett esophagus.

Germ cell line—The germ cell line produces gametes (ova in females and sperm in males). "Germ cell line" needs to be distinguished from "germline," which usually denotes the common genetic heritage of all of the cells in the body, determined at conception.

Germline mutation—A germline mutation is one that is found in every cell of the body. It should not be confused with mutations in the germ cell lineage (the lineage that gives rise to sperm and ova).

Gestational trophoblastic disease—A somewhat euphemistic term that covers a variety of neoplasms and proliferative lesions of trophoblast cells as well as the abnormal tissues that often precede the development of trophoblastic neoplasms. The lesions included as gestational trophoblastic diseases are: hydatidiform mole (complete and partial types), invasive mole, choriocarcinoma, placental site trophoblastic tumor, and epithelioid trophoblastic tumor. These lesions all arise from the extraembryonic lineage of a conceptus and have no relationship to germ cell tumors that yield a neoplastic trophoblastic component.

Glass slide—Anatomic pathologists render diagnoses on glass slides containing stained thin sections (4 to 8 micrometer) of tissues. The pathologist places each slide on the stage of the microscope. The stage lies between the light source and the objective lenses. The pathologist peers through the two eyepieces of the microscope (all modern diagnostic microscopes are binocular), viewing every cell on every prepared slide before rendering a diagnosis.

Glial cells—Glial cells are supportive cells that occur exclusively in the central nervous system (CNS). They surround the CNS neurons and produce myelin. They develop exclusively from the neuroectoderm. The peripheral nervous system, developed from the neural crest, cannot produce glial cells. The cell in the peripheral nervous system that serves a similar purpose as the glial cell is the Schwann cell. Gliomas (tumors of glial cells) are exclusively of neuroectodermal origin. Schwannomas are exclusively of neural crest origin. A reactive increase in glial cells is called gliosis.

Glioblastoma multiforme (GBM)—A poorly differentiated and aggressive brain tumor of astrocyte lineage. The suffix "blastoma" is somewhat misleading because the tumor cells do not have cytologic or biologic features of blastemal cells that develop into the cells of the brain. Cytologically, the cells of glioblastoma multiforme can be large and do not often resemble the small round blue cells found in other "blastomas" (neuroblastoma, retinoblastoma, and medulloblastoma). GBM is the most common form of primary brain cancer (the most common form of brain cancer is "secondary" cancer that has metastasized to the brain from another organ). Despite its aggressive growth, GBM only rarely invades outside the CNS.

Glucagonoma—A glucagon-producing tumor arising from gut-derived endocrine cells found in the pancreas or elsewhere in the alimentary tract. Glucagon is a hormone that opposes insulin. Not surprising, glucagonomas can produce diabetes and an otherwise rare condition known as necrolytic migratory erythema.

Glycolysis—Glycolysis is a cellular mechanism for generating ATP (adenosine triphosphate, the energy-carrying molecule in cells) through glucose metabolism, and not by oxygen-dependent respiration. The microenvironment of cancers is relatively hypoxic, and the rate of glycolysis is high in many cancers. The Warburg hypothesis, described in the 1920s, held that cancer cells are different from normal cells because their metabolism is glycolytic. This shift from respiration to glycolytic metabolism accounts for the neoplastic phenotype. As noted throughout the book, cancer cells differ from normal cells in many ways. Few scientists today would suggest that a single aberration in tumor metabolism could account for the malignant phenotype.

Grading—The process of assigning a diagnosed tumor specimen to one of several biologic categories based on the examination of morphologic features. Often, a tumor's grade is determined by the extent of nuclear atypia. A tumor with minimal nuclear atypia might be assigned a "low grade," indicating an expected indolent behavior. Another tumor with marked nuclear atypia might be labeled a "high-grade" tumor with an expected aggressive clinical course. Sometimes grading is not based on nuclear atypia. In the case of prostate cancer, grading is based on glandular size, shape, and distribution and is assigned a number (Gleason score) between 2 and 10. Prostatic adenocarcinomas with Gleason scores of 2 through 4 are expected to have an indolent behavior. Prostatic adenocarcinomas with Gleason scores of 8 through 10 are expected to have an aggressive clinical course. See staging and diagnosis.

Grant—A grant is an award of money to a researcher in the hope that he or she will work productively. Most grants are awarded after reviewing a grant application, and in this case, the funders expect that the researcher will work on the project described in the application. The grantee's only real responsibility is to use the grant funds to support his or her work. The most important distinction between a research grant and a research contract is that the contract specifies a deliverable product, whereas the grant does not.

H&E (Hematoxylin & Eosin)— The most common stain used in pathology laboratories. With H&E staining, the nuclei of cells are blue, and the cytoplasm is pink. Without staining, cells are colorless (except for those cells that are pigmented, such as melanocytes), and the anatomic parts of the cell (nucleus, cell membrane, nucleolus, brush borders, etc.) cannot be easily distinguished. The development of histologic staining techniques in the 19th century was among the most important advances in medical science. Hematoxylin is a complex stain extracted from the logwood tree (Haematoxylum campechianum), which grows primarily in South America (Figure G.3). Hematoxylin stains the nuclei of cells blue, having an affinity for acidic nuclear proteins. Eosin is a red counterstain that provides a pink or red color to cytoplasm, proteins, and some granular organelles.

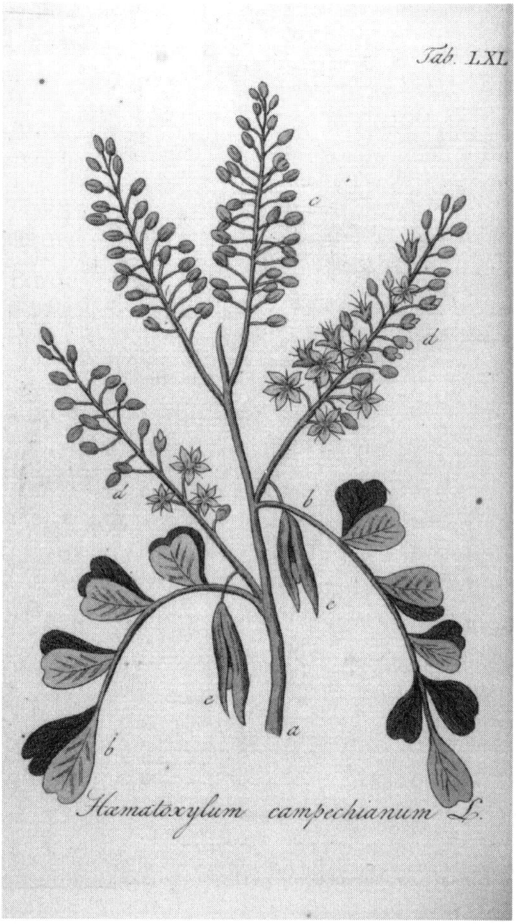

FIGURE G.3 The logwood tree, *Haematoxylum campechianum*.

Hamartoma—A hamartoma is a localized overgrowth of a tissue resulting in the formation of a tumorous mass. A hamartoma can be thought of as a malformation that happens to result in a mass. Hamartomas can be quite large (e.g., splenic hamartomas), but they usually reach a maximum size and stop growing. After they have reached their maximal size, hamartomas sometimes shrink. When a hamartoma exhibits progressive growth, it is likely that another neoplasm (benign or malignant) has developed from within the hamartoma. See choristoma, tuberous sclerosis, Cowden syndrome, and proteus syndrome.

Hamartoma of breast—A benign, circumscribed lesion of the breast that contains adipose tissue, stroma, and glands. A capsule is often present. It is seen in some cases of Cowden syndrome.

Haploinsufficiency—Occurs when one of two alleles of a required gene is inactivated, and the other allele does not express sufficient quantities of the gene product to sustain normal cellular functionality. Haploinsufficiency may result in cancer if one copy of a tumor suppressor gene is inactivated and the other copy cannot provide sufficient functionality to suppress tumorigenesis (326, 327).

Hematogenous dissemination—Cancers metastasize through the body via two routes: lymphatic dissemination and hematogenous dissemination. In general, lymphatic dissemination, which uses lymphatic drainage vessels, is slow and regional. Hematogenous dissemination (through blood vessels) is fast and distant. Tumors that disseminate via a hematogenous route tend to produce metastatic lesions that occur in the lung, brain, bone, and spleen (Figure G.4). Sarcomas are more prone to hematogenous dissemination than are tumors of endodermal or ectodermal origin, although there are many exceptions to this generalization.

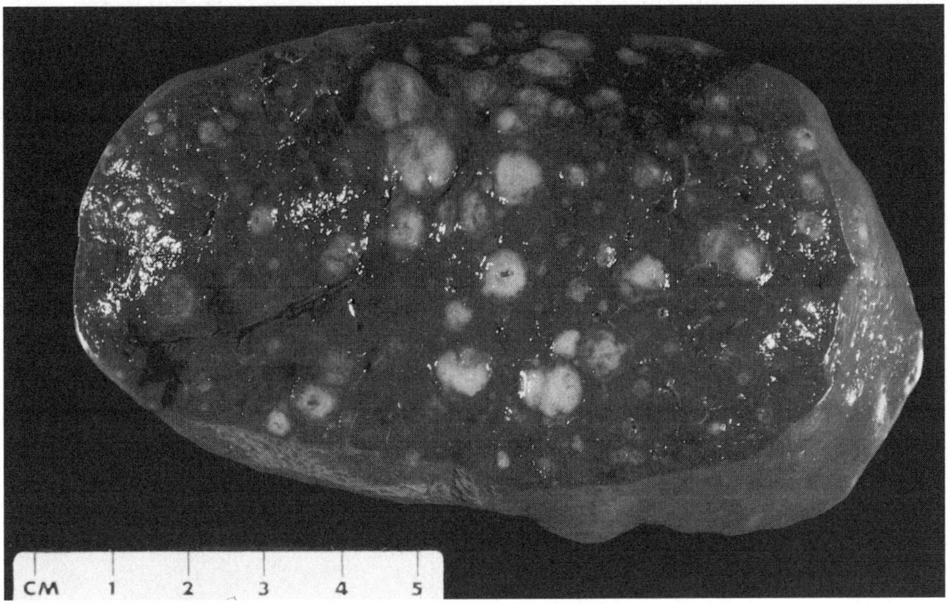

FIGURE G.4 Multiple nodules of metastatic cancer in the spleen.

Hepatocellular carcinoma—A malignant neoplasm of hepatocytes. Worldwide, hepatocellular carcinoma is the third greatest cause of cancer deaths, accounting for 611,000 deaths annually, representing 8.7% of total number of humans killed by cancer. Animal models of hepatocellular carcinoma and precancer lesions for hepatocellular carcinoma indicate that after initiation with a carcinogen, many small nodules will grow if the liver is stimulated to divide with a partial hepatectomy (29). Most of these early nodules regress, but some continue to become cancers. In humans, liver precancers are probably represented by nodules in cirrhotic livers that contain foci of highly dysplastic hepatocytes (328).

Hepatoma—Same as hepatocellular carcinoma. See hepatocellular carcinoma.

Heterogeneity—See tumor heterogeneity.

Heterogeneous data—A pathology report and a histopathology image are very different types of data. One is a narrative text containing diagnostic terms along with clinical, demographic, and administrative information. The other is a binary file relating color and position to pixels. These are examples of heterogeneous data that are tied by a medical relationship. A digital representation of a bone x-ray may have a relationship to an operative note, a bill for hospital services, or an accident report. In these instances, the data are likely to have very different types of representations and formats and may have relationships that were not anticipated by the persons who created the data formats for each type of collected information.

Hierarchy—Traditionally, a hierarchy is a system in which groups are arranged, top to bottom, by rank. The classification of all organisms on earth is a hierarchical classification. Every organism belongs in a class, and every class (except the top class) has one superclass. The top class is called the root. Today, the concept of a hierarchy has been expanded to accommodate ontologies in which a class may have more than one superclass (i.e., multiple inheritance is permitted). Ontologies are represented graphically by directed acyclic graphs. The practical implication of this change is that modern hierarchies are much more complex than traditional hierarchies.

Histology—Histology is the study of tissue organization. The study is accomplished with a microscope and with stained thinly sliced sections of tissues mounted on glass slides.

Histologic examination—All anatomic pathologists and many types of biologists spend countless hours examining tissues under a microscope. They study the different cells in the tissue and the relationship among the cells and the changes in cells that occur after a biological event. In the case of pathologists, the biological event is the occurrence of a disease. Most human diseases can be diagnosed through careful histological examination of samples of involved tissues.

Histogenesis—Biologists use the term histogenesis to indicate the process whereby undifferentiated germ layers produce all the differentiated cells of the body (about 200). Pathologists use the term "histogenesis" to indicate the particular cell type from which a neoplasm arises. For example, the histogenesis of chronic myelocytic leukemia is a myeloid stem cell, and the histogenesis of neuroblastoma is a primitive neural crest cell.

Histone—The major protein in chromatin. Histone deacetylation causes histones to tighten around DNA, reducing normal transcription by blocking transcription factors from attaching to their normal target sites.

HNPCC (human nonpolyposis colon cancer)—Also known as Lynch syndrome. An inherited defect in DNA mismatch repair leads to microsatellite instability and an increased risk of several different types of cancer, particularly colon cancer. The syndrome can be caused by mutations in several different genes, including MLH1, MSH2, MSH6, PMS, and PMS2. The cancers often occur before the age of 50 years (somewhat younger than the age of occurrence of sporadically occurring tumors).

Hodgkin lymphoma—A type of lymphoma that is distinctively different from most other lymphomas (Figure G.5) (List G.1).

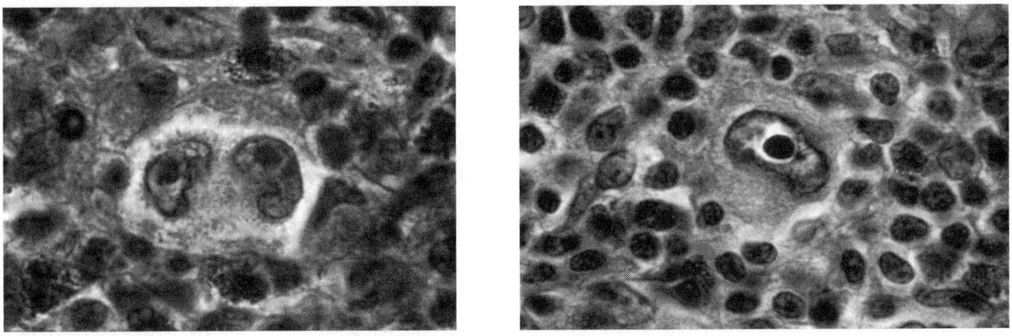

(a) **(b)**

FIGURE G.5 Hodgkin lymphoma. Figure A shows a characteristic Reed-Sternberg cell (center). This large, malignant lymphocyte is binucleated. Each bean-shaped nucleus has a single, large nucleolus. Figure B has a central, large Hodgkin-type cell, which has some features of a Reed-Sternberg cell, but with a single nucleus. Figures A and B have a mixture of normal and abnormal lymphocytes and a sprinkling of eosinophils.

Homeotic gene—A homeotic gene controls normal embryologic development. A mutation in a homeotic gene can change the tissue (from one type to another type) found in an anatomic location. In rare instances, a tumor may seem to change its lineage. For example, a gastric adenocarcinoma may develop a component of choriocarcinoma (229). Nobody knows how this occurs, but it is possible that a mutation in a homeotic gene is involved.

List G.1 Distinctive features of Hodgkin lymphoma.

Bimodal age distribution (15–35 years and over 55 years)

Very high cure rate (over 90%)

Most lymphomas have a characteristic lymphoid cell type and a characteristic stage of development. Hodgkin lymphoma has no specifically identified progenitor lymphoid cell type or stage of lymphoid cell development. Histologically, many different cell types are present in Hodgkin lymphoma, including a large atypical cell known as a Reed-Sternberg cell.

Stepwise anatomic spread of tumor from its primary site of origin. Most other lymphomas are widely disseminated from an early phase of disease.

Homolog—A gene related to another gene through descent from a DNA sequence in a common ancestor. Two genes in the same organism can be homologous to each other if they both descended from the same gene in an ancestor.

Homozygosity—Occurs when both alleles of a gene are the same.

Human nonpolyposis colon cancer—See HNPCC.

Hydatidiform mole—Also called a gestational mole. A hydatidiform mole is a malformed conceptus that consists of placental villi and a small, variable amount of malformed embryonic tissue. Although only one quarter of choriocarcinomas occur with hydatidiform moles, each occurrence of a hydatidiform mole is associated with a significant risk of trophoblastic neoplasia. There are two types of hydatidiform mole: complete mole, and partial mole. See complete mole and partial mole.

IEN—See intraepithelial neoplasm.

Imaginal disc—Focal collections of cells in an insect larva that give rise to adult structures (such as wings and antennae).

Immunosuppression—The body's immune defenses can be suppressed with a variety of medications, including cyclosporine and prednisone. Immune suppression can be very useful in transplant medicine, in which it is important to suppress the body's normal tendency to reject a foreign tissue. Immune suppression may also be useful to control diseases in which the immune system is overexpressed (e.g., rheumatoid arthritis and systemic lupus erythematosus). Sometimes the immune system is weakened by disease. AIDS is the best known immune suppressing disease.

Imprinting—In mammals, after erasure of the germ cell's epigenomic programming (early in embryogenesis), about 100 genes are inactivated through epigenomic modification. If the germ cell matures into a sperm or ova that is fertilized, a haploid set of imprinted genes is passed to the conceptus, and the offspring has a phenotype determined, in small part, by the maternal and paternal patterns of imprinting. An example is the NOEY2 tumor suppressor gene, which is imprinted (inactivated) in females and which contributes to some cases of breast and ovarian cancers (329). See loss of imprinting.

Inflammatory pseudotumor—Old terminology for inflammatory myofibroblastic tumor.

Inflammatory myofibroblastic tumor—Inflammatory myofibroblastic tumor can arise from multiple locations and is composed of spindle cells and inflammatory cells, suggesting a chronic inflammatory condition confined to a location. These lesions that were once considered nonneoplastic are now considered true neoplasms of myofibroblasts. The inflammatory component, which masks its neoplastic identity, is preserved in the current terminology, "inflammatory myofibroblastic tumor." Recent work has identified a rearrangement of the ALK (anaplastic lymphoma kinase) gene (63).

Inherited retinoblastoma—Inherited retinoblastomas are often bilateral and often occur in children. Nonfamilial retinoblastomas are almost always unilateral and occur in an older age group. The reason for this difference is that children born with an inherited retinoblastoma syndrome have a germline inactivation of a tumor suppressor (the RB gene). To develop cancer, they need only acquire one additional mutation inactivating the alternate allele. Because only one additional mutational event is required in the inherited syndrome, the tumors occur often (two tumors in a single patient) and early. In nonfamilial cases of retinoblastoma, two mutational events (one for each allele) must occur sometime during the life of the patient, and this usually takes time (accounting for the late onset) and occurs only rarely (accounting for only one occurrence of a tumor). The theoretical underpinnings of these observations were made by Knudsen and preceded the discovery of tumor suppressor genes and the RB gene mutation (330).

In situ—Literally, in its place. When referring to a cancer, *in situ* implies that the cancer has not metastasized or invaded surrounding tissues.

Initiation—The mysterious events that occur when a cell is exposed to a carcinogen. Changes that occur in the descendants of initiated cells may lead to the eventual clonal growth of a neoplasm.

Insulinoma—See islet cell adenoma.

Intraepithelial neoplasm (IEN)—A specific class of precancers that derive from an epithelial mucosa or epithelial lining that is delimited by an anatomically distinct basement membrane lining. Often, the term "intraepithelial neoplasia" is used interchangeably with the term "precancer." This is simply wrong because the term precancer includes many types of lesions that are not intraepithelial (List G.2).

Throughout this book, we use the term "precancer" to include the intraepithelial neoplasias as well as other identifiable lesions that precede the development of malignant neoplasms.

List G.2 Examples of precancers that are not intraepithelial neoplasms.

- Myelodysplasias and myeloproliferative disorders
- Lymphoproliferative lesions
- Nodular epithelial lesions for which there is no surrounding basement membrane, including preneoplastic nodules of liver, cortical adenomas of kidney, C-cell proliferative lesions that precede medullary carcinoma of thyroid
- Dysplastic nevi which may have intraepithelial and dermal components
- Noninvasive neoplasms of nonepithelial origin such as intratubular germ cell neoplasms
- The identifiable precursors for sarcomas

Intramembranous bone—Bone that arises by calcification of a preshaped mesenchymal matrix (e.g., the cranial bones). All intramembranous bones in humans are of neural crest origin. Intramembranous bone formation is synonymous with dermal bone formation. See enchondral bone.

Intramucosal—See lamina propria and intraepithelial neoplasia.

Invasion—The ability of tumor cells to move through normal tissues. All tumors that can metastasize can also invade; however, there are a few tumors that invade but virtually never metastasize (e.g., basal cell carcinoma of skin and most tumors of the brain).

Invasive mole—Also called invasive hydatidiform mole and previously called chorioadenoma destruens. Occasionally, a hydatidiform mole will invade into the myometrium. The invading tissue consists of molar villi. The presence of invading molar villi and the occasional co-occurrence with morphologically similar noninvasive molar tissue help to distinguish an invasive mole from a choriocarcinoma. The term "invasive mole" should not be confused with the totally unrelated lesion, malignant melanoma arising from a mole. See gestational trophoblastic disease.

Islet cell adenoma—A tumor arising from the endoderm-derived endocrine cells distributed as small epithelial islets throughout the pancreas. These tumors can secrete a variety of different hormones, including insulin, gastrin, glucagon, and vasoactive peptide. Most often, the term islet cell adenoma refers to an insulin-producing tumor (sometimes called an insulinoma).

Karyotype—From the Greek root *karyon*, meaning nucleus, the karyotype is an expression of the chromosomal complement of a cell. Chromosomes can be examined after physically spreading a cell when it is in mitosis and staining the chromosomes (with Giemsa or some other technique). The spread chromosomes can be counted and compared with a normal set of spread chromosomes. Abnormal chromosomes are described using a syntax approved by the International System for Human Cytogenetic Nomenclature (ISCN). See translocation.

Keratoacanthoma—A small tumor of skin composed of atypical squamous cells that grows around a keratinous crater. Keratoacanthomas are unusual neoplasms because they appear suddenly (with no known precursor lesion), grow extremely rapidly until they reach their maximal size, and then regress through a process of terminal differentiation in which the squamous cells produce an enlarging keratin core that replaces the dividing cell population and eventually sloughs. The entire process usually lasts just a few months. The atypical squamous cells in the growing lesion are morphologically indistinguishable from the cells in a squamous cell carcinoma (331).

Kuttner tumor—Chronic sclerosing sialadenitis, a nonneoplastic inflammatory condition of salivary glands. It can be mistaken grossly and histologically for carcinoma arising from the salivary gland. The term "Kuttner tumor" is an example wherein a misnomer (the lesion is not a neoplasm) has endured.

Lamina propria—The loose fibrovascular tissue between the mucosal epithelium and the muscularis mucosa (the mucosal muscle band). See submucosa.

Latency (of tumors)—The period between exposure to a carcinogen (the initiating event) and the emergence of a detectable neoplasm is the latency period.

Leser-Trelat—A paraneoplastic syndrome characterized by the appearance of new seborrheic keratoses or an increase in the size of existing seborrheic keratoses as a consequence of a cancer somewhere in the body. The sign of Leser-Trelat is particularly associated with patients who have melanoma. Other paraneoplastic syndromes of skin include acanthosis nigricans (hyperpigmented and hyperplastic band of skin) and eruptive acrochordons (benign skin tags). Ellis and colleagues have speculated that all three of these paraneoplastic syndromes are caused by TGF-α (transforming growth factor-α) produced by tumor cells (77). See paraneoplastic syndrome.

Leukemia—Neoplasms of blood cell origin that manifest as circulating malignant cells. The first symptoms of leukemia may include anemia, infection, fever, hemorrhage, inappropriate clotting, immune dysregulation, and disturbances of the normal function of organs. It is unusual for leukemias to present as tumor masses. Leukemias are always considered malignant (i.e., there are no benign leukemias), although some leukemias have a long, indolent course characterized by the simple increase in number of cells that closely resemble their normal counterparts (e.g., chronic lymphocytic leukemia).

Lhermitte-Duclos—A syndromic variant of Cowden disease, characterized by a hamartomatous overgrowth of enlarged ganglion cells which replace the granular cell layer and Purkinje cells of the cerebellum. This results in global hypertrophy of the cerebellum, with enlarged, coarse gyri.

Li-Fraumeni—A rare inherited syndrome that predisposes individuals to a variety of different cancers, including rhabdomyosarcoma, soft-tissue sarcomas, breast cancer, brain tumors, osteosarcoma, leukemia, adrenocortical carcinoma, lymphoma, lung adenocarcinoma, melanoma, gonadal germ cell tumors, prostate carcinoma, and pancreatic carcinoma. It is caused by a mutation in the p53 tumor suppressor gene.

Lineage infidelity—There are two major lineages of lymphoid cells: T and B. Each cell type is functionally distinct and each derives from a different fetal organ (B cells derive from the bone marrow in humans or the bursa of Fabricius in birds, and T cells derive from the thymus). Occasionally, a tumor of T cells will develop B-cell markers or vice versa.

Liver cancer—Although there are many different neoplasms that can arise from the liver, the term "liver cancer" generally signifies hepatocellular carcinoma.

LOH—See loss of heterozygosity.

Loss of heterozygosity—If one of the two alleles of a gene is inactivated, only one of the alleles will be expressed, and this is referred to as loss of heterozygosity.

Loss of imprinting (LOI)—Loss of an epigenetic suppression of an allele that occurs in germ cells during imprinting. LOI results in the expression of the allele. The first discovered instance of LOI was the overexpression of insulin-like growth factor-2 (IGF-2) in Wilms tumor. In this case, a mutation disrupts the H19 imprinting control region that would normally silence the maternally inherited IGF2 allele.

Lung cancer—Any of the dozens of different malignant tumors that arise from lung tissue. As commonly used, the term "lung cancer" refers to the common human cancers that arise from the bronchi of the lung: squamous carcinoma, adenocarcinoma, and small cell carcinoma. See bronchogenic carcinoma.

Lymphoma—Any neoplasm composed of neoplastic lymphocytes. All lymphomas are considered malignant (i.e., there are no benign lymphomas and there are no generally accepted lymphoid precancers). By putting all lymphomas into the same category (malignant lymphoma), tumors of lymphoid origin display a wide variety of perplexing biological diversity (including spontaneous regression, unexpected indolence, phenotypic transformation) that should properly be assigned to defined stages of development (e.g., precancer stage, benign tumor, hamartoma, and cancer).

Lymphoproliferative lesions—Although some people include lymphomas in the class of lymphoproliferative lesions, as used herein, the term is restricted to sustained hyperplasias of one or more lymphoid cell types that lack some of the features that would be expected in a fully developed lymphoid malignancy. The lymphoproliferative lesions would include the prelymphomas (lesions that sometimes progress to lymphomas). The lymphoproliferative lesions are the lymphoid equivalent of myeloproliferative lesions. Although there is enormous disagreement over which conditions should be included in this category, the list of lymphoproliferative lesions usually includes angioimmunoblastic lymphadenopathy with or without dysproteinemia, hairy cell leukemia, posttransplant lymphoproliferative disease, *Helicobacter*-responsive MALTomas, x-linked lymphoproliferative disease, Epstein-Barr virus–associated lymphoproliferative disease, and the histiocytoses of childhood.

Lynch syndrome—see HNPCC.

Malformation—A malformation is a disorder of normal growth or development that usually occurs in utero or during childhood. The term malformation is sometimes applied to focal growth abnormalities that may occur in adults (such as small angiomas and vascular anomalies that can occur in the skin and intestines of older adults) (Figure G.6). When the malformation involves an identifiable subpopulation of cells that grows to form a tumor-like mass, it is generally called a hamartoma. Confusion arises in cases where the tumor-like mass consists of a well-organized tissue composed of many different cell types. For example, arteriovenous malformations (AVMs) can grow quite large, but they are composed of large, functioning walled vessels. They are different from the congenital hemangiomas composed of small capillaries. We tend to think of the AVM as a malformation (not a type of neoplasm) and of the infantile hemangioma as a hamartoma (a neoplasm).

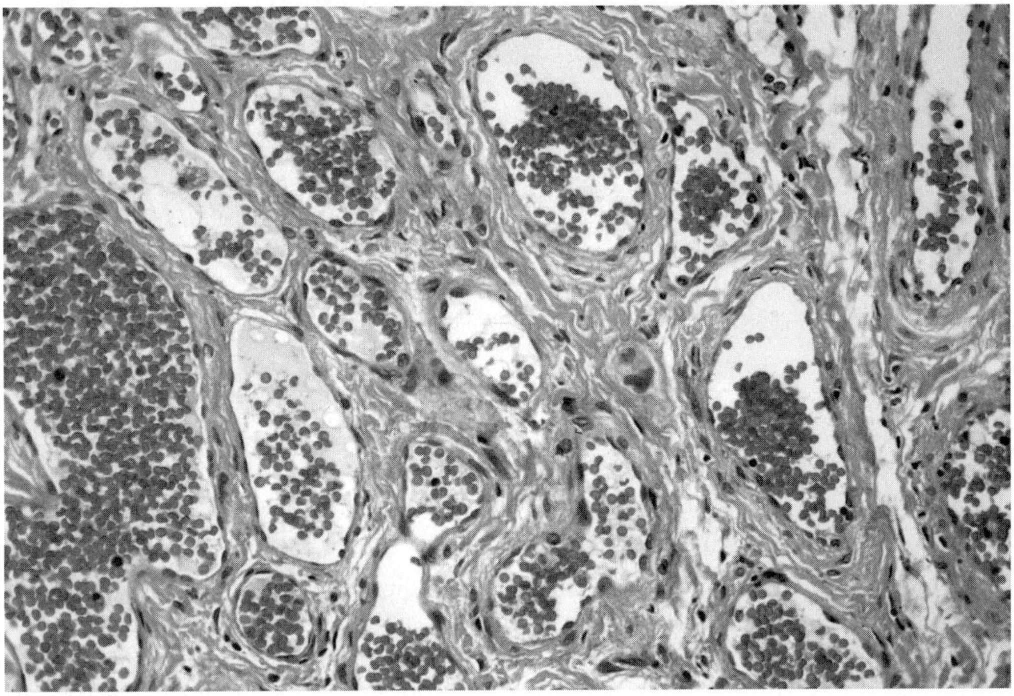

FIGURE G.6 Hemangioma. Multiple vascular spaces, filled with red blood cells, characterize this tumor.

Malign—One might assume that any tumor that is not benign must be malign. Actually, tumors are either benign or malignant (not malign), and I am not sure why the word "malign" is never used in the cancer literature to describe non-benign tumors. See malignant.

Malignancy—Synonym for malignant tumor.

Malignant—If a tumor's typical untreated biological course is to kill its host, it is called a malignant tumor. For the most part, malignant tumors are tumors that are capable of progressive growth and invasion. Most malignant tumors are also capable of metastasizing.

Malignant melanoma—Synonym for melanoma (Figure G.7). In the special case of melanoma, there is no lesion that takes the name, "benign melanoma." A benign neoplasm of melanocytes is called a nevus. The single word "melanoma" is always assumed to be a shortened form for "malignant melanoma."

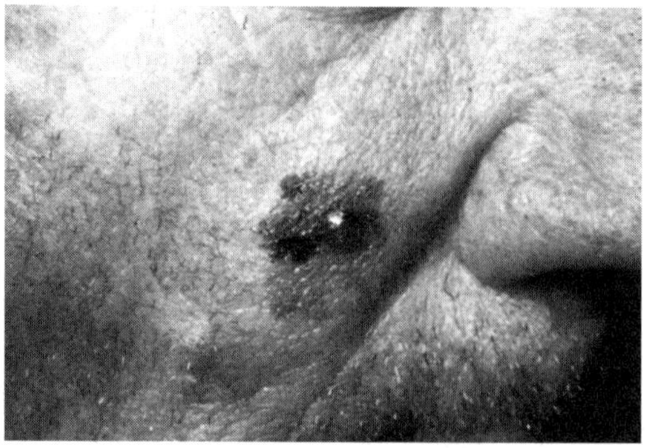

FIGURE G.7 Clinical appearance of a malignant melanoma. The lesion is mottled, with light and dark areas, raised and flat areas, and an irregular border. Benign nevi are symmetric and have a uniform appearance.

Malignant phenotype—Refers to the general morphologic and biologic properties of cancers (persistent growth, invasion, metastasis, and cytologic atypia) or to the properties of a specific cancer.

Malignant process—any biological process that results in life-threatening disease or death. It should not be confused with malignancy.

MALToma—Mucosa-associated lymphoid tissue lymphoma. Most of the gastrointestinal tract and the salivary glands are lined by mucosal lymphoid tissue that lacks the architecture of lymph nodes. Basically, the lymphoid tissue that lines the gut mucosa consists of aggregates of lymphocytes, with some lymphoid follicles. Most of the cells are specialized B lymphocytes (monocytoid B lymphocytes) that share many features with marginal zone B lymphocytes found in lymph nodes of marginal type. Tumors that arise from these MALT tissues are called MALTomas. MALTomas may occur in the stomach after chronic infection with *Helicobacter pylori.*

Medullary—Literally, in the middle. The term is used to mean many different things depending on its particular anatomic context. Thus, a medullary tumor of thyroid and a medullary tumor of breast are unrelated tumors. In addition, a medullary tumor of the thyroid is not related to a papillary carcinoma of thyroid, even though they both arise inside the thyroid gland. The marrow of bones is often referred to as the medulla, and histiocytic medullary reticulosis is a lymphoma involving bone marrow. Furthermore, terms that use a suffix derived from "medullary" may have no predictable meaning. A medulloblastoma of the brain is not a tumor composed of medulloblasts nor does it typically arise from the medulla oblongata. A medulloblastoma is a brain tumor that usually appears near the midline of the CNS axis, almost always within the cerebellum. This just goes to show that in medicine a single word may have many different meanings that cannot be adequately described within a definition.

Medullary thyroid carcinoma—A tumor of the thyroid arising from C cells, which derive from the neural crest. C cells (and, in many cases, medullary thyroid carcinoma cells) produce calcitonin, a hormone involved in calcium homeostasis.

Melanoma—A malignant tumor derived from melanocytes.

Mesoderm—The embryonic germ layer sandwiched between the ectoderm and endoderm.

Metaplasia—Metaplasia is the growth of cells in an anatomic site that is not the normal site of growth for the cells. Metaplasia is a process of gradual replacement of one cell type for another. Metaplasia usually follows chronic toxicity. In general, the metaplastic cells are better able to withstand the toxin than the cells that are being replaced. An example of metaplasia is squamous metaplasia of the bronchus in which ciliated epithelial cells that are chronically exposed to cigarette smoke are slowly replaced by squamous cells. The normal lung contains no squamous cells. All squamous cells found in the lung result from bronchial metaplasia. It is inferred that every bronchogenic squamous cell carcinoma emerges from a focus of squamous metaplasia.

Metaplastic breast carcinoma—Foci of chondrosarcoma or fibrosarcoma or osteosarcoma may sometimes arise in breast carcinomas. When this occurs, the tumors are called metaplastic breast carcinoma, indicating that the tumor contains elements that are not usually present in the tissue of origin (241). The likely source of metaplastic elements in breast carcinomas is myoepithelial differentiation. Breast ducts contain epithelial duct cells and adjacent myoepithelial cells, both of ectodermal origin. Myoepithelial cells are capable of mesenchymal differentiation and are the presumed source of rare sarcomatous components in breast cancers.

Metastasis—Metastasis occurs when cells from a tumor grow at a location that is noncontiguous with the primary growth site of the tumor. The ability to metastasize is acquired late in the life of a tumor (after it has acquired the ability to grow progressively and invade). Many biological steps are required for successful metastasis: A tumor cell must invade through the wall of a lymph vessel or a blood vessel; once inside the vessel, the tumor cell must be carried away from the primary tumor to another site within the vascular system, where it may reside for some indeterminate time as a cell or as a small group of cells living a precarious life pressed against the endothelium of the vessel. The endovascular tumor cells must invade through the vascular endothelium into the surrounding tissue. The invading cell or cells must begin to grow. The growing colony must adjust to its new microenvironment and must eventually attract the growth of vessels from which it can derive oxygen and other external nutrients. Because surgical oncologists have become proficient at excising the tumor from its primary site of growth, most people who die of cancer will die from their metastases (not the primary cancer).

Metathesaurus—A thesaurus formed by combining multiple independent thesauruses. The U.S. National Library of Medicine's UMLS (Unified Medical Language System) is a metathesaurus composed of over a hundred different medical nomenclatures. See thesaurus.

MGUS (Monoclonal Gammopathy of Undetermined Significance)—MGUS is a monoclonal increase in plasma cells with a resultant spike in circulating levels of the specific antibody synthesized by the clonal plasma cells. MGUS is actually a common condition found in the older population and may occur in about 1% of the population over 70 years of age. Progression of MGUS to multiple myeloma is infrequent, with a conversion of about 1% to 2% per year. Because MGUS occurs in an older population, the chance of MGUS progressing to myeloma within the lifespan of the patient is quite low. MGUS is an example of a precancer that is not an intraepithelial neoplasm. See IEN.

Microsatellite—Microsatellites (also known as simple sequence repeats [SSRs]) are DNA sequences that consist of repeating units of 1 to 4 base pairs. Microsatellites are inherited and polymorphic. Within a population there may be wide variation in the number of repeats at a chosen microsatellite locus. See microsatellite instability.

Microsatellite instability (MSI)—An indicator of replication errors in neoplasms. MSI results from deficiencies in mismatch repair (usually caused by mutations in the mismatch repair genes). Many kinds of cancer have microsatellite instability. It is present in almost all colon tumors in HNPCC (hereditary nonpolyposis colorectal cancer).

Minisatellite—A unique or nearly-unique DNA sequence, 10 to 100 bases in length, that reoccurs at more than 1,000 locations in the human genome. Minisatellites are often found proximal to translocation breakpoints.

Mitochondria—Mitochondria are self-replicating, DNA-containing organelles within cells. Every mitochondrion in a human body is descended from mitochondria contained in the maternal oocyte. For this reason, mitochondria have a purely maternal lineage. Mitochondria are specialized prokaryotes that evolved into obligate intracellular symbiotes in animals. Cellular ATP synthesis occurs inside mitochondria. The number of mitochondria in a human cell can vary from one to several thousand.

Mitosis—The phase of the cell cycle in which the duplicated chromosomes condense and separate. The chromosomes are duplicated during S phase. During mitosis, the cell splits (Figure G.8).

Mitotic—Relating to mitosis. See mitosis.

Molecular signature—Soon after the advent of gene expression microarrays, it seemed likely that we could find gene expression profiles that are unique for each different cancer and for biological categories within each cancer type. For instance, there might be a gene expression profile that is characteristic of squamous cell carcinomas of lung, and within that profile, there might be a set of expression values that specify squamous cell carcinomas that are likely to metastasize. Despite encouraging early work, we have no comprehensive collection of signatures for the different cancers of man. Currently, we do have a few gene expression tests that refine certain subsets of breast cancers and lymphomas. Based on the limited success of a molecular signature approach, it seems unlikely that expression profiling will satisfactorily determine the identity and biology of the majority of tumors.

Monophyletic—If all of the members of a class have developed from a common ancestor and if this constraint extends to every class in the classification, the classification is monophyletic. Neither the Tree of Life (Classification of terrestrial organisms) nor the Developmental Classification of Neoplasms is monophyletic.

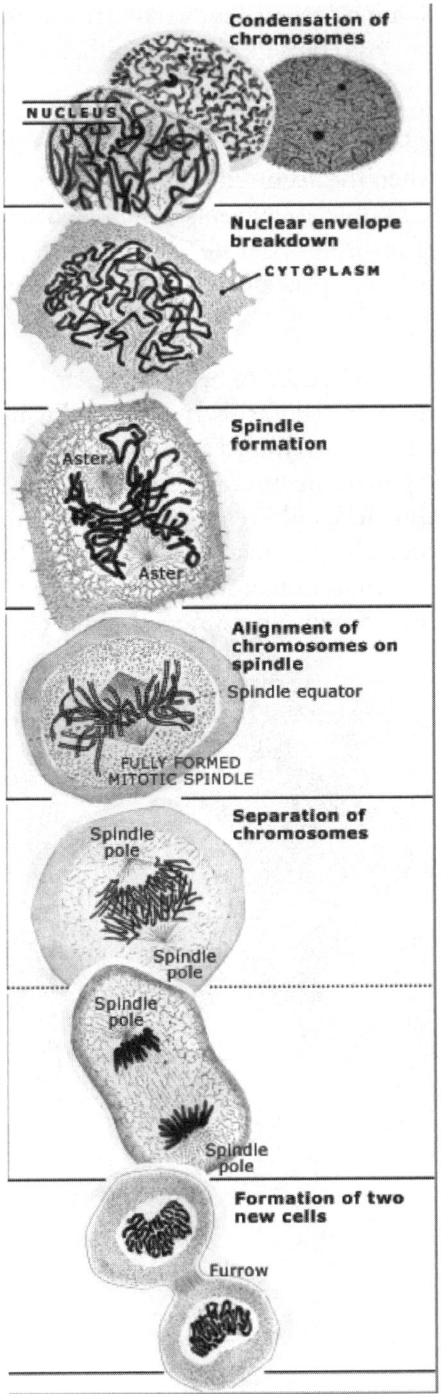

FIGURE G.8 Mitosis. Reproduced from Flemming, W., 1879. *Archiv fur Mikroskopische Anatomie.*

Mucosa—Consists of the lamina propria (connective tissue between the surface epithelium and the muscularis mucosa) plus the surface epithelial layer.

Multidrug resistance—Sometimes when a cancer stops responding to a chemotherapeutic drug, it simultaneously develops resistance to other drugs. This phenomenon of multidrug resistance occurs when the acquired method of resistance is a general property that is effective against a variety of external agents. Examples of a multidrug resistance mechanism are decreased rate of absorption of external molecules into cells, higher rate of deactivation of drug by smooth endoplasmic reticulum, increased synthesis of intracellular targets for toxic agents, and so forth.

Multiple myeloma—A tumor composed of neoplastic plasma cells (Figure G.9). It often occurs in multiple locations and usually arises from the bone marrow. Plasma cells secrete immunoglobulins and multiple myelomas are typically characterized by a single immunoglobulin produced by all of the tumor cells and producing a sharp immunoglobulin peak on serum protein electrophoresis. The production of an immunoglobulin molecule uniquely identifies each myeloma and indicates that the tumor is clonal, that the tumor is differentiated (not a collection of stem cells), and that the clonal progenitor cell was committed to producing a unique immunoglobulin. See MGUS.

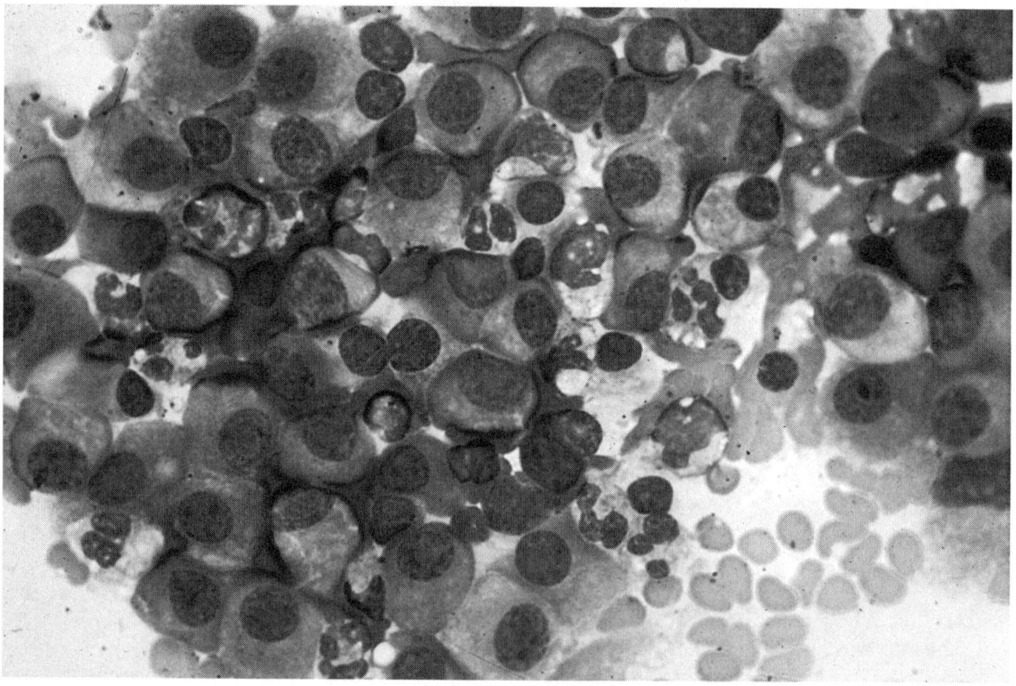

FIGURE G.9 Multiple myeloma. Plasma cells dominate. The neoplastic plasma cells have abundant cytoplasm, an eccentric, round nucleus with fine, granular chromatin, and an area of cytoplasmic clearing adjacent to the nucleus (the perinuclear huff). Neutrophils, red blood cells, and hematopoietic precursor cells are also noted.

Multipotent stem cell—Cells that can produce differentiated cells of more than one cell type. It is synonymous with pluripotent stem cell. See totipotent stem cell.

Muscularis mucosae—In most ectodermal and endodermal structures, the thin, occasionally discontinuous muscle layer that separates the lamina propria from the submucosa.

Mutagen—A substance that can produce inherited alterations in the genome of a cell. Many carcinogens are mutagens, and many, if not all, mutagens are carcinogens. The close relationship between mutagenesis and the initiating events in carcinogenesis provide the rationale for the Ames test for carcinogens. See Ames test.

Mutator phenotype—Cancer cells often have many different mutations. It has been hypothesized that the progenitor cell of cancers has a mutator phenotype that heightens the likelihood that a tumor cell will acquire new mutations. Cancer may ensue when the acquired mutations confer a malignant phenotype.

Mycosis fungoides—A lymphoma of T lymphocytes that occurs in the skin. When malignant mycosis fungoides cells are found in the blood, the constellation of clinical findings is often called Sezary syndrome. Despite its name, mycosis fungoides is not caused by a fungus.

Myelodysplasia—See myelodysplastic syndrome.

Myelodysplastic syndrome—The myelodysplasias (formerly known as preleukemias) consist of several closely related diseases characterized by anemia and pancytopenia, disordered myeloid growth and maturation, the appearance of circulating blast cells, chromosomal abnormalities, and frequent progression to leukemia. The myelodysplastic entities are (listed as increasing grades of dysplasia) refractory anemia, refractory anemia with ringed sideroblasts, refractory anemia with excess blasts, and refractory anemia with excess blasts in transformation. Chronic myelomonocytic leukemia is sometimes included with the myelodysplasias. Transitions commonly occur from one level of dysplasia to the next higher grade of dysplasia. Refractory anemia with excess blasts in transformation, as its name implies, often develops into acute myelogenous leukemia. Myelodysplasias are examples of potentially lethal precancers. Often, patients die directly from their myelodysplasia without ever becoming leukemic. Causes of death can include chronic anemia, clotting disorders (from thrombocytopenia or thrombocytosis), infection (due to neutropenia), or iron overload (due to dyserythropoietic hemolysis). Myelodysplasias occur most often in older people and sometimes follow the occurrence of aplastic anemia. See myeloproliferative disorder, aplastic anemia, and dyserythropoiesis.

Myelofibrosis—Characterized by a sustained hyperplasia of extramedullary hematopoietic cells (myeloid growth outside of the bone marrow). The pathogenesis of myelofibrosis is obscure and somewhat controversial. It is thought that extramedullary hematopoiesis results from a primary fibrosing disease of the bone marrow. A specific JAK2 mutation is found in more than half of patients with myelofibrosis (332).

Myeloproliferative disorder—Myeloproliferative disorders are neoplastic or hyperplastic conditions characterized by the sustained growth of a stem cell subpopulation of myeloid origin that produces an increase in one or more types of circulating blood cells (such as platelets or red blood cells or granulocytes). The myeloproliferative disorders include chronic eosinophilic leukemia, chronic myelocytic leukemia, chronic myelomonocytic leukemia, chronic neutrophilic leukemia, essential thrombocythemia, hypereosinophilic syndrome, juvenile myelomonocytic leukemia, polycythemia vera, primary myelofibrosis, and systemic mastocytosis. The list includes disorders of diverse etiologies. Most of these disorders have characteristic mutations: BCR-ABL in chronic myelogenous leukemia; JAK2 mutations in polycythemia vera, essential thrombocythemia and primary myelofibrosis; c-KIT mutation in systemic mastocytosis; rearrangements of PDGFRB in chronic eosinophilic leukemia and chronic myelomonocytic leukemia; and RAS/PTPN11/NF1 mutations in juvenile myelomonocytic leukemia (169). The myeloproliferative disorders show us that any myeloid stem cell disorder that yields increased numbers of circulating differentiated cells will yield a hematologic disease that is similar to leukemia.

Myoepithelial cell—A ductal cell that has an epithelial morphology, but with functionality more akin to muscle (including contractility and responsiveness to acetylcholine). The physiologic role of myoepithelial cells is to squeeze ducts. Most myoepitheliomas are of ectodermal origin and only rarely develop from endodermal tissues. Myoepithelial cells are the presumed source of sarcomatous changes in rare metaplastic carcinomas of breast and provide the "mixed" appearance of pleomorphic adenoma of salivary glands. See metaplastic breast carcinoma and pleomorphic adenoma.

Myometrium—The muscular wall of the uterus.

Necropsy—A postmortem examination of a nonhuman animal. The term "autopsy" is never used for nonhuman postmortem examinations for a reason that relates to the derivation of the term autopsy, meaning to examine oneself. When we examine nonhuman animals, we are not examining ourselves (our own species).

Neoplasm—A general term that includes all focal growths composed of genetically abnormal cells. It includes benign and malignant tumors, precancers, and hamartomas. It excludes tissues that grow reactively as part of a normal physiologic process, such as hyperplasias and inflammatory lesions.

Neoplastic regression—See regression.

Neural crest—The neural crest is an embryonic germ layer that arises from specialized cells that lie between the neural tube and the ectoderm. In this book, neural crest is considered another embryonic layer on par with ectoderm, endoderm, mesoderm, neuroectoderm, and trophectoderm. Only the neural crest and germ cells are capable of giving rise to cell types that are produced by the other embryonic layers (e.g., neurons and osteocytes).

Neurectoderm—Alternate, less popular, spelling of neuroectoderm. See neuroectoderm.

Neurocristopathy—A disease of cells that derive from the neural crest. Examples are MEN2 (multiple endocrine neoplasm syndrome type 2), aganglionic diseases of the gastrointestinal tract, and neurofibromatosis. Several descriptive but embryologically nonspecific terms overlap with neurocristopathy. For instance, phacomatoses are syndromes that produce abnormalities and hamartomas of the nervous system, skin, and eye. Some of the phacomatoses would be more appropriately classified as neurocristopathies. The same is true for several of the inherited tumorous or hamartomatous syndromes described as genodermatoses.

Neuroectoderm—The embryologic layer that gives rise to the central nervous system. In the two authoritative embryologic classifications (Nomina Embryologica and Nomina Embryologica Veterinaria), the neuroectoderm is considered a derivative of the ectoderm and has been categorized as a subdivision of ectodermal organs. In the Developmental Classification, class neuroectoderm is considered a germ layer on par with class ectoderm. The reason for this is that tumors of Class Neuroectoderm do not inherit properties from Class Ectoderm.

Neurofibromatosis—There are several inherited conditions associated with neurofibromas (a benign neoplasm of Schwann cells). Common use assigns "neurofibromatosis" to type 1 neurofibromatosis, or Von Recklinghausen disease, which is associated with multiple cafe au lait spots (large epidermal nevi) plus neurofibromas. The so-called central form of neurofibromatosis, now called neurofibromatosis type 2, is characterized by bilateral acoustic schwannomas and meningiomas. The neurofibromatoses are prototypical neural crest disorders (neurocristopathies) wherein affected tissues derive from neural crest. See neurocristopathy.

Neutrophils—Circulating blood cells are either red (red blood cells) or white (white blood cells). More accurately, blood cells are either red or they are unpigmented. The red cells greatly outnumber the white cells, but there are a large variety of circulating white cells, dominated by lymphocytes, platelets, monocytes, and granulocytes (neutrophils, eosinophils, and basophils). The granulated cells can be distinguished by the dyes that can be absorbed by the different types of granules. Eosinophils retain a pink dye. Basophils retain a blue dye, and neutrophils do not retain dyes of either color. The most numerous type of white cell is the neutrophil, the most important defender against virulent microorganisms.

Nevus—A hamartomatous lesion of melanocytic cells, usually a mixture of melanocytes and so-called nevocellular cells. Almost all nevi occur in the skin, the normal site of melanocytic growth. Nevi can be identified as small (usually several millimeters across), brown spots. Most people have nevi. They can be found at birth (congenital nevi) or they can develop at any time of life.

Node-negative cancer—The term "node-negative cancer" is misleading. The "nodes" referred to in the term are lymph nodes. The term does not indicate that the cancer lacks lymph nodes. When a cancer achieves the ability to metastasize, the first destination for metastatic tumor is often the regional lymph nodes near the primary tumor. When the primary cancer is excised, the surgeon may remove the local lymph nodes that drain the tumor. These lymph nodes are examined by a pathologist for the presence of metastatic cancer foci. If the pathologist sees a focus of tumor that has metastasized to a lymph node, the lymph node will be labeled "positive" (for metastatic cancer). Every lymph node is examined, and if none of the removed lymph nodes are found to have cancer foci, the cancer is called a "node-negative cancer." If any of the lymph nodes contain cancer, the number of positive lymph nodes is counted along with the total number of resected lymph nodes, and these numbers appear in the surgical pathologist's report as part of the staging for the cancer. The pathologist does not inspect every part of every lymph node resected by the surgeon. Typically, the pathologist cuts the lymph node along its length and inspects one slice through the middle of the lymph node. A small focus of tumor nestled in an eccentric position in the node may not be detected by this examination. Also, single cells that have metastasized to the lymph node may not be noticed by the pathologist; therefore, the term node-negative cancer does not actually guarantee that cancer cells have not reached local lymph nodes. Regardless, traditional pathologic examination of lymph nodes has proven to be a valuable staging procedure and a useful therapeutic predictor. See hematogenous dissemination, sentinel lymph node, and occult cancer.

Nomenclature—A list of all of the significant words used within a knowledge domain. A nomenclature differs from a dictionary. The latter includes definitions for all of the words. A nomenclature is also different from a taxonomy. The latter includes all of the named instances of the objects in a classification. A nomenclature is also different from an ontology, which is a list of classes and their relations to one another. See taxonomy and ontology.

Non-Hodgkin lymphoma—For somewhat obscure reasons, the medical profession has divided lymphoid neoplasms into two categories: Hodgkin lymphoma and non-Hodgkin lymphoma. Non-Hodgkin lymphoma comprises every lymphoid neoplasm other than Hodgkin lymphoma. I know of several incidents in which patients were solemnly informed that their lymph node biopsy came back as "non-Hodgkin lymphoma." Patients always misunderstand this diagnosis, thinking, "Whew! I don't know what Hodgkin lymphoma is, but it doesn't sound good and I'm very glad I don't have it!" To my knowledge, "non-Hodgkin lymphoma" is the only diagnosis that informs patients what they don't have and declines to specify what they do have. See Hodgkin lymphoma.

Nonhomologous—Two very similar proteins, genes, and organs can be nonhomologous because homology requires that both objects derive from a common ancestral object. Because similar objects can be nonhomologous, it is usually a bad idea to build a biological classification based entirely on similarities among objects. See homolog.

Nontoxic cancer chemotherapy—Until the last few years, all cancer chemotherapy was designed to be toxic to human cells. The underlying rationale for using toxic drugs is that tumor cells are especially sensitive to the toxicity of cancer drugs. If we dose patients with enough toxic drug to kill all their tumor cells, a sufficient number of normal cells will survive to sustain life. In most cases, toxic cancer drugs preferentially kill rapidly dividing cells, such as cancer cells, gastric lining cells, and bone marrow stem cells. Nontoxic drugs are aimed at inhibiting pathways in tumor cells that play a key role in the malignant phenotype (cell division, invasion or metastasis). Because these new drugs are not designed to kill cells, they are often referred to as nontoxic chemotherapeutic agents. The term "nontoxic" is somewhat misleading because all medications have unwanted side effects, and thus are toxic. Still, the new generation of nontoxic chemotherapeutic agents is vastly safer and more tolerable than the preceding generation of cytotoxic agents.

Nuclear atypia—In normal tissues, the nuclei of any cell type will have a uniform morphology. Each nucleus will have roughly the same size, shape, and texture (distribution of chromatin). In tumor tissues, the nuclei of malignant cells are almost always characterized by aberrations from normal. Malignant nuclei are almost always larger than the normal nuclei of the same cell type. A malignant colon carcinoma cell will be larger, on average, than the nuclei of normal mucosal lining cells of the colon. This is true for every cancer of humans. Even small cell carcinoma of the lung, a tumor named for the small size of its tumor cells, has nuclei that are considerably larger than the nuclei of normal bronchial epithelial cells. Simply knowing the nuclear sizes of a population of cells is often sufficient to make a cytologic diagnosis of malignancy. Cancer nuclei are characterized by nonuniformity. The size of a nucleus will vary from tumor cell to tumor cell. The same holds for every other feature of the malignant nucleus. Tumor nuclei will not be uniformly rounded. Some nuclei will be round; others will have small indentations in the nuclear edge. Others will be smooth but show convexity. The area of convexity may change from cell to cell, sometimes facing into the center of the cell and sometimes facing the nearest cytoplasmic border. The nuclear texture will vary greatly from focus to focus within the nucleus and between one nucleus and the next. Areas of chromatin clumping (focal hyperchromaticity) will alternate with areas of nuclear hypochromaticity (lack of color). The coarse texture of chromatin from malignant cells will be different from the fine chromatinic texture of normal cells. Cancer cells will have increased numbers of nucleoli. In theory, every transcriptionally active chromosome is entitled to a nucleolus. Light microscopic examination of stained human cells seldom reveals more than a half dozen of these organelles. In cancer cells, there are increased numbers of nucleoli, and the number varies from one malignant nucleus to the other. Whereas normal nucleoli are uniform in size and round, the nucleoli of malignant cells tend to vary in size and may show angularities and assorted noncircular geometries. These principles of cancer cytology can be described in a few paragraphs. Mastering these secrets to the point of rendering reliable diagnoses may require years of training and the utmost professional concentration and patience.

Nucleolar organizing region (NOR)—NORs are associated with a particular locus in each chromosome. They help organize the nucleolus, and several NORs from different chromosomes may cooperate to organize a single nucleolus. NORs contain a fibrillar protein and can be easily identified in cells with a silver-based stain with affinity for NOR proteins.

Nucleolus—A circumscribed area of the nucleus that contains proteins and RNA and that synthesizes ribosomal RNA. The ribosomal RNA in the nucleolus must be exported through nuclear pores to the cytoplasm, where it serves in the transcription of mRNA to protein. A single nucleus usually contains one or several nucleoli. The presence of an increased number of nuclei with abnormal shapes (deviating from the normally round shape) and sizes all varying from cell to cell is a cytologic features often observed in cancers.

Nucleus—In the cells of eukaryotes and their descendants, a membrane-bound organelle containing the cell's chromatin, which packages the cell's genome. See anucleate cells.

Occult cancer—Oncologists use the term "occult cancer" to represent a number of different but related concepts. An occult cancer may be a metastatic or primary tumor that is not medically detectable. The term is sometimes used to describe a tumor of unknown origin. If a patient presents with multiple metastases in the brain and lung but with no known primary source for the metastatic foci, the primary tumor may be referred to as occult (i.e., hidden). Sometimes the term is used as a synonym for a dormant cancer focus that is not growing. The most common current use of the term "occult cancer" refers to circulating cancer cells or cancer cells that are found as microfoci (small clumps of just a few cells) in lymph nodes. Sensitive tumor markers tested in blood samples and tissue samples allow us to visualize small numbers of cancer cells from cancer patients in clinical remission. For the most part, we simply do not know the biological significance of occult tumor cells. We cannot predict whether these cells are likely to grow into metastatic foci or whether they will lie dormant forever or whether they will die. Understanding the fate of occult cancer cells is a very important goal of cancer researchers (333).

Odontogenic keratocyst—A neoplasm arising within jaws in close proximity to teeth and composed of a squamous cell-lined cystic structure that can invade into adjacent bone. This lesion occurs sporadically or as part of Gorlin syndrome (multiple basal cell carcinoma syndrome). Multiple keratocysts are typical of Gorlin syndrome. Gorlin syndrome is characterized by a germline mutation in the PTCH gene. PTCH gene mutations are also found in sporadic (i.e., nonsyndromic) basal cell carcinomas and odontogenic keratocysts. The presence of PTCH mutations in a cyst is further evidence that at least some examples of squamous cell-lined cysts are true neoplasms.

Oligoastrocytoma—A brain tumor composed of two morphologic cell types, one resembling oligodendrocytes and the other resembling astrocytes. Despite the bipartite morphology, these tumors have the same developmental lineage (glial cells of Class Neuroectoderm). The tumor cells with oligodendroglial morphology and the cells of astrocytic morphology both share many of the same genetic alterations (334).

Oligodendrocyte—Oligodendrocytes (oligodendroglial cells) are roughly the neuroectodermal equivalent of the neural crest Schwann cell. Oligodendroglial cells wrap around axons. The cell membrane of oligodendroglial cells contains myelin, and the wrapped oligodendrocyte forms the myelin sheath that protects the axon and aids in the conduction of the axonal impulse. One of the most important distinctions between oligodendroglial cells and Schwann cells is that a single oligodendroglial cell may wrap around many neuronal axons, whereas a single Schwann cell can wrap around only one neuronal axon. The other major neuroectodermal glial cell is the astrocyte. See astrocytes.

Oligodendroglioma—A glioma derived from cells that resemble oligodendrocytes. Oligodendrogliomas frequently have a combined loss of most or all of the short arm of chromosome 1 (1p) and most or all of the long arm of chromosome 19 (19q). Oligodendrogliomas with this combination of chromosome markers have a better prognosis and a better response to treatment than other gliomas. This favorable treatment response is seen in well-differentiated oligodendrogliomas and in anaplastic (poorly differentiated) oligodendrogliomas.

Oncogene—Normal genes that, when mutated or otherwise altered to a more active form, or overexpressed, contribute toward a neoplastic phenotype within a specific range of cell types. The normal form of the oncogene is called the proto-oncogene. The altered, more active form of the gene is called the oncogene. Some retroviruses contain oncogenes and can cause tumors by adding their oncogene to the host genome. See tumor suppressor gene and activated oncogene.

Oncology—The branch of medicine devoted to treating cancer.

Ontogeny—The development of the adult organism from embryonic structures.

Ontology—An ontology is a rule-based grouping of classes in a knowledge domain. Ontologies support queries and logical inferences pertaining to classes. They can be used to test data for logical consistency and can be designed to discover relationships between different classes of data. See classification.

Oral cancer—Cancer of the mouth. The mouth includes oral mucosa, salivary glands, teeth, tongue and glottal appendages, tonsils, and palate and their associated connective tissues.

Pap smear (full name, Papanicolaou smear)—A cytologic specimen obtained by scraping or brushing the junction between the endocervix and the ectocervix, spreading the sampled cells onto a glass slide, and staining the cells with a special histologic reagent (the Papanicolaou stain) that allows the cytologist to visualize subtle alterations in cell morphology. A typical Pap smear contains about 20,000 cells, and every cell must be inspected to rule out dysplasia and other pathologic abnormalities (Figures G.10 and G.11). A large cytology laboratory can handle hundreds of thousands of Pap smears in a year and can screen samples of cells obtained from virtually any anatomic site and from any body fluid. In the last half of the 20th century, the Pap smear cytologic evaluation led to at least a 70% drop in the number of deaths from cervical cancer in every country that fully deployed the test.

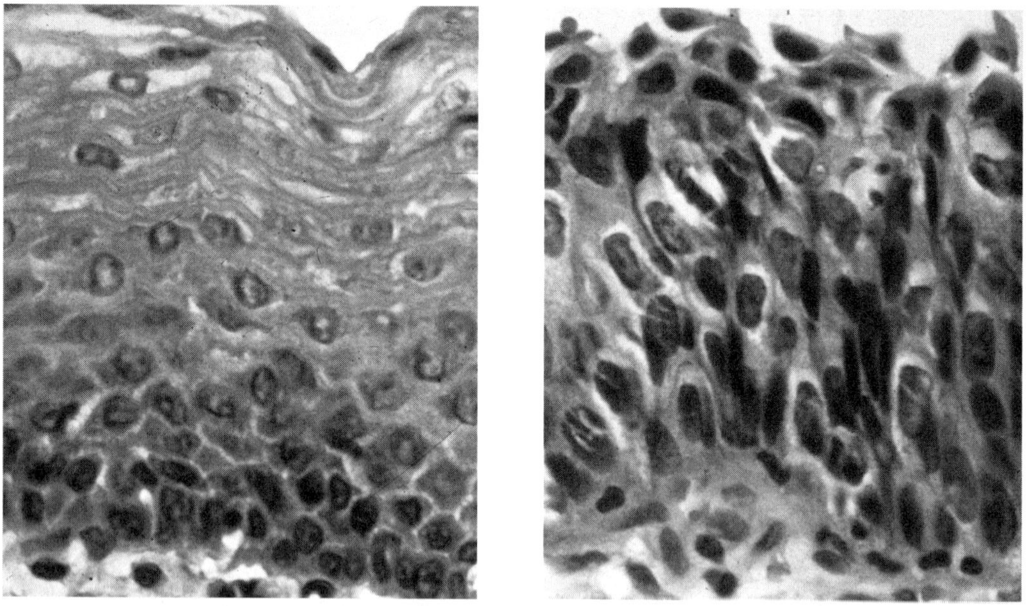

(a) (b)

FIGURE G.10 Figure A. Normal cervical epithelium. Figure B. Carcinoma in situ of the cervix. The atypical, malignant cells have large, dark nuclei. Atypical cells occupy the full thickness of the mucosa, but do not invade below the epithelial lining.

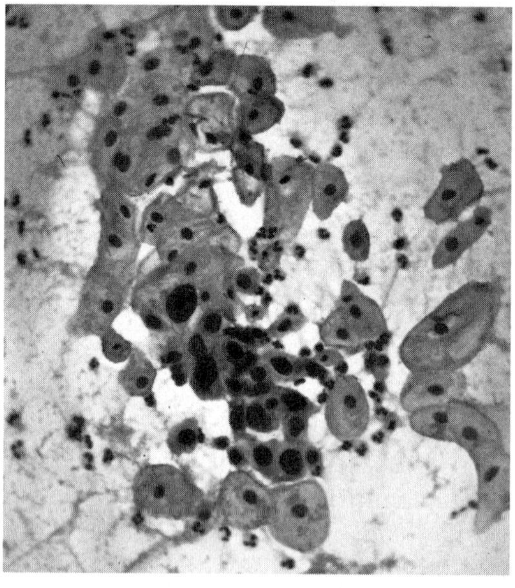

FIGURE G.11 Pap smear. In the middle of the Pap smear is a clump of cells with large, dark, irregular nuclei. These cells are consistent with a high grade squamous intraepithelial lesion. These dysplastic cells may have arisen from a lesion similar to that seen in Figure G.10.B. Surrounding the dysplastic cells are normal squamous cells. These cells have small, dark, round, central nuclei and abundant flat cytoplasm. Scattered neutrophils are noted.

Paraneoplastic syndrome—A constellation of clinical findings caused by a cancer but not produced by the destructive growth of the tumor (i.e., not due to tumor growth or invasion or metastasis). Most paraneoplastic syndromes are caused by humoral factors released by the tumor that have a specific effect on one or more normal tissues. Examples of paraneoplastic syndromes are Eaton-Lambert syndrome, acanthosis nigricans, carcinoid syndrome, hyperthermia of malignancy, night sweats of malignancy, sign of Leser-Trelat, hypoglycemia, limbic encephalopathy, anemia of malignancy, wasting disease of malignancy, autoimmune hemolytic anemia, opsoclonic-myoclonic ataxia, and others. See Leser-Trelat.

Paraphyletic classification—A class is paraphyletic if it does not contain every descendant member. A pure cladistic classification (the kind of classification that is easiest to understand because it is created purely through ancestry and not be similarity) prohibits paraphyletic groupings. The Developmental Lineage Classification and Taxonomy of Neoplasms contains paraphyletic groups. The neural crest class is paraphyletic because tumors of the skull bones are omitted from the class of neural crest-derived tumors, although skull bones derive from neural crest. See polyphyletic.

Parathyroid adenoma—Neoplastic enlargement of (usually) one parathyroid gland. A benign parathyroid adenoma has a morphology that is identical to that of parathyroid hyperplasia. A clinically useful distinction between parathyroid adenoma and parathyroid hyperplasia is that most or all of the parathyroid glands are enlarged in parathyroid hyperplasia. Usually, only one gland is involved by parathyroid adenoma. See parathyroid hyperplasia.

Parathyroid hyperplasia—The parathyroid glands are small structures located in, on, or near the thyroid glands. The number of parathyroid glands in each person may vary, but most people have at least four glands. The purpose of the parathyroid gland is to secrete parathyroid hormone (PTH), which regulates calcium metabolism in the body. Increased secretion of PTH can result in hypercalcemia. In parathyroid hyperplasia, all or most of the parathyroid glands are enlarged. See parathyroid adenoma.

Parthenogenesis—Reproduction from an ovum without fertilization by a sperm. This occurs in plants, invertebrates, and some vertebrates, but does not occur in mammals, a class of animals that uses genomic imprinting. Mammalian imprinting may interfere with parthenogenesis. There are about 80 to 100 genes in mammals which are imprinted, and many of these genes are vital to normal embryogenesis and fetal development. In parthenogenesis, all genes have maternal derivation, and the organism loses the benefit of paternal imprinting, which may be required for the normal development of a human embryo (270, 271).

Partial mole—There are two types of hydatidiform mole: complete mole and partial mole. A partial mole has mostly nonedematous placental villi and a variable amount of malformed embryonic tissue. A partial mole, like a complete mole, can develop into an invasive mole or a choriocarcinoma, but the incidence of conversion is much smaller (about 20% for complete mole and about 3% for partial mole). Partial moles are triploid and usually formed by the dispermic fertilization of an ovum, which produces two sets of paternal chromosomes and one set of maternal chromosomes.

Pathogenesis—The biologic events that lead to the expression of a disease. Carcinogenesis is a special instance of pathogenesis.

Pathologist—A professional who studies diseases in the expectation that a deep understanding of disease processes will lead to the cure or eradication of the disease. Today, the term is sometimes reserved (inappropriately) for specially trained medical doctors who render diagnostic services in a hospital. The term "pathologist" actually includes PhDs, veterinarians, and other professionals who devote their lives to understanding diseases. See anatomic pathologist and clinical pathologist.

Pathway—The term "pathway" means something different today than it meant a few decades ago. According to traditional thinking, a pathway was a sequence of biochemical reactions, involving a specific set of enzymes and substrates that produced a chemical product. The classic pathway was the Krebs cycle. It was common for students to calculate the output of the cycle (in moles of ATP) based on stoichiometric equations employing known amounts of substrate. Today, in the field of cancer research, pathways operative in tumors usually involve linked actions involving receptors, activators, enzymes, and sites in macromolecules. Their products are often a cellular action, rather than a chemical product. These pathways may not always occur in a specific organelle. The pathways may interact with other pathways, and their direction and biological consequences may be complex and variable among different cell types or the different physiological status of one cell type. Still, the term "pathway" is a convenient conceptual device to organize classes of molecules that interact with a generally defined set of partner molecules to produce a consistent range of biological actions.

PEComa—PEComas (perivascular epithelioid cell tumors) are rare, usually benign lesions that comprise a morphologically diverse family of neoplasms including angiomyolipoma, lymphangioleiomyomatosis, clear cell "sugar" tumors, and clear cell myomelanocytic tumors (335). In addition to their common origin, from a perivascular mesenchymal cell, the cells of these tumors have clear cytoplasm and immunostain positively with HMB-45.

Perinatal carcinogenesis—A process that begins during gestation, or soon after birth, that leads to cancer. The term is usually restricted to a process that begins with an external exposure to a carcinogen and would exclude inherited or germline carcinogenesis. In general, newborns are much more sensitive to carcinogens than are adult organisms. See transplacental carcinogenesis.

Phenotype—The observable properties of a cell, organ, or organism. The concept of a phenotype exists in contrast to the concept of a genotype. The phenotype expresses the sum of environmental, genetic, and developmental influences. The term "cancer phenotype" usually refers to the minimal properties of growth, persistence, invasion and metastasis that characterize virtually all cancers.

Philadelphia chromosome—Almost all patients with chronic myelogenous leukemia have a characteristic translocation that is called the Philadelphia chromosome. Parts of chromosomes 9 and 22 translocate and form a fusion gene composed of a part of the BCR (breakpoint cluster region) gene and a part of the ABL (Abelson Leukemia) gene (Figure G.12). The fused gene creates a protein product with several functions, including tyrosine kinase activity. We know that the Philadelphia chromosome appears in a stem cell because in chronic myelogenous leukemia, cells of hematopoietic lineages, in addition to the granulocytic lineage, will contain the same Philadelphia chromosome. This is an example where an activated oncogene pervades a stem cell line but preferentially produces tumors of a single differentiated cell type within the lineage (neutrophils). Not surprisingly, the Philadelphia chromosome can be found in leukemias other than chronic myelocytic leukemia. It occurs in about 25% of cases of acute lymphoblastic leukemia occurring in adults, in about 5% of cases of acute lymphoblastic leukemia occurring in children, and in some cases of acute myelogenous leukemia.

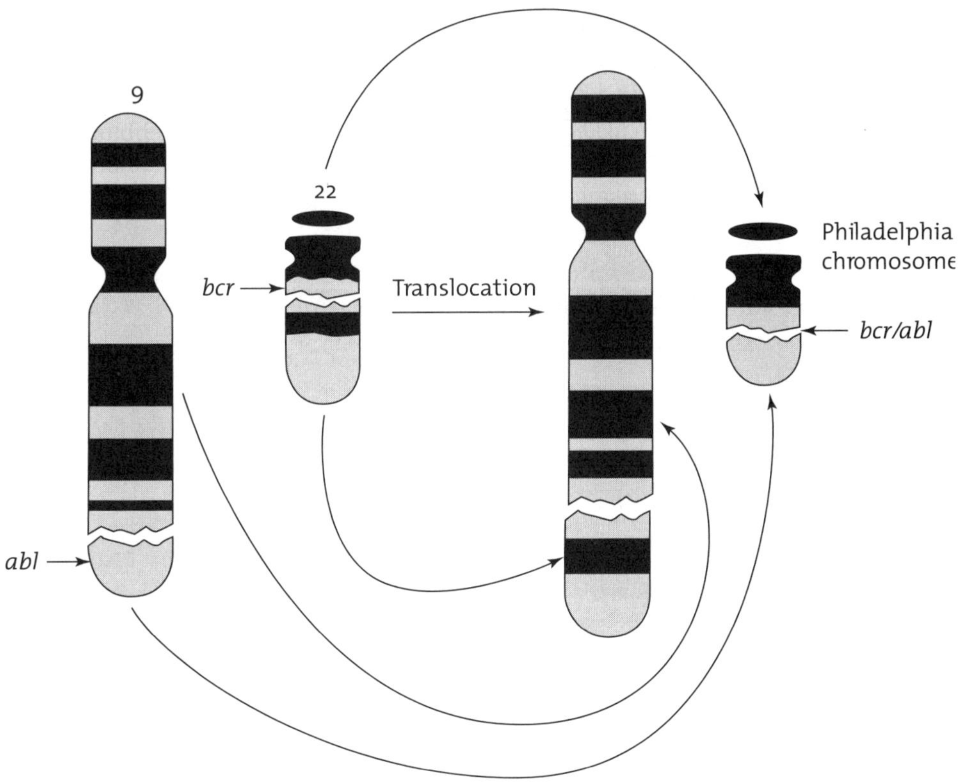

FIGURE G.12 Philadelphia chromosome. A reciprocal translocation between chromosomes 9 and 22 is seen.

Phylogeny—Classifications based on ancestral lineage.

Pilomatrixoma—A benign skin tumor. The putative histogenesis of pilomatrixoma is the hair matrix.

Placental site trophoblastic tumor (PSTT)—From the stem cell trophoblast, several different kinds of differentiated trophoblastic cells develop. These are the villous syncytiotrophoblasts, the anchoring trophoblasts, and the invasive intermediate trophoblasts. Several different lesions can develop from the intermediate trophoblasts, including exaggerated placental site, placental site nodule, epithelioid trophoblastic tumor, and PSTT. It seems that PSTT can occur after either molar or nonmolar pregnancies.

Pleiotypic—The occurrence of multiple different responses from a single stimulus. Something that causes a pleiotypic response is pleiotrophic.

Pleomorphic adenoma—A benign salivary gland tumor composed of a mixture of epithelial cells, some forming glands, some secreting copious mucus, and a myxoid component containing round or spindle cells. This tumor, once called a mixed tumor, is now considered to be the result of a salivary ductal cell that differentiates along several sublineages, including ductal, acinar, and myoepithelial. Over 80% of pleomorphic adenomas occur in the parotid gland, and nearly 80% of benign parotid gland tumors are pleomorphic adenomas.

Plesiomorphy—A trait that is shared by some classes but not others and that derives from a very primitive ancestor. Plesiomorphic traits have limited value in establishing the phylogenetic relationships among the descendant classes of the primitive ancestor that first developed the trait.

Pluripotent stem cell—A cell that can produce any of the differentiated cells of endoderm, ectoderm, and mesoderm but cannot produce the extraembryonic cells of the placenta, such as trophoblasts and cannot produce germ cells. See totipotent stem cell and multipotent stem cell.

Polygene—A gene that alone does not affect phenotype but can operate in a group of nonallelic genes to influence a phenotypic trait. In cancer, as mutations accumulate, polygene mutations may influence the malignant phenotype.

PNET (primitive neuroectodermal tumor)—Includes several malignant tumors of childhood that are composed of small, blue cells, all derived from neural crest. The word "neuroectodermal" appears in the name, rather than the more accurate term "neural crest," possibly because early developmental biologists thought that the neural crest grew from the neural tube (part of the neuroectoderm). In this book, the neural crest and the neuroectoderm are treated as two distinct embryonic layers.

Polyphyletic—A class is polyphyletic if some of the members have a different most recent ancestor than other members of the same class.

Postmitotic—Refers to cells that will never divide again. Many of the differentiated cells of the body are post-mitotic. Once they have finished their physiologic purpose, they are sloughed off into the gut (e.g., fully differentiated colonic goblet cells) or the air (e.g., fully differentiated squamous cells of the skin), or they are decomposed by the reticuloendothelial system (e.g., red blood cells).

Precancer—Precancers are the lesions from which cancers derive. Precancers have some of the properties of cancers, but not all of those properties. Precancers have some properties that cancers lack. Eliminating precancers cures the patient of the cancer that might have eventually developed. Two of the most interesting properties of precancers (lacking in cancers) are their propensity for spontaneous regression and the ease with which they can be prevented or treated.

Precancer regression—See regression.

Precancerous condition—A precancerous condition is a pathologic process that predisposes a person to the development of a precancer. It differs from the term "precancer." The latter refers to a clone of cells from which a cancer emerges. An example of a precancerous condition would be cirrhosis, which carries a high risk of developing cancers and precancers. An example of a precancer is "atypical nodule of liver," a lesion often found in cirrhotic livers from which hepatocellular carcinomas often arise.

Predictive markers—Same as predictor. See predictor.

Predictor—In medical practice, a predictor indicates the likelihood of a particular clinical response to a particular choice of treatment. In other words, the term "predictor" is a shortened version of the more complete term, "response predictor." The term "predictive test" is imprecisely used as a synonym for "screening test." It is best to distinguish "screening" test from "predictive" test. A "screening test" divides a population into a group that is likely to have the disease and another group that is unlikely to have the disease. The determination of the presence or absence of disease in individuals is accomplished with a "diagnostic test," not with a "predictive test" or a "screening test."

Premalignancy—Same as precancer. See precancer.

Premalignant condition—Same as precancerous condition. See precancerous condition.

Prognosis—The likelihood of recovery from a disease. A prognostic marker estimates the likelihood that the patient will recover. Prognostic markers have very little clinical value, and many people use the term "prognostic marker" when they really mean to say "predictive marker." See predictor.

Prognostic test—A prognostic test is used to determine the likelihood that a person with a certain disease and a certain set of circumstances (stage of disease, age of patient, concomitant conditions) will recover. When you think about it, a prognostic test is has very limited value. Doctors and patients more often need to know the likelihood that a patient will respond to one treatment versus another treatment option, and for this, they need a "predictive test." The term "prognostic test" is often confused with "predictive test."

Progression—Cancer progression is the acquisition of additional properties of the malignant phenotype over time. Progression is achieved through a variety of mechanisms (e.g., genetic instability, epigenetic instability, and aberrant cell death regulation) and results in the emergence of heterogeneous subclones that have growth advantages over other cells descended from the same progenitor. When a lesion grows larger and larger without changing its genotype of phenotype, this is simple growth and should not be confused with progression. Expansile growth without progression is typical of benign lesions. The direct result of tumor progression is tumor heterogeneity. See tumor heterogeneity.

Prolactinoma—An adenoma of the pituitary gland that secretes prolactin, a hormone that stimulates breast milk production. A large percentage of pituitary adenomas produce very small amounts of prolactin, but in most cases, the concentration of prolactin produced is insufficient to produce any clinical symptoms of hyperprolactinemia.

Promoter sites—DNA sites near (usually upstream from) the site where RNA polymerase binds to DNA and initiates transcription. Transcription factors bind to regions containing promoter sites.

Promotion—A poorly understood phenomenon whereby tissues treated with a carcinogenic initiating agent and followed by a period of latency in which cancers do not arise, are rapidly induced to produce cancers after a subsequent treatment with another agent (called the promoter). Many cancer researchers think in terms of an initiation–promotion step in carcinogenesis, wherein initiation involves a mutation of a gene or genes that confer an early cancer genotype in a cell, and promotion is a hyperplasiogenic event (an event that causes an increase in cell division and cell number). The assumption is that after promotion there is a clonal population of cells endowed with at least some of the genotype found in precancerous cells. The term "promotion" is not related to "promoter sites."

Proteus syndrome—A rare, congenital, and apparently sporadic disease characterized by the overgrowth of many different tissues. Findings include epidermal nevus, disproportionate overgrowth, and a variety of benign and malignant tumors.

Proto-oncogene—An oncogene that has not been activated. See oncogene.

Pseudoepitheliomatous hyperplasia—Sometimes hyperplastic growth of epithelial tissues can be exuberant and disorganized, simulating cancer. Pseudoepitheliomatous growth of squamous mucosa is commonly found with an underlying granular cell tumor. It is easy for an unwary pathologist to mistakenly diagnose "squamous cell carcinoma" when a biopsy, intended for an intraoral submucosal granular cell tumor, captures the mucosal surface instead.

PSTT—See placental site trophoblastic tumor.

Rank—In hierarchical biological nomenclatures, classes are given ranks. The biggest problem with ranking is that it cannot keep up with the complexity of the classification. The early Linnaean classification made due with a handful of ranks (species, genus, order, class, and kingdom). With the splitting and addition of branches to the classification, we have now about 21 ranks, including family, phylum, division, domain, and new prefixes (super, sub, infra). These rankings are unnecessary. Each class is a subclass of only one direct ancestor. This means that the only necessary rankings in a classification are class and subclass. The location of a class in a classification can be determined by naming its direct ancestor (i.e., the class for which it is a subclass). In computational terms, this would be the basis by which a graph traversal program could construct the hierarchy and produce a graphic representation of the hierarchy (i.e., a dendrogram). Throughout this book, we use a rankless system of description, employing the terms class and subclass and naming the ancestor class, as appropriate.

Regression—The general rule for the natural progression of cancer is unrelenting growth followed by invasion, metastasis, and death. This is not always the case. Sometimes neoplasms simply stop growing and even disappear (regress). There are several different mechanisms for tumor regressions: a population of neoplastic cells may not have acquired all of the biological properties needed for sustained growth, the likely mechanism for precancer regression; the population of tumor cells is sufficiently immunogenic to stimulate a host response that eliminates the primary cancer (sometimes seen in small primary melanomas); the tumor differentiates to a nondividing population of cells (for example keratoacanthoma and some neuroblastomas); the growth advantage of the tumor, compared to normal cells, is lost when a growth stimulus is removed from the host environment (for example, hormone-responsive tumors, some cases of *Helicobacter*-induced MALTomas, possibly immunostimulated lymphomas); and virally induced tumors for which viral replication ceases (e.g., tumors occurring and regressing in transiently immunosuppressed patients).

Regulatory elements—The transcription (of DNA into RNA) is regulated by DNA sequences that bind with specialized proteins (such as transcription factors). DNA regulatory elements are given names based on their biological and their specific sequences (e.g., promoters and enhancers).

Renal cell carcinoma—The term renal cell carcinoma is a general term that includes all of the different morphologic and cytogenetic subtypes of cancers that arise from the renal epithelial tubule. These include clear-cell (conventional) renal carcinoma, papillary renal carcinoma, and chromophobe renal carcinoma. Tumors that arise from other parts of the kidney, such as tumor arising from the interstitial cells, the collecting ducts, the transitional lining cells, and primitive tumors and sarcomas arising from the kidney are called by their exact names and are not included under the term "renal cell carcinoma." All tumors arising from the kidney, including all of the renal cell carcinomas, belong to Class Mesoderm.

Rhabdoid tumor—A very rare tumor that occurs in children and develops from either of several different embryonic layers (neuroectoderm and mesoderm). Rhabdoid tumors are characterized by biallelic loss of the INI1 tumor suppressor gene (Chapter 15).

RNA splicing—See alternative RNA splicing.

Screening test—Screening is done on a population of people who have no symptoms of the disease that is being tested. The screening population may consist of everyone (population screening), or it may consist of a subpopulation that is at greatest risk for a particular disease (e.g., a particular age group, a particular gender, or a particular ethnicity). The screening test may involve looking for a particular risk marker (e.g., high cholesterol) or may involve looking for a disease marker (e.g., hypertension). A diagnostic test is different from a screening test because it is performed on persons with a sign or symptom of disease for the purpose of determining whether the symptom is due to a particular disease. In general, test validation criteria (sensitivity, specificity, and predictive values) are different for screening tests than for diagnostic tests. See predictor.

Schwannoma—A tumor composed of neoplastic Schwann cells. Schwann cells have a neural crest origin. The presence of schwannomas may indicate an inherited neurocristopathy. See neurocristopathy and neurofibromatosis.

Seminoma—A tumor of differentiated male germ cells arising almost exclusively in the testis. The female equivalent of the seminoma is the ovarian dysgerminoma.

Sentinel lymph node—When cancers metastasize, particularly cancers that derive from ectoderm and endoderm, their first site of metastasis is often the regional lymph nodes that drain the primary site of the tumor. Within the regional lymph nodes, tumors tend to metastasize to the lymph node that most actively drains the tumor, which is called the sentinel lymph node. The sentinel lymph node is considered the biologically most likely location to contain metastatic cancer. If the sentinel lymph node is negative for metastatic cancer, there is a reasonable chance that the other regional lymph nodes will be negative and that the tumor has not metastasized to distant sites. Surgeons sometimes opt to remove the sentinel lymph node in the region of a tumor, sparing resection of the other lymph nodes in the region. The sentinel lymph node can be identified by a perioperative procedure in which a marker is injected into the tissue near the tumor. The lymph node absorbing the greatest amount of marker is the sentinel node.

Sign of Leser-Trelat—See Leser-Trelat.

Similarity classification—Classifying by similarity is a simple approach to classification in which a variety of features are observed for all of the members of a domain, and members are grouped together based on the total number of features that they have in common. This was one of the first methods used to classify living organisms. This produced groupings that combined all animals that fly (birds, bats, bumblebees), swim (fish, cephalopods, lampreys), or walk (humans, bears, penguins). This approach cannot take into account the dissimilarities within a single entity during various stages of its life (e.g., caterpillar and butterfly, tadpole and frog, Barrett esophagus and invasive carcinoma of esophagus). The resulting classifications did not really add to our understanding of organisms and had value only as identification schemes. Classifying neoplasms by their gene expression profiles is just another example of similarity classification.

Skin cancer—Any cancer arising from the epidermis, adnexa or integument (dermis and subcutis). Basal cell carcinoma and squamous cell carcinoma account for the greatest number of skin tumors. Melanoma, although less frequent, accounts for the greatest number of deaths due to skin tumors.

Solid tumors—Tumors that grow as an identifiable mass of cells. Leukemia is an example of a non-solid (sometimes called liquid) tumor. Some tumors present as malignant effusions in body fluids (such as peritoneal or pleural fluids). Malignant effusions may also be considered non-solid or liquid tumors.

Somatic mutation—The word "soma" refers to something that occurs to the body or to the cells that compose the body. A somatic mutation is an acquired mutation that occurs in any of the cells of the body other than the fertilized ovum (zygote). Cancer, with the exception of some cancers occurring in inherited syndromes, is a disease that results from mutations in somatic cells (i.e., somatic mutations). Cells that are not part of the tumor will not have the somatic mutations found in the tumor cells. Tumors that arise from acquired mutations in somatic cells are sometimes called "sporadic tumors."

Species—Organisms that descended from a common ancestral species and that interbreed. Always refers to the plurality of members, and I suppose this is the reason that only the plural form is used (i.e., species, not specie).

Spitz nevus—A type of benign nevus found most often in children and young adults and notable for histologic and gross features in common with malignant melanoma. The tumor contains atypical melanocytes, has high cellularity, and can appear in clusters on the skin (simulating superficial dissemination). It is helpful to know the age of the patient (malignant melanomas rarely occur in children and Spitz nevi are uncommon in adults past 40). Histologic features that would discourage a diagnosis of melanoma include symmetry (melanomas usually have asymmetric growth pattern) and a reduction of cell size (to a nevoid morphology in Spitz nevi) in the deeper parts of the lesion. Still, the distinction between benign Spitz nevus and malignant melanoma can be a difficult diagnostic decision for a pathologist.

Spontaneous regression—See regression.

Sporadic tumor—A tumor that occurs in a population without a predisposing inherited cause. The term applied to any tumor is subject to re-evaluation. If a patient develops breast cancer and is a member of a family where many of the women develop breast cancer, the patient is said to have familial breast cancer (i.e., nonsporadic). If the basis of the high breast cancer rate among family members is found to be the presence of BRCA and the patient is found to lack the BRCA mutation, then the tumor is presumed to be sporadic in that patient, even though her sisters and cousins may have inherited their predisposition to breast cancer. If additional studies show that the patient inherited a completely different gene (from her mother or her father) that predisposed her to breast cancer, then the tumor shifts from sporadic to inherited. See BRCA.

Staging—The process of determining the extent of tumor at the time of diagnosis. This usually involves determining the size of the tumor, the extent of local growth of the tumor, determining whether the tumor has metastasized to regional lymph nodes, and determining whether the tumor has metastasized to distant sites. If lymph nodes are involved by tumor, it is important to know the number of lymph nodes involved by tumor; and if tumor has metastasized to distant sites, it is important to know which sites are involved and the extent of involvement at each site. Staging information is used, along with grading information, to choose the best treatment option available to the patient. See grading.

Stewart-Treves syndrome—Conditions that block lymph drainage (such as the dissection and removal of regional lymph nodes during breast cancer surgery) may result in edema in the involved tissues, along with reactive proliferation in small lymphatic vessels. Occasionally, an angiosarcoma may arise from the lymphatic vessels growing within this tissue. The syndrome of lymphatic blockage followed by angiosarcoma is called Stewart-Treves syndrome.

Submucosa—The connective tissue that lies between the muscularis mucosa and the muscularis propria (Figure G.13).

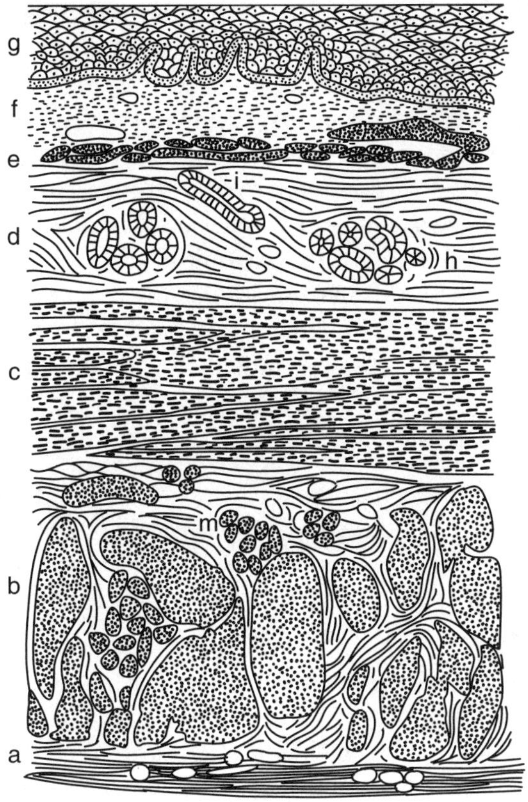

FIGURE G.13 Drawing of submucosa: b, c = muscularis propria, d = submucosa, e = lamina propria.

Sui generis—From the Latin, meaning one of a kind. In the Developmental Classification of Neoplasms, the only *sui generis* taxon is the atypical teratoid rhabdoid tumor.

Suppressor gene—See tumor suppressor gene.

Syngeneic—Two organisms are syngeneic if they belong to the same species and if they are genetically compatible for purposes of tissue transplantation.

Syndrome—A constellation of diseases and distinct physical findings that appear together. Syndromes are often inherited, but they can be acquired as well.

Synonymous SNPs—Different SNPs (single nucleotide polymorphisms) that have the same transcriptional meaning because the genetic code has some codon triplet ambiguity. For example, guu, guc, gua, gug will all code for the amino acid valine (when translated in the same reading frame). These triplets are synonymous though their sequences are nonidentical.

Synteny (from syn, along, and tainia, band)—An ordering of a group of genes along the chromosome that is the same among related species. During evolution, chromosomal rearrangement occurs naturally over time. Closely related species will tend to preserve the order of sets of homologous genes. The disruption of synteny often indicates divergent evolution. In cancer research, synteny between human and mouse is used to find mouse chromosomal regions that correspond to human chromosomal regions. Human oligodendrogliomas are often characterized by losses of chromosomes 1p, 19q. Regions in mouse syntenic to 1p are found on murine chromosomes 3 and 4 and regions on mouse syntenic to 19q are found on murine chromosome 7. In an experimental mouse oligodendroglioma model, no losses were found in either of these syntenic chromosomal regions, indicating that the cytogenetic marker for human oligodendroglioma is absent from the mouse tumor (336).

Taxa—Plural form of taxon.

Taxon—The list of instances in a particular class within a classification. In the classification of living organisms, this would correspond to all of the organisms that fall into a genus. In the Developmental Lineage Classification of Neoplasms, a taxon would be all of the named tumors that belong to a particular subclass of tumors, for example, the subclass of neural crest tumors that have a melanocytic origin.

Taxonomy—The list of every instance in an entire classification, along with its assigned class. You can think of the classification as a scaffold to hold organized objects and the taxonomy is the collection of named objects that fills the scaffold. The Developmental Lineage Classification of Neoplasms contains a taxonomy of about 6,000 different tumors comprising nearly all of the tumors of mankind. The taxonomy fills a classification of about 40 classes. Each tumor in the taxonomy has, on average, over 20 synonyms, yielding a neoplasm nomenclature of over 130,000 neoplastic terms. See nomenclature and classification.

Teilum tumor—See endodermal sinus tumor.

Telomeres—Cells that contain linear DNA (such as eukaryotes) have a problem not encountered in cells with circular DNA (such as bacteria). DNA replication in one of the two strands of linear DNA is discontinuous and requires the availability of template sequences of DNA to initiate replication fragments. Because DNA is not built as a tandem collection of complete replicative subunits, the last fragment of DNA (corresponding to the tip of a chromosome) is incomplete, and the template sequence of the last piece cannot be replicated. After each mitosis, a short template fragment of DNA is lost from the tip of every chromosome in eukaryotic cells. Chromosomes are built with a long padding sequence, consisting of repetitive DNA, at the chromosome tips, and this sequence is called the telomere. In normal cells, the telomere becomes shorter and shorter with each cell replication. Eventually (usually after about 50 to 100 mitoses), the telomere is consumed, and the cell ceases further replication. The phenomenon of telomere exhaustion is called senescence and is thought to play an important role in the cellular aging process. Some cells can restore their telomeres with an enzyme, telomerase. Some cancer cells have higher than normal concentrations of telomerase. The ability of some cancer cells to restore their telomeres contributes to the ability of cancer cells to divide without limit, a phenomenon sometimes called cancer cell immortality. Telomerase insufficiency has been suggested as a possible cause of bone marrow failure (337) and spontaneous regression of some tumors (45).

Teratogenesis—The process leading to a developmental malformation. Teratogenesis is a phenomenon that acts quickly on a population of cells, and the major mechanisms of teratogenesis seems to be tissue-specific toxicity involving embryonic or fetal cells that have a required role in normal development. Although a carcinogen can be a teratogen, the mechanisms of carcinogenesis and teratogenesis are probably unrelated.

Thesaurus—A thesaurus is a collection of terms, with all of the synonymous or closely related terms joined under a single entry. In modern medical nomenclatures, the thesaurus lists terms along with a concept code. All of the synonymous terms are given the same concept code. For example, hepatic cancer, hepatocellular carcinoma, HCC, and malignant hepatoma might be assigned the same seven-digit code. Terms found in, or extracted from, a medical text can be assigned the concept code that applies to the term in the text. Concept code assignments in medical text permit computers to collect reports that contain any desired concept code, regardless of the choice of term used in any text record.

Tissue blocks—See archived tissue blocks.

Tissue organization field theory (TOFT)—See field theory.

Totipotent stem cell—A stem cell that can give rise to any tissue of an organism, including extraembryonic tissues, tissues of the earliest embryo, and germ cells. A totipotent stem cell is different from a pluripotent stem cell, which cannot produce extraembryonic cells and cannot produce germ cells.

Transcription factor—A protein that binds to specific sequences of DNA and activates or represses the rate of transcription of a gene. See promoter sites.

Transdifferentiation—A biological phenomenon whereby a cell changes its morphology and functionality from one type of cell to another. If a retinal photoreceptor were to become a ductal cell lining a tear duct, that would be an example of transdifferentiation. Most people restrict usage of the word "transdifferentiation" to changes that exceed normal modes of differentiation. During renal organogenesis, when a metanephric mesodermal cell condenses around a ureteric bud to produce a renal epithelial cell, that might not be considered "transdifferentiation" because it is the normal and obligatory pathway by which the kidney is created. For the purposes of this book, the word "transdifferentiation" refers to a process by which a cell becomes a type of cell that lies outside its normal developmental lineage.

Transformation (in chemical carcinogenesis)—Refers to the tissue changes that occur in lines of cultured cells that seem to mimic the acquisition of the malignant phenotype in carcinogenesis. Cell lines may, after treatment with a carcinogen, change their behavior. The

List G.3 Karyotypes for some neoplasms that have a simple Robertsonian translocation.

t(2;5)(p23;q35)—Anaplastic large cell lymphoma
t(8;14)—Burkitt lymphoma (c-myc)
t(9;22)(q34;q11)—Philadelphia chromosome, CML, ALL
t(11;14)—mantle cell lymphoma (Bcl-1)
t(11;22)(q24;q11.2-12)—Ewing sarcoma
t(14;18)(q32;q21)—follicular lymphoma (Bcl-2)
t(17;22)—dermatofibrosarcoma protuberans
t(15;17)—acute promyelocytic leukemia
t(1;12)(q21;p13)—acute myelogenous leukemia
t(9;12)(p24;p13)—chronic myelocytic leukemia, acute lymphocytic leukemia
t(X;18)(p11.2;q11.2)—synovial sarcoma

cells may start to grow faster, and they may pile higher in the tissue culture flask (loss of density-dependent growth inhibition). The cells may gain the ability to grow in soft agar. Cells that can grow in soft agar are said to have anchorage-independent growth. Cells that can grow in a tissue culture flask but cannot grow in soft agar are said to have anchorage-dependent growth. Anchorage-independent growth is considered to be analogous to invasive growth seen in cancers. Transformation occurs spontaneously in cultured animal cells, but the rate of transformation increases after carcinogen exposure. For this reason, cell transformation is sometimes used as an in vitro model for carcinogenesis. Transformation has several other meanings depending on biological discipline. See transformation (in microbiology).

Transformation (in microbiology)—To microbiologists, "transformation" is one means of introducing genes into an organism, resulting in the acquisition of some new biological feature. Virologists use the term "transformation" to mean the acquisition of an altered cellular phenotype after viral infection.

Translational research—This term includes all biomedical research that attempts to find new therapies, devices, techniques, or tests that can be used in a clinical setting. Much of translational research is aimed at deriving benefit from the many advances in basic biomedical research achieved in the past decade.

Translocation—Numerous cancers contain reciprocal translocations in which parts of chromosomes are exchanged between two noncomplementary chromosomes (i.e., nonpaired chromosomes) (List G.3). The translocations are expressed by a "t" (for translocation), followed by a parenthetic list of the chromosomes involved, followed by another parenthetic list of the parts of the chromosomes exchanged. In the list, the second parenthesized expression uses a "p" to indicate the short arm of the chromosome, a "q" for the long arm, and numbers to indicate bands, and a decimal point to separate a band number from its subband number. See karyotype.

Transplacental carcinogenesis—The induction of a cancer in an offspring through exposure of the mother to a carcinogen during pregnancy. In experimental rodent models, transplacental carcinogenesis results in tumors appearing in adult offspring. The kinds of tumors that appear are typically "adult" type cancers (i.e., epithelial, differentiated cancers). In the case of humans, the best studied example of transplacental carcinogenesis is DES exposure, which results in malformations and tumors of the genital tract in young adult women. See perinatal carcinogenesis.

Transposon—Transposons, also called transposable elements, are repeated gene sequences scattered throughout chromosomes that can replicate themselves and move to different locations. They may contain regulatory elements that can modify gene expression. Transposons may play a role in the altered expression of genes in cancer cells (338).

Trilateral retinoblastoma—The occurrence of bilateral retinoblastomas is sometimes followed by the occurrence of a pineoblastoma, and this is referred to as trilateral retinoblastoma. The pineal gland has an evolutionary anlage identical to that of the eye. The difference is that neuroectoderm-derived photoreceptors of the pineal gland take up residence deep within a midline recess in the brain, whereas the photoreceptors of the eyes developed as paired external structures. The evolved pineal gland, like the evolved eyes, reacts to filtered sunlight, albeit any light reaching the pineal is highly filtered. In response to light cessation, the pineal gland releases melatonin, a hormone that influences circadian rhythms (e.g., sleep). The same germline mutation that leads to bilateral retinoblastomas may occasionally cause a pineoblastoma, with all three tumors sharing a common lineage and organ homology.

Trophectoderm—The trophectoderm is formed in the blastocyst stage and gives rise to the placenta and the amniotic membranes. Because these tissues are not part of the developed animal, they are referred to as extraembryonic tissue. All gestational trophoblastic neoplasia arises from the trophectoderm.

Trophoblast—The trophectoderm produces one general type of epithelial cell, called the trophoblast.

Tuberous sclerosis—Also known as tuberous sclerosis complex (TSC) and formerly known as Bourneville disease. Tuberous sclerosis can be caused by one of at least two different genetic lesions at two different cytogenetic loci: TSC1 on chromosome 9q34 encoding hamartin and TSC2 on chromosome 16p13.3 encoding tuberin. Both proteins are tumor-suppressor genes. The syndromes produced by either gene are essentially identical to one another. In keeping with other tumor syndromes caused by alterations in tumor suppressor genes, the resulting neoplasms involve cells from more than one developmental lineage. The neoplasms are hamartomas or benign neoplasms, ash-leaf spots, shagreen patches, periungual fibromas, so-called adenoma sebaceum (actually facial angiofibromas), retinal hamartomas ("mulberry lesions"), cardiac rhabdomyomas, renal angiomyolipomas, and subependymal giant cell astrocytomas (47).

Tumor heterogeneity—The direct result of tumor progression is tumor heterogeneity. A cell that has acquired a new genetic aberration will produce a new growing subclone within a clonal neoplasm if the new aberration confers a survival advantage. Tumor heterogeneity is almost always a bad prognostic indicator because the new subclones will tend to have phenotypic properties that permit them to grow under conditions that prevail in an enlarging tumor. These will likely consist of enhanced anaerobic energy metabolism (a lessened dependence of respiration), enhanced invasiveness (providing growing cells with a mechanism to avoid being crushed by neighbor cells), increased growth rate (the ability to grow when other clones have slowed or stopped), greater incidence of metastasis (the ability to seed to distant sites after invading through the walls of lymphatics or blood vessels), and shortened dormancy (shorter time for cells seeded in distant anatomic sites to grow as metastatic subclones).

Tumor progression—See progression.

Tumor suppressor gene—When a tumor suppressor gene is absent, cellular events that either express the malignant phenotype or that lead to the development of a malignant phenotype can proceed at a faster rate or with a higher likelihood. Tumor suppressor genes have several properties that separate them from oncogenes. First, none of the known tumor suppressor genes occurs on the x or y chromosomes (i.e., they are all autosomal). Tumor suppressor genes have a biological effect when they are inactive (a mutation that increases the activity of a tumor suppressor gene will not lead to cancer). A single active copy of a tumor suppressor gene is usually sufficient to maintain a normal cellular phenotype (i.e., both copies of a tumor suppressor gene are inactivated to increase a cell's risk of acquiring a cancer phenotype). Loss of tumor suppressors alone does not result in cancer cells (additional events must occur). Known tumor suppressors work through alterations in the control of cell division, the control of DNA repair, or the control of apoptosis.

Tylosis—A rare, inherited disease associated with epidermolytic hyperkeratosis of palms and soles (so-called palmoplantar keratoderma) and with a high incidence of esophageal carcinoma.

Ubiquitination—Ubiquitin is a protein found in all eukaryotic cells (hence its name) that attaches to other proteins. Ubiquitination is thought to be a process that tags proteins that need to be broken down and removed from cells.

Ulcerative colitis—A chronic inflammatory condition of the colon that predisposes affected patients to colon cancer.

Undifferentiated tumor—As a result of tumor progression, cancer cells tend to lose features found in their normal cell counterparts. A tumor is said to be undifferentiated if it lacks all such features. If the cells in a tumor lack most (but not all) of these differentiated features, the tumor is said to be poorly differentiated. Likewise, tumors may be called moderately differentiated or well differentiated based on their deviance from normal cells. The term "undifferentiated tumor" is generally meant to confer a poor prognosis because tumors generally become "undifferentiated" as the result of tumor progression. We should remember that "undifferentiation" is a morphologic assessment and that so-called undifferentiated tumors may employ pathways that are highly conserved for their cell lineage, enhancing the vulnerability of cells to therapeutic agents targeted to lineage-specific pathways. In addition, primitive tumors of childhood are, ab initio, "undifferentiated". Because these tumors are never differentiated, they do not go through the biological process that leads a differentiated cell to an undifferentiated state.

Uniparental disomy—This occurs when the embryo receives a chromosome pair or a paired partial chromosome from one parent and no contribution from the other parent. Uniparental disomy is frequently found in myelodysplastic syndrome. The most extreme form of uniparental disomy occurs in hydatidiform moles.

Validation—Reproducibility is when you get the same result (e.g., a diagnostic test result or an observed experimental value) over and over when you perform the test. Validation is when you draw the same conclusion over and over.

Von Hippel Lindau (VHL)—VHL is a dominantly inherited familial cancer syndrome predisposing to a variety of malignant and benign neoplasms, including retinal angiomata, liver hemangiomas, pancreatic cysts, bilateral papillary cystadenoma of the epididymis, bilateral papillary cystadenomas of the broad ligament, renal angiomyolipoma, renal cell carcinoma, renal cysts, cerebellar hemangioblastoma, spinal cord hemangioblastoma, adrenal hemangioma, polycythemia, pheochromocytoma, paraganglioma, and adenocarcinoma of the ampulla of Vater. There are at least four clinical variants of VHL: VHL type 1—renal carcinoma and hemangioblastoma, VHL type 2A—hemangioblastoma and pheochromocytoma, VHL type 2B—renal carcinoma and pheochromocytoma, and VHL type 2C—pheochromocytoma only (81). VHL is caused by the VHL tumor suppressor gene that, when inactive, cannot serve its normal function of inhibiting transcription elongation (339, 340).

Von Recklinghausen disease—See neurofibromatosis.

Warburg hypothesis—See glycolysis.

Warthin tumor—Also known as papillary cystadenoma lymphomatosum, Warthin tumor is a benign tumor of salivary glands that most often arises in the parotid gland. Warthin tumors are often bilateral (left and right parotid glands) or multiple (more than one tumor per gland). The cell of origin of the Warthin tumor seems to be a variant of duct cell (the cells lining the salivary gland ducts) in which there is a proliferation of mitochondria, producing oncocytes (cells in which the cytoplasm is filled with proliferating mitochondria, characterized by enlarged cells with prominent, somewhat foamy cytoplasm).

In addition to neoplastic oncocytes, the Warthin tumor is characterized by an infiltrate of reactive (nonneoplastic) lymphocytes and numerous cystic spaces. Recently, it has been suggested that the mitochondria of oncocytes proliferate due to mutations of mitochondrial DNA and that cigarette smoke contributes to the mutation rate of mitochondria (341). See mitochondria.

X chromosome—The female sex chromosome. Genetically normal human females have paired x chromosomes. Genetically normal human males have one x chromosome and one y chromosome.

Xeroderma pigmentosum—An inherited disorder in which affected persons cannot efficiently repair DNA damage produced by ultraviolet light. Persons with xeroderma pigmentosum must avoid sunlight. Otherwise, they develop skin cancers such as basal cell carcinoma, squamous cell carcinoma, and melanoma.

Yolk sac tumor—See endodermal sinus tumor.

References

1. Willis RA. *Borderland of Embryology and Pathology.* London: Butterworth, 1958. In the mid 20th century, many pathologists believed that neoplasms were pathologic forms of tissue embryogenesis. This excellent and scholarly book discusses the similarities between embryologic growth and neoplastic growth.
2. Foulds L. *Neoplastic Development.* New York: Academic Press, 1969.
3. Bailar JC, Gornik HL. Cancer undefeated. *N Engl J Med* 336:1569–1574, 1997.
4. Esteller M. The necessity of a human epigenome project. *Carcinogenesis* 27:1121–1125, 2006.
5. Helmboldt CF, Jungherr EL, Caparo AC. Pulmonary adenomatosis in the chinchilla. *Am J Vet Res* 19:270–276, 1958.
6. Berman JJ, Murray RJ, Lopez-Plaza IM. Widespread posttracheostomy atypia simulating squamous cell carcinoma: a case report. *Acta Cytologica* 35:713–716, 1991.
7. Benvenuti S, Arena S, Bardelli A. Identification of cancer genes by mutational profiling of tumor genomes. *FEBS Lett* 579:1884–1890, 2005. The authors state that progression results by the serial accumulation of genetic accidents in somatic cells.
8. Fialkow PJ. Use of genetic markers to study cellular origin and development of tumors in human females. *Adv Cancer Res* 1972;15:191–226, 1972.
9. Engels K, Oeschger S, Hansmann ML, Hillebrand M, Kriener S. Bone marrow trephines containing lymphoid aggregates from patients with rheumatoid and other autoimmune disorders frequently show clonal B-cell infiltrates. *Hum Pathol* 38:1402–1411, 2007. The authors show that simply demonstrating clonality in a lymphoid lesion is not unequivocal evidence of lymphoma.
10. Elias H, Sherrick JC. *Morphology of the Liver.* New York: Academic Press, 1969.
11. Berman JJ, Moore GW. The role of cell death in the growth of preneoplastic lesions: a Monte Carlo simulation model. *Cell Proliferation* 25:549–557, 1992. Regression of precancerous lesions is common. This article examines the hypothesis that

early lesions operate under the identical growth kinetics of "late" lesions (neoplasms), but kinetic features favoring continuous growth in established lesions tend to favor extinction of lesions composed of small numbers of cells. Growth simulations of early lesions demonstrated that small increments in the intrinsic cell loss probability in even the earliest progenitors of malignancy can strongly influence the subsequent development of neoplasia from initiated foci.

12. Berman JJ, Moore GW. Spontaneous regression of residual tumor burden: prediction by Monte Carlo simulation. *Anal Cell Pathol* 4:359–368, 1992.
13. Moore GW, Berman JJ. Cell growth simulations that predict polyclonal origins for "monoclonal" tumors. *Cancer Lett* 60:113–119, 1991.
14. Shen Q, Lantvit DD, Lin Q, et al. Advanced rat mammary cancers are growth hormone dependent. *Endocrinology* 148:4536–4544. 2007.
15. Lyman CP, Fawcett DW. The effect of hibernation on the growth of sarcoma in the hamster. *Cancer Res* 14:25–28, 1954.
16. Orbo A, Rise CE, Mutter GL. Regression of latent endometrial precancers by progestin infiltrated intrauterine device. *Cancer Res* 66:5613–5617, 2006.
17. Cardiff RD, Munn RJ, Galvez JJ. The tumor pathology of genetically engineered mice: a new approach to molecular pathology. In: Fox JG, Davisson MT, Quimby FW, Barthold SW, Newcomer CE, Smith AL, eds. *The Mouse in Biomedical Research: Experimental Biology and Oncology* (2nd ed.). New York: Elsevier, 2006:581–622.
18. Kimura ET, Nikiforova MN, Zhu Z, et al. High prevalence of BRAF mutations in thyroid cancer: genetic evidence for constitutive activation of the RET/PTC-RAS-BRAF signaling pathway in papillary thyroid carcinoma. *Cancer Res* 63:1454–1457, 2003.
19. Soares P, Trovisco V, Rocha AS, et al. BRAF mutations and RET/PTC rearrangements are alternative events in the etiopathogenesis of PTC. *Oncogene* 22:4578–4580, 2003.
20. Cohen Y, Xing M, Mambo E, et al. BRAF mutation in papillary thyroid carcinoma. *J Natl Cancer Inst* 95:625–627, 2003.
21. Melillo RM, Castellone MD, Guarino V, et al. The RET/PTC-RAS-BRAF linear signaling cascade mediates the motile and mitogenic phenotype of thyroid cancer cells. *J Clin Invest* 115:1068–1081, 2005.
22. Sansom OJ, Clarke AR. p53 null mice: damaging the hypothesis? *Mutat Res* 452:149–162, 2000.
23. Sjoblom T, Jones S, Wood LD, et al. The consensus coding sequences of human breast and colorectal cancers. *Science* 314:268–274, 2006.
24. Lengauer C, Kinzler KW, Vogelstein B. Genetic instability in colorectal cancers. *Nature* 386:623–627, 1997.

25. Wang Z, Cummins JM, Shen D, et al. Three classes of genes mutated in colorectal cancers with chromosomal instability. *Cancer Res* 64:2998–3001, 2004.

26. Michor F, Iwasa Y, Vogelstein B, Lengauer C, Nowak MA. Can chromosomal instability initiate tumorigenesis? *Semin Cancer Biol* 15:43–49, 2005.

27. Nouzova M, Holtan N, Oshiro MM, et al. Epigenomic changes during leukemia cell differentiation: analysis of histone acetylation and cytosine methylation using CpG island microarrays. *J Pharmacol Exp Ther* 311:968–981, 2004.

28. Flynn PJ, Miller WJ, Weisdorf DJ, Arthur DC, Brunning R, Branda RF. Retinoic acid treatment of acute promyelocytic leukemia: in vitro and in vivo observations. *Blood* 62:1211–1217, 1983.

29. Solt D, Medline A, Farber E. Rapid emergence of carcinogen-induced initiated hepatocytes in liver carcinogenesis. *Am J Pathol* 88:595–618, 1977.

30. Gestl SA, Leonard TL Biddle JL, Debies MT, Gunther EJ. Dormant WNT-initiated mammary cancer can participate in reconstituting functional mammary glands. *Mol Cell Biol* 27:195–207, 2007.

31. Guess HA, Hoel DG. The effect of dose on cancer latency period. *J Environ Pathol Toxicol* 1:279–286, 1977.

32. Oller AR, Rastogi P, Morgenthaler S, Thilly WG. A statistical model to estimate variance in long term low dose mutation assays: testing of the model in a human lymphoblastoid mutation assay. *Mutat Res* 216:149–161, 1989.

33. Cranor C. Scientific inferences in the laboratory and the law. *Am J Public Health* 95:S121–S128, 2005.

34. National Academy of Sciences. *Toxicity Testing: Strategies to Determine Needs and Priorities.* Washington, DC: National Academy Press, 1984.

35. Matsushima T, Matsumoto H, Shirai A, Sawamura M, Sugimura T. Mutagenicity of the naturally occurring carcinogen cycasin and synthetic methylazoxymethanol conjugates in *Salmonella typhimurium. Cancer Res* 39:3780–3782, 1979.

36. Pierce JP, Natarajan L, Caan BJ, et al. Influence of a diet very high in vegetables, fruit, and fiber and low in fat on prognosis following treatment for breast cancer: the Women's Healthy Eating and Living (WHEL) randomized trial. *JAMA* 298:289–298, 2007.

37. de Meester C, Gerber GB. The role of cooked food mutagens as possible etiological agents in human cancer. A critical appraisal of recent epidemiological investigations. *Rev Epidemiol Sante Pub* 1995;43:147–161, 1995.

38. Mathers CD, Shibuya K, Boschi-Pinto C, Lopez AD, Murray CJL. Global and regional estimates of cancer mortality and incidence by site: I: application of regional cancer survival model to estimate cancer mortality distribution by site. *BMC Cancer* 2:36, 2002.

39. Bonnette SG, Kittrell FS, Stephens LC, Meyn RE, Medina D. Interactions of apoptosis, proliferation and host age in the regression of the mouse mammary preneoplasia, TM3, carrying an unusual mutation in p53. *Carcinogenesis* 20:1715–1720, 1999.

40. Brash DE, Ponten J. Skin precancer. *Cancer Surv* 32:69–113, 1998.

41. Moretti S, Spallanzani A, Pinzi C, Prignano F, Fabbri P. Fibrosis in regressing melanoma versus nonfibrosis in halo nevus upon melanocyte disappearance: could it be related to a different cytokine microenvironment? *J Cutan Pathol* 34:301–308, 2007.

42. Razack AH. Bacillus Calmette-Guérin and bladder cancer. *Asian J Surg* 30:302–309, 2007.

43. Evans AE, Gerson J, Schnaufer L. Spontaneous regression of neuroblastoma. *Natl Cancer Inst Monogr* 44:49–54, 1976.

44. Aterman K, Schueller EF. Maturation of neuroblastoma to ganglioneuroma. *Am J Dis Child* 120:217–222, 1970.

45. Bodey B. Spontaneous regression of neoplasms: new possibilities for immunotherapy. *Expert Opin Biol Ther* 2:459–476, 2002.

46. Chieco-Bianchi L, Colombatti A, Collavo D, Sendo F, Aoki T, Fischinger PJ. Tumor induction by murine sarcoma virus in AKR and C58 mice. *J Exp Med* 140:1162–1179, 1974

47. *Holland-Frei Cancer Medicine.* Kufe D, Pollock R, Weichselbaum R, Bast R, Gansler T, Holland J, Frei E, eds. Ontario, Canada: BC Decker, 2003.

48. Fung YK, Crittenden LB, Fadly AM, Kung HJ. Tumor induction by direct injection of cloned v-src DNA into chickens. *Proc Natl Acad Sci USA* 80:353–357, 1983.

49. Hacein-Bey-Abina S, Von Kalle C, Schmidt M, et al. LMO2-associated clonal T cell proliferation in two patients after gene therapy for SCID-X1. *Science* 302:415–419, 2003.

50. Shachaf CM, Kopelman AM, Arvanitis C, et al. MYC inactivation uncovers pluripotent differentiation and tumour dormancy in hepatocellular cancer. *Nature* 431:1112–1117, 2004.

51. Giuriato S, Ryeom S, Fan AC, et al. Sustained regression of tumors upon MYC inactivation requires p53 or thrombospondin-1 to reverse the angiogenic switch. *Proc Natl Acad Sci USA* 103:16266–16271, 2006.

52. Biggs PJ, Chapman P, Lakhani SR, Burn J, Stratton MR. The cylindromatosis gene (cyld1) on chromosome 16q may be the only tumour suppressor gene involved in the development of cylindromas. *Oncogene* 12:1375–1377, 1996.

53. Zhao R, Xing S, Li Z, et al. Identification of an acquired JAK2 mutation in polycythemia vera. *J Biol Chem* 280:22788–22792, 2005.

54. Chan EF, Gat U, McNiff JM, Fuchs E. A common human skin tumour is caused by activating mutations in beta-catenin. *Nat Genet* 21:410–413, 1999.

55. Kandasamy J, Smith A, Diaz S, Rose B, O'Brien C. Heterogeneity of PLAG1 gene rearrangements in pleomorphic adenoma. *Cancer Genet Cytogenet* 177:1–5, 2007.

56. Fletcher JA, Gibas Z, Donovan K, et al. Ovarian granulosa-stromal cell tumors are characterized by trisomy 12. *Am J Pathol* 138:515–520, 1991.

57. Behboudi A, Winnes M, Gorunova L, et al. Clear cell hidradenoma of the skin—a third tumor type with a t(11;19)-associated TORC1-MAML2 gene fusion. *Genes Chromosome Cancer* 43:202–205, 2005.

58. Sciot R, Akerman M, Dal Cin P, et al. Cytogenetic analysis of subcutaneous angiolipoma: further evidence supporting its difference from ordinary pure lipomas: a report of the CHAMP Study Group. *Am J Surg Pathol* 21:441–444, 1997.

59. Iwabuchi H, Sakamoto M, Sakunaga H, et al. Genetic analysis of benign, low-grade, and high-grade ovarian tumors. *Cancer Res* 55:6172–6180, 1995.

60. Oliveira AM, Perez-Atayde AR, Inwards CY, et al. USP6 and CDH11 oncogenes identify the neoplastic cell in primary aneurysmal bone cysts and are absent in so-called secondary aneurysmal bone cysts. *Am J Pathol* 165:1773–1780, 2004.

61. Mabuchi T, Shimizu M, Ino H, et al. PRKAR1A gene mutation in patients with **cardiac myxoma**. *Int J Cardiol* 102:273–277, 2005.

62. Seidman JD, Berman JJ, Hitchcock CL, et al. DNA analysis of cardiac myxomas: flow cytometry and image analysis. *Hum Pathol* 22:494–500, 1991.

63. Bridge JA, Kanamori M, Ma Z, et al. Fusion of the ALK gene to the clathrin heavy chain gene, CLTC, in inflammatory myofibroblastic tumor. *Am J Pathol* 159:411–415, 2001.

64. Donna A, Betta PG. Mesodermomas: a new embryological approach to primary tumours of coelomic surfaces. *Histopathology* 5:31–44, 1981.

65. Cools J, DeAngelo DJ, Gotlib J, et al. A tyrosine kinase created by fusion of the PDGFRA and FIP1L1 genes as a therapeutic target of imatinib in idiopathic hypereosinophilic syndrome. *N Engl J Med* 348:1201–1214, 2003.

66. Leithner A, Weinhaeusel A, Zeitlhofer P, et al. Evidence of a polyclonal nature of myositis ossificans. *Virchows Arch* 446:438–441, 2005.

67. Californial Cancer Registry. Data standards and quality control memorandum DSQC 2000-12. http://ccrcal.org/PDF/memo2000-12.pdf. (Accessed July 12, 2008).

68. Berman JJ. Borderline ovarian tumor workshop, Bethesda, Maryland, August 27–28, 2003. *Hum Pathol* 35:907–909, 2004.

69. Gordon R. *Great Medical Disasters.* New York: Dorset Press, 1986:155–160. These are lively essays on the history of medical catastrophes, including the black death, influenza, the long history of scurvy, yellow fever, and others.

70. Kolata G. Reversing Trend, Big Drop Is Seen in Breast Cancer. *New York Times,* December 15, 2006.

71. Berman JJ, Albores-Saavedra J, Bostwick D, et al. Precancer: a conceptual working definition: results of a consensus conference. *Cancer Detect Prev* 30:387–394, 2006.

72. Wabinga HR. Pattern of cancer in Mbarara, Uganda. *East Afr Med J* 79:193–197, 2002.

73. Nze-Nguema F, Sankaranarayanan R, Barthelemy M, et al. Cancer in Gabon, 1984–1993: a pathology registry-based relative frequency study. *Bull Cancer* 83:693–696, 1996.

74. Palatianos GM, Cintron JR, Narula T, et al. Father of modern cytology: a 30-year commemorative. *J Fla Med Assoc* 79:837–838, 1992.

75. Bergstrom R, Sparen P, Adami HO. Trends in cancer of the cervix uteri in Sweden following cytological screening. *Br J Cancer* 81:159–166, 1999.

76. Anttila A, Pukkala E, Soderman B, et al. Effect of organised screening on cervical cancer incidence and mortality in Finland, 1963–1995: recent increase in cervical cancer incidence. *Int J Cancer* 83:59–65, 1999.

77. Ellis DL, Kafka SP, Chow JC, et al. Melanoma, growth factors, acanthosis nigricans, the sign of Leser-Trelat, and multiple acrochordons: a possible role for alpha-transforming growth factor in cutaneous paraneoplastic syndromes. *N Engl J Med* 317:1582–1587, 1987.

78. Sioutos N, de Coronado S, Haber MW, Hartel FW, Shaiu WL, Wright LW. NCI Thesaurus: a semantic model integrating cancer-related clinical and molecular information. *J Biomed Inform* 40:30–43, 2007.

79. Brownstein MH, Mehregan AH, Bikowski JBB, Lupulescu A, Patterson JC. The dermatopathology of Cowden's syndrome. *Brit J Derm* 100:667–673, 1979.

80. Schrager CA, Schneider D, Gruener AC, Tsou HC, Peacocke M. Clinical and pathological features of breast disease in Cowden's syndrome: an underrecognized syndrome with an increased risk of breast cancer. *Hum Pathol* 29:47–53, 1998.

81. OMIM™. Online Mendelian Inheritance in Man™. http://www.ncbi.nlm.nih.gov/omim/ (accessed July 12, 2008). OMIM is a marvelous and free database of the genetic diseases of mankind. OMIM lists every known inherited condition in humans. Each condition is annotated with biologic and clinical descriptions in a detailed textual narrative that includes a listing of relevant citations. There are nearly 17,000 conditions described in OMIM.

82. PubMed. http://www.ncbi.nlm.nih.gov/sites/entrez. PubMed home is the host site of Entrez, the search engine sponsored by the National Library of Medicine that conducts text-based searches over the major NCBI (National Center for Bioinformatics) databases, including PubMed and OMIM.

83. Jina Y, Heo K, Han W, Lim H, Wang M. Morphological and genetic characteristics of *Nicotiana langsdorffii, N. glauca,* and its hybrid. *EXCLI Journal* 4:25–33, 2005.

84. Baker NE. Patterning signals and proliferation in *Drosophila* imaginal discs. *Curr Opin Genet Dev* 17:287–293, 2007.

85. Woodhouse E, Hersperger E, Shearn A. Growth, metastasis, and invasiveness of *Drosophila* tumors caused by mutations in specific tumor suppressor genes. *Dev Genes Evol* 207:542–550, 1998.

86. Hariharan IK, Bilder D. Regulation of imaginal disc growth by tumor-suppressor genes in *Drosophila. Ann Rev Genet* 40:335–361, 2006.

87. Tuberous sclerosis complex in flies too? A fly homolog to TSC2, called gigas, plays a role in cell cycle regulation. Posted July 27, 2000, http://www.ncbi.nlm.nih.gov/books/bv.fcgi?rid=coffeebrk.chapter.25. (Accessed July 12, 2008.)

88. Tsonis PA, Eguchi G. Abnormal limb regeneration without tumor formation in adult newts directed by carcinogens 20-methylcholanthrene and benzo(a)pyrene. *Dev Growth Differ* 24:183–190, 1982.

89. Breedis C. Induction of accessory limbs and of sarcoma in the newt (*Triturus viridescens*) with carcinogenic substances. *Cancer Res* 12:861–866, 1952.

90. Prehn RT. Regeneration versus neoplastic growth. *Carcinogenesis* 18:1439–1444, 1997.

91. Anver MR. Amphibian tumors: a comparison of anurans and urodeles. *In Vivo* 6:435–437, 1992.

92. Khudoley VV. Tumor induction by carcinogenic agents in anuran amphibian *Rana temporaria. Arch Geschwulstforsch* 47:385–399, 1977.

93. Sinnhuber RO, Hendricks JD, Wales JH, Putnam GE. Neoplasms in rainbow trout, a sensitive animal model for environmental carcinogenesis. *Ann NY Acad Sci* 298:389–408, 1977.

94. Wales JH, Sinnhuber RO, Hendricks JD, Nixon JE, Eisele TA. Aflatoxin B1 induction of hepatocellular carcinoma in the embryos of rainbow trout (*Salmo gairdneri*). *J Natl Cancer Inst* 60:1133–1139, 1978.

95. Bailey GS, Williams DE, Hendricks JD. Fish models for environmental carcinogenesis: the rainbow trout. *Environ Health Perspect* 104(Suppl 1):5–21, 1996.

96. Spitsbergen JM, Kent ML. The state of the art of the zebrafish model for toxicology and toxicologic pathology research: advantages and current limitations. *Toxicol Pathol* 31(Suppl):62–87, 2003.

97. Kelsh RN, Eisen JS. The zebrafish colourless gene regulates development of non-ectomesenchymal neural crest derivatives. *Development* 127:515–525, 2000.

98. Rowley JD. Most genes involved in translocations have homologs in other organisms. *Annu Rev Genet* 32:495–519, 1998.

99a. Robanus-Maandag E, Dekker M, van der Valk M, et al. p107 is a suppressor of retinoblastoma development in pRb-deficient mice. *Genes Dev* 12:1599–1609, 1998.

99. Nikitin AY, Lee WH. EArly loss of the retinoblastoma gene is associated with impaired growth inhibitory innervation during melanothroph carcinogenesis in Rb+/– mice. *Genes Dev* 10:1870–1879, 1996.

100. Williams BO, Remington L, Albert DM, Mukai S, Bronson RT, Jacks T. Cooperative tumorigenic effects of germline mutations in Rb and p53. *Nat Genet* 7:480–484, 1994.

101. Rajagopalan H, Bardelli A, Lengauer C, Kinzler KW, Vogelstein B, Velculescu VE. Tumorigenesis: RAF/RAS oncogenes and mismatch-repair status. *Nature* 418:934, 2002.

102. Dawe C. Phylogeny and oncogeny. In: *National Cancer Institute Monograph* 31. Bethesda, MD: U.S. National Cancer Institute, 1968:12.

103. Christie AE, Kutz-Naber KK, Stemmler EA, et al. Midgut epithelial endocrine cells are a rich source of the neuropeptides APSGFLGMRamide (Cancer borealis tachykinin-related peptide Ia) and GYRKPPFNGSIFamide (Gly1-SIFamide) in the crabs *Cancer borealis, Cancer magister* and *Cancer productus. J Exp Biol* 210(Pt 4):699–714, 2007.

104. Syder AJ, Karam SM, Mills JC, Ippolito JE, Ansari HR, Farook V, Gordon JI. A transgenic mouse model of metastatic carcinoma involving transdifferentiation of a gastric epithelial lineage progenitor to a neuroendocrine phenotype. *Proc Natl Acad Sci USA* 101:4471–4476, 2004.

105. Murgia C, Pritchard JK, Kim SY, Fassati A, Weiss RA. Clonal origin and evolution of a transmissible cancer. *Cell* 126:477–87, 2006.

106. Saif MW. Pancreatoblastoma. *JOP (Online Journal of Pancreas)* 8:55–63, 2007. URL: www.joplink.net.

107. Cuilliere P, Lazure T, Bui M, et al. Solid adenoma with exclusive hepatocellular differentiation: a new variant among pancreatic benign neoplasms? *Virchows Arch* 441:519–522, 2002.

108. Wells JM, Esni F, Boivin GP, et al. Wnt/beta-catenin signaling is required for development of the exocrine pancreas. *BMC Dev Biol* 7:4, 2007.

109. Anderson FJ. *An Illustrated History of the Herbals.* iUniverse, 1999. http://books.iuniverse.com/viewbooks.asp?isbn=1583481141&page=15. Dioscorides wrote *Materia Medica* in 60 AD, one of the earliest and most influential taxonomies of botanical medicine. This scholarly work is recommended as an early source of thoughtful discussion on the topic of taxonomic organization.

110. Margulis L, Schwartz KV. Five kingdoms: an illustrated guide to the phyla of life on earth, third edition. New York: W.H. Freeman, 1998.

111. Burke AP, Sobin LH, Federspiel BH, Shekitka KM, Helwig EB. Goblet cell carcinoids and related tumors of the vermiform appendix. *Am J Clin Pathol* 94:27–35,

1990. This article describes the clinical biology of goblet cell carcinoid of appendix, which happens to be completely different from the clinical biology of plain old carcinoid tumor of appendix. The lesson learned is that small variations in the name of a tumor may have important clinical consequences; therefore, nomenclatures must strive to include and distinguish all the variant names of clinically distinct neoplasms.

112. National Cancer Institute. Director's challenge: toward a molecular classification of tumors. http://*grants*.nih.gov/*grants*/guide/rfa-files/RFA-CA-98-027.html. Posted January 20, 1999; Accessed Jul 12, 2008.

113. Rosai J. The continuing role of morphology in the molecular age. *Mod Pathol* 14:258–260, 2001. Juan Rosai, a leading pathologist, describes the clinical value of morphologic classification of disease.

114. Rosenwald A, Wright G, Leroy K, et al. Molecular diagnosis of primary mediastinal B cell lymphoma identifies a clinically favorable subgroup of diffuse large B cell lymphoma related to Hodgkin lymphoma. *J Exp Med* 198:851–862, 2003.

115. Schuetz AN, Yin-Goen Q, Amin MB, et al. Molecular classification of renal tumors by gene expression profiling. *J Mol Diagn* 7:206–218, 2005.

116. Zhou X, Temam S, Oh M, Pungpravat N, Huang BL, Mao L, Wong DT. Global expression-based classification of lymph node metastasis and extracapsular spread of oral tongue squamous cell carcinoma. *Neoplasia* 8:925–932, 2006.

117. Cawley GC, Talbot NL. Gene selection in cancer classification using sparse logistic regression with Bayesian regularization. *Bioinformatics* 22:2348–2355, 2006.

118. Yu J, Yu J, Almal AA, et al. Feature selection and molecular classification of cancer using genetic programming. *Neoplasia* 9:292–303, 2007.

119. Mayr E. *The Growth of Biological Thought: Diversity, Evolution and Inheritance.* Cambridge: Belknap Press, 1982. Ernst Mayr (1905–2005) wrote dozens of books and hundreds of article on a variety of subjects in biology. His most enduring work has been on the subject of evolution and biological classification. This, in my opinion, is one of the most important books written in the field of biology because its treatment of the subject of biological classification has enormous value in medical research.

120. Diamandopoulos GT, Meissner WA. Neoplasia. In: Kissane JM, ed. *Anderson's Pathology.* St. Louis: Mosby, 1985:518–520. Diamandopoulos wrote, "Since there are almost limitless varieties of tumors, a complete table of classification would require many pages. Any shortened version is not only necessarily incomplete but also likely to be confusing."

121. Argani P, Antonescu CR, Illei PB, et al. Primary renal neoplasms with the ASPL-TFE3 gene fusion of alveolar soft part sarcoma: a distinctive tumor entity previously included among renal cell carcinomas of children and adolescents. *Am J Pathol* 159:179–192, 2001.

122. Panagopoulos I, Tiziana Storlazzi C, Fletcher CD, et al. The chimeric FUS/CREB3l2 gene is specific for low-grade fibromyxoid sarcoma. *Genes Chromosomes Cancer* 40:218–228, 2004.

123. Lai CYL, Chan KYK, Khoo U, et al. Analysis of gestational trophoblastic disease by genotyping and chromosome in situ hybridization. *Mod Pathol* 17:40–48, 2004.

124. French CA, Kutok JL, Faquin WC, et al. Midline carcinoma of children and young adults with NUT rearrangement. *J Clin Oncol* 22:4135–4139, 2004.

125. Ioannidis JPA. Is molecular profiling ready for use in clinical decision making? *Oncologist* 12:301–311, 2007.

126. Abrikossoff A. Uber Myome, ausgehend von der quergestreiften willkurlichen Muskulatur. *Virchows Arch Pathol Anat* 260:215–233, 1926.

127. Gray SH, Gruenfeld GE. Myoblastoma. *Am J Cancer* 30:699–708, 1937.

128. Fisher ER, Wechsler H. Granular cell myoblastoma: a misnomer: electron microscopic and histochemical evidence concerning its Schwann cell derivation and nature (granular cell schwannoma). *Cancer* 15:936–954, 1962.

129. Sobel HJ, Schwarz R, Marquet E. Light and electron-microscopic study of the origin of granular cell myoblastoma. *J Pathol* 109:101–111, 1973.

130. Berman JJ, Rice JM, Strandberg J. Granular cell variants in a rat schwannoma: evidence of neurogenic origin of granular cell tumor (myoblastoma). *Vet Pathol* 15:725–731, 1978.

131. Mukai M. Immunohistochemical localization of S-100 protein and peripheral nerve myelin proteins (P2 protein, P0 protein) in granular cell tumors. *Am J Pathol* 112:139–146, 1983.

132. National Cancer Institute. SEER *Cancer Statistics Review* 1975–2004. Pages 30–31, Table 1–4. http://seer.cancer.gov/csr/1975_2004/results_merged/sect_01_overview.pdf.

133. U.S. Orphan Drug Act of 1983. http://www.fda.gov/orphan/oda.htm.

134. Clark WH Jr, Hood AF, Tucker MA, Jampel RM. Atypical melanocytic nevi of the genital type with a discussion of reciprocal parenchymal-stromal interactions in the biology of neoplasia. *Hum Pathol* 29(1 Suppl 1):S1–S24, 1998.

135. Makretsov N, He M, Hayes M, et al. A fluorescence in situ hybridization study of ETV6-NTRK3 fusion gene in secretory breast carcinoma. *Genes Chromosomes Cancer* 40:152–157, 2004. The authors conclude that ETV6-NTRK3 fusion gene is a specific genetic alteration in secretory breast carcinoma.

136. Antonioli DA. Gastrointestinal autonomic nerve tumors: expanding the spectrum of gastrointestinal stromal tumors. *Arch Pathol Lab Med* 113:831–8333, 1989.

137. Berman J, O'Leary TJ. Gastrointestinal stromal tumor workshop. *Hum Pathol* 32:578–582, 2001.

138. O'Leary T, Berman JJ. Gastrointestinal stromal tumors: answers and questions. *Hum Pathol* 33:456–458, 2002.
139. Fletcher CD, Berman JJ, Corless C, et al. Diagnosis of gastrointestinal stromal tumors: a consensus approach. *Int J Surg Pathol* 10:81–89, 2002.
140. Druker BJ. Perspectives on the development of a molecularly targeted agent. *Cancer Cell* 1:31–36, 2002.
141. Chen YB, Rahemtullah A, Hochberg E. Primary effusion lymphoma. *Oncologist* 12:569–576, 2007.
142. Royds JA, Iacopetta B. p53 and disease: when the guardian angel fails. *Cell Death Differ* 13:1017–1026, 2006.
143. Marsh DJ, Theodosopoulos G, Martin-Schulte K, et al. Genome-wide copy number imbalances identified in familial and sporadic medullary thyroid carcinoma. *J Clin Endocrinol Metab* 88:1866–1872, 2003.
144. Vezzosi D, Bennet A, Caron P. Recent advances in treatment of medullary thyroid carcinoma. *Ann Endocrinol (Paris)* 68:147–153, 2007.
145. Blanquet V, Turleau C, Gross-Morand MS, Senamaud-Beaufort C, Doz, F, Besmond C. Spectrum of germline mutations in the RB1 gene: a study of 232 patients with hereditary and nonhereditary retinoblastoma. *Hum Mol Genet* 4:383–388, 1995.
146. Johnson RL, Rothman AL, Xie J, et al. Human homolog of patched, a candidate gene for the basal cell nevus syndrome. *Science* 272:1668–1671, 1996.
147. Gerber C, von Hochstetter AR, Schuler G, Hofmann V, Rosenthal C. Penis carcinoma in a young chimney sweep: case report 200 years following the description of the first occupational disease. *Schweiz Med Wochenschr* 125:1201–1205, 1995.
148. Sharp GB. The relationship between internally deposited alpha-particle radiation and subsite-specific liver cancer and liver cirrhosis: an analysis of published data. *J Radiat Res* 43:371–380, 2002.
149. Wagoner JK. Toxicity of vinyl chloride and poly(vinyl chloride): a critical review. *Environ Health Perspect* 52:61–66, 1983.
150. Hill RB, Anderson RE. Pathologists and the autopsy. *Am J Clin Pathol* 95(Suppl):42, 1991. The authors extoll the virtues of autopsies.
151. Herbst AL, Ulfelder H, Poskanzer DC. Association of maternal stilbestrol therapy and tumor appearance in young women. *N Engl J Med* 284:878–881, 1971.
152. Herbst AL, Scully RE, Robboy SJ. The significance of adenosis and clear-cell adenocarcinoma of the genital tract in young females. *J Reprod Med* 15:5–11, 1975.
153. Strohsnitter WC, Noller KL, Hoover RN, et al. Cancer risk in men exposed in utero to diethylstilbestrol. *J Natl Cancer Inst* 93:545–551, 2001.
154. Rose G. Sick individuals and sick populations. *Int J Epidemiol* 14:32–38, 1985.
155. Morales Surez-Varela MM, Llopis Gonzalez A, Marquina Vila A, Bell J. Mycosis fungoides: review of epidemiological observations. *Dermatology* 201:21–28, 2000.

156. Thurman RG, Scholz R. Mixed function oxidation in perfused rat liver. *Fed Proc* 28:902, 1969.

157. Berman JJ. Cell proliferation and the aetiology of hepatocellular carcinoma. *J Hepatol* 7:305–309, 1988.

158. Craddock VM. Liver carcinomas induced in rats by single administration of dimethylnitrosamine after partial hepatectomy. *J Natl Cancer Inst* 47:899–907, 1971.

159. Vesselinovitch SD, Koka M, Rao KVN, Mihailovich N, Rice JM. Perinatal multicarcinogenesis by ethylnitrosourea in mice. *Cancer Res* 37:1822–1828, 1977.

160. Johnson PJ, Williams R. Cirrhosis and the aetiology of hepatocellular carcinoma. *J Hepatol* 4:140–147, 1987.

161. Niederan C, Fischer R, Sonnenberg A, Stremmel W, Teampisch HL, Strohmeyer G. Survival and causes of death in cirrhotic and non-cirrhotic patients with primary hemochromatosis. *N Engl J Med* 313:1256–1262, 1985.

162. Bomford A, Williams R. Long-term results of venosection therapy in idiopathic hemochromatosis. *Q J Med* 45:611–623, 1976.

163. Dunsford HA, Sell S, Chisari FV. Hepatocarcinogenesis due to chronic liver cell injury in hepatitis B virus transgenic mice. *Cancer Res* 50:3400–3407, 1990.

164. National Cancer Institute. Surveillance Epidemiology End Results (SEER). http://seer.cancer.gov/.

165. Auerbach AD, Allen RG. Leukemia and preleukemia in Fanconi anemia patients: a review of the literature and report of the International Fanconi Anemia Registry. *Cancer Genet Cytogenet* 51:1–12, 1991.

166. Auerbach AD. Fanconi anemia and leukemia: tracking the genes. *Leukemia* 6(Suppl 1):1–4, 1992.

167. German J. Bloom's syndrome: incidence, age of onset, and types of leukemia in the Bloom's Syndrome Registry. In: Bartsocas CS, Loukopoulos D, eds. *Genetics of Hematological Disorders.* Washington, DC: Hemisphere Publishers, 1992:241–258.

168. Snyder R, Witz G, Goldstein BD. The toxicology of benzene. *Environ Health Perspect* 100:293–306, 1993.

169. Tefferi A, Gilliland DG. Oncogenes in myeloproliferative disorders. *Cell Cycle* 6:550–566, 2007.

170. Li JP, D'Andrea AD, Lodish HF, Baltimore D. Activation of cell growth by binding of Friend spleen focus-forming virus gp55 glycoprotein to the erythropoietin receptor. *Nature* 343:762–764, 1990.

171. Ruscetti SK, Janesch NJ, Chakraborti A, Sawyer ST, Hankins WD. Friend spleen focus-forming virus induces factor independence in an erythropoietin-dependent erythroleukemia cell line. *J Virol* 64:1057–1062, 1990.

172. Pinkston JA, Cole P. Cigarette smoking and Warthin's tumor. *Am J Epidemiol* 144:183–187, 1996.

173. Shamanin V, zur Hausen H, Lavergne D, et al. Human papillomavirus infections in nonmelanoma skin cancers from renal transplant recipients and nonimmunosuppressed patients. *J Natl Cancer Inst* 88:802–811, 1996.

174. Dessoukey MW, Omar MF, Abdel-Dayem H. Eruptive keratoacanthomas associated with immunosuppressive therapy in a patient with systemic lupus erythematosus. *J Am Acad Dermatol* 37(3 Pt 1):478–480, 1997.

175. Guitart J, McGillis ST, Bergfeld WF, Tuthill RJ, Bailin PL, Camisa C. Muir-Torre syndrome associated with alpha 1-antitrypsin deficiency and cutaneous vasculitis: report of a case with exacerbation of a cutaneous neoplasm during immunosuppressive therapy. *J Am Acad Dermatol* 24(5 Pt 2):875–877, 1991. Describes a keratoacanthoma developed after immunosuppressive therapy for necrotizing vasculitis.

176. Murphy SC, Lowitt MH, Kao GF. Multiple eruptive dermatofibromas in an HIV-positive man. *Dermatology* 190:309–312, 1995.

177. Cohen PR. Multiple dermatofibromas in patients with autoimmune disorders receiving immunosuppressive therapy. *Int J Dermatol* 30:507–508, 1991.

178. Jensen P, Clausen OP, Geiran O, et al. Cutaneous complications in heart transplant recipients in Norway 1983–1993. *Acta Derm Venereol* 75:400–403, 1995.

179. Kammula US, Marincola FM. Cancer immunotherapy: is there real progress at last? *BioDrugs* 11:249–260, 1999.

180. Prehn RT. Immunomodulation of tumor growth. *Am J Pathol* 77:119–122, 1974. The author wrote that "immunologic surveillance, as originally conceived, probably does not exist."

181. Tomar SL, Winn DM, Swango PA, Giovino GA, Kleinman DV. Oral mucosal smokeless tobacco lesions among adolescents in the United States. *J Dent Res* 76: 1277–1286, 1997.

182. de Visscher JG, Grond AJ, Otter R, van der Waal I. Lip carcinoma: a review. *Ned Tijdschr Tandheelkd* 109:391–395, 2002.

183. Neville BW, Damm DD, Weir JC, Fantasia JE. Labial salivary gland tumors. *Cancer* 61:2113–2116, 1988.

184. Ries LAG, Smith MA, Gurney JG, Linet M, Tamra T, Young JL, Bunin GR, eds. Cancer Incidence and Survival among Children and Adolescents: United States SEER Program 1975–1995, National Cancer Institute, SEER Program. NIH Pub. No. 99-4649. Bethesda, MD, 1999. Graphs from this resource were reprinted under rights specified on the SEER monograph front-page, "All material in this report is in the

public domain and may be reproduced or copied without permission; citation as to source, however, is appreciated."

185. Cohn SL. Diagnosis and classification of the small round-cell tumors of childhood. *Am J Pathol* 155:11–15, 1999.
186. Antonescu CR, Nafa K, Segal NH, Dal Cin P, Ladanyi M. EWS-CREB1: a recurrent variant fusion in clear cell sarcoma: association with gastrointestinal location and absence of melanocytic differentiation. *Clin Cancer Res* 12:5356–5362, 2006.
187. Kho AT, Zhao Q, Cai Z, et al. Conserved mechanisms across development and tumorigenesis revealed by a mouse development perspective of human cancers. *Genes Dev* 18:629–640, 2004.
188. The Leukemia and Lymphoma Society. Leukemia Facts & Statistics. http://www.leukemia-lymphoma.org/.
189. Anderson LM, Bhalchandra A, Diwan BA, Nicola T, Fear NT, Roman E. Critical windows of exposure for children's health: cancer in human epidemiological studies and neoplasms in experimental animal models. *Environ Health Perspect* 108(Suppl 3):573–594, 2000.
190. Wakeford R. The risk of childhood cancer from intrauterine and preconceptional exposure to ionizing radiation. *Environ Health Perspect* 103:1018–1025, 1995.
191. Mole RH. Childhood cancer after prenatal exposure to diagnostic x-ray examinations in Britain. *Br J Cancer* 62:152–168, 1990.
192. Bove KE, McAdams AJ. Multifocal nephroblastic neoplasia. *J Natl Cancer Inst* 61:285–294, 1978.
193. Beckwith JB, Kiviat NB, Bonadio JF. Nephrogenic rests, nephroblastomatosis, and the pathogenesis of Wilms' tumour. *Pediatr Pathol* 10:1–36, 1990.
194. Friedmann AS, Malott KA, Memoli VA, Pai SI, Yu XM, North WG. Products of vasopressin gene expression in small-cell carcinoma of the lung. *Br J Cancer* 69:260–263, 1994.
195. Weissglas M, Schamhart D, Lowik C, Papapoulos S, Vos P, Kurth KH. Hypercalcemia and cosecretion of interleukin-6 and parathyroid hormone related peptide by a human renal cell carcinoma implanted into nude mice. *J Urol* 153:854–857, 1995.
196. Johnson RL, Hartmann WH. The thyroid. In: Silverberg SG, ed. *Principles and Practice of Surgical Pathology.* New York: Wiley, 1983:1415–1441.
197. Skogseid B, Larsson C, Lindgren PG, et al. Clinical and genetic features of adrenocortical lesions in multiple endocrine neoplasia type 1. *J Clin Endocrinol Metab* 75:76–81, 1992.
198. Hanahan D, Weinberg RA. The hallmarks of cancer. *Cell* 100:57–70, 2000.
199. Corless CL, Schroeder A, Griffith D, et al. PDGFRA mutations in gastrointestinal stromal tumors: frequency, spectrum and in vitro sensitivity to imatinib. *J Clin Oncol* 23:5357–5364, 2005.

200. Sutovsky H, Gazit E. The von Hippel-Lindau tumor suppressor protein is a molten globule under native conditions: implications for its physiological activities. *J Biol Chem* 279:17190–17196, 2004.

201. Sidhar SK, Clark J, Gill S, et al. The t(X;1)(p11.2;q21.2) translocation in papillary renal cell carcinoma fuses a novel gene PRCC to the TFE3 transcription factor gene. *Hum Mol Genet* 5:1333–1338, 1996.

202. Argani P, Fritsch M, Kadkol SS, Schuster A, Beckwith JB, Perlman EJ. Detection of the ETV6-NTRK3 chimeric RNA of infantile fibrosarcoma/cellular congenital mesoblastic nephroma in paraffin-embedded tissue: application to challenging pediatric renal stromal tumors. *Modern Pathol* 13:29–36, 2000.

203. Brandi ML, Gagel RF, Angeli A, et al. Guidelines for diagnosis and therapy of MEN type 1 and type 2. *J Clin Endocrinol Metab* 86:5658–5671, 2001.

204. Sasaki H, Zlatescu MC, Betensky RA, Ino Y, Cairncross JG, Louis DN. PTEN is a target of chromosome 10q loss in anaplastic oligodendrogliomas and PTEN alterations are associated with poor prognosis. *Am J Pathol* 159:359–367, 2001.

205. Moore JL, Rush LM, Breneman C, Mohideen MPK, Cheng KC. Zebrafish genomic instability mutants and cancer susceptibility. *Genetics* 174:585–600, 2006.

206. McDonald JM, Colman H, Perry A, Aldape K. Molecular and clinical aspects of 1//19q loss in oligodendrogloioma. In: Zhang W, Fuller GN, eds. *Genomic and Molecular Neuro-Oncology.* Sudbury, MA: Jones and Bartlett Publishers, 2004: 17–29.

207. Ohgaki H, Eibl RH, Schwab M, et al. Mutations of the p53 tumor suppressor gene in neoplasms of the human nervous system. *Mol Carcinog* 8:74–80, 1993.

208. Bogler O, Cavenee WK. Methylation and genomic damage in gliomas. In: Zhang W, Fuller GN, eds. *Genomic and Molecular Neuro-Oncology.* Sudbury, MA: Jones and Bartlett Publishers, 2004:3–16.

209. Costello JF, Fruhwald MC, Smiraglia DJ, et al. Aberrant CpG-island methylation has non-random and tumour-type-specific patterns. *Nat Genet* 24:132–138, 2000.

210. Esteller M, Corn PG, Baylin SB, Herman JG. A gene hypermethylation profile of human cancer. *Cancer Res* 61:3225–3229, 2001.

211. Herman JG, Latif F, Weng Y, et al. Silencing of the VHL tumor-suppressor gene by DNA methylation in renal carcinoma. *Proc Natl Acad Sci USA* 91:9700–9704, 1994.

212. Laird PW, Jackson-Grusby L, Fazeli A, et al. Suppression of intestinal neoplasia by DNA hypomethylation. *Cell* 81:197–205, 1995.

213. Atallah E, Kantarjian H, Garcia-Manero G. The role of decitabine in the treatment of myelodysplastic syndromes. *Expert Opin Pharmacother* 8:65–73, 2007.

214. Atallah E, Garcia-Manero G. Use of hypomethylating agents in myelodysplastic syndromes. *Clin Adv Hematol Oncol* 5:544–552, 2007.

215. Weisenberger DJ, Siegmund KD, Campan M, et al. CpG island methylator phenotype underlies sporadic microsatellite instability and is tightly associated with BRAF mutation in colorectal cancer. *Nat Genet* 38:787–793, 2006.

216. Toward a Comprehensive Genomic Analysis of Cancer—NCI/NHGRI Workshop: Questions and Answers. http://www.cancer.gov/cancertopics/factsheet/HumanCancerGenomeProjectFAQs.

217. Greenman C, Stephens P, Smith R, et al. Patterns of somatic mutation in human cancer genomes. *Nature* 446:153–158, 2007.

218. O'Driscoll L, McMorrow J, Doolan P, et al. Investigation of the molecular profile of basal cell carcinoma using whole genome microarrays. *Mol Cancer* 5:74, 2006.

219. Sorek R, Dror G, Shamir R. Assessing the number of ancestral alternatively spliced exons in the human genome BMC. *Genomics* 7:273, 2006.

220. Bolande RP. Developmental pathology. *Am J Pathol* 94:623–683, 1979.

221. Bolande RP. Models and concepts derived from human teratogenesis and oncogenesis in early life. *J Histochem Cytochem* 32:878–884, 1984.

222. Bolande RP. Neurocristopathy: its growth and development in 20 years. *Pediatr Pathol Lab Med* 17:1–25, 1997.

223. *Nomina Embryologica Veterinaria.* Prepared by the International Committee on Embryological Veterinary Nomenclature. Knoxville, Tennessee, 2003. http://www.wava-amav.org/downloads/nev_2006.pdf. (Accessed July 12, 2008.)

224. *Nomina Anatomica* (5th ed.). Prepared by the International Anatomical Nomenclature Committee. Baltimore: Williams and Wilkins, 1980.

225. Shibata D, Tavare S. Counting divisions in a human somatic cell tree: how, what and why? *Cell Cycle* 5:610–614, 2006.

226. Frumkin D, Wasserstrom A, Kaplan S, et al. Genomic variability within an organism exposes its cell lineage tree. *PLoS Comput Biol* 1:e50, 2005.

227. Young HM. Embryological origin of interstitial cells of Cajal. *Microsc Res Tech* 47:303–308, 1999.

228. Ritter MR, Reinisch J, Friedlander SF, Friedlander M. Myeloid cells in infantile hemangioma. *Am J Pathol* 168:621–628, 2006.

229. Liu AY, Chan WY, Ng EK, et al. Gastric choriocarcinoma shows characteristics of adenocarcinoma and gestational choriocarcinoma: a comparative genomic hybridization and fluorescence in situ hybridization study. *Diagn Mol Pathol* 10:161–165, 2001.

230. Houghton J, Stoicov C, Nomura S, et al. Gastric cancer originating from bone marrow-derived cells. *Science* 306:1568–1571, 2004.

231. Theise ND, Nimmakayalu M, Gardner R, et al. Liver from bone marrow in humans. *Hepatology* 32:11–16, 2000.

232. Bryder D, Rossi DJ, Weissman IL. Hematopoietic stem cells: the paradigmatic tissue-specific stem cell. *Am J Pathol* 169:338–346, 2006.

233. Kucia M, Reca R, Jala VR, Dawn B, Ratajczak J, Ratajczak MZ. Bone marrow as a home of heterogenous populations of nonhematopoietic stem cells. *Leukemia* 19:1118–1127, 2005.

234. Grove JE, Bruscia E, Krause DS. Plasticity of bone marrow-derived stem cells. *Stem Cells* 22:487–500, 2004.

235. Cho YM, Choi J, Lee OJ, Lee HI, Han DJ, Ro JY. SMARCB1/INI1 missense mutation in mucinous carcinoma with rhabdoid features. *Pathol Int* 56:702–706, 2006.

236. Donner LR, Wainwright LM, Zhang F, Biegel JA. Mutation of the INI1 gene in composite rhabdoid tumor of the endometrium. *Hum Pathol* 38:935–939, 2007.

237. Roberts CW, Galusha SA, McMenamin ME, Fletcher CD, Orkin SH. **Haploinsufficiency** of Snf5 (integrase interactor 1) predisposes to malignant rhabdoid tumors in mice. *Proc Natl Acad Sci USA* 97:13796–13800, 2000.

238. Kwon CS, Wagner D. Unwinding chromatin for development and growth: a few genes at a time. *Trends Genet* 23:403–412, 2007.

239. Okita K, Ichisaka T, Yamanaka S. Generation of germline-competent induced pluripotent stem cells. *Nature* 448:313–317 2007.

240. Morozov A, Lee SJ, Zhang ZK, Cimica V, Zagzag D, Kalpana GV. INI1 induces interferon signaling and spindle checkpoint in rhabdoid tumors. *Clin Cancer Res* 13:4721–4730, 2007.

241. Fowler MH, Fowler J, Ducatman B, Barnes L, Hunt JL. Malignant mixed tumors of the salivary gland: a study of loss of heterozygosity in tumor suppressor genes. *Mod Pathol* 19:350–355, 2006.

242. Mitelman F, Johansson B, Mertens F. Fusion genes and rearranged genes as a linear function of chromosome aberrations in cancer. *Nat Genet* 36:331–334, 2004.

243. Wang E, Lichtenfels R, Bukur J, et al. Ontogeny and oncogenesis balance the transcriptional profile of renal cell cancer. *Cancer Res* 64:7279–7287, 2004.

244. Fujii H, Yoshida M, Gong ZX, et al. Frequent genetic heterogeneity in the clonal evolution of gynecological carcinosarcoma and its influence on phenotypic diversity. Cancer Res 60:114–120, 2000.

245. Hubalek M, Ramoni A, Mueller-Holzner E, Marth C. Malignant mixed mesodermal tumor after tamoxifen therapy for breast cancer. *Gynecol Oncol* 95:264–266, 2004.

246. Komoto M, Tominaga K, Nakata B, Takashima T, Inoue T, Hirakawa K. Complete regression of low-grade mucosa-associated lymphoid tissue (MALT) lymphoma in the gastric stump after eradication of *Helicobacter pylori*. *J Exp Clin Cancer Res* 25:283–285, 2006.

247. Hu-Lieskovan S, Zhang J, Wu L, Shimada H, Schofield DE, Triche TJ. EWS-FLI1 fusion protein up-regulates critical genes in neural crest development and is responsible for the observed phenotype of Ewing's family of tumors. *Cancer Res* 65:4633–4644, 2005.

248. Pomeroy SL, Tamayo P, Gaasenbeek M, et al. Prediction of central nervous system embryonal tumour outcome based on gene expression. *Nature* 415:436–442, 2002.

249. Mak BC, Yeung RS. The tuberous sclerosis complex genes in tumor development. *Cancer Invest* 22:588–603, 2004.

250. Trainor PA, Melton KR, Manzanares M. Origins and plasticity of neural crest cells and their roles in jaw and craniofacial evolution. *Int J Dev Biol* 47:541–553, 2003.

251. Berman JJ, Rice JM. Odontogenic tumours produced in Fischer rats by a single intraportal injection of methylnitrosourea. *Arch Oral Biol* 1980; 25: 213–220. Twenty-nine odontogenic tumors were found in 134 rates injected only once with methylnitrosourea (0.4 millimole/kg body weight) via the portal vein within 48 hours after partial hepatectomy; 26 tumors were odontomas and 3 were ameloblastomas.

252. Berman JJ. Methylnitrosourea induction of rat odontomas. *Tox Pathol* 16:309, 1988. Ameloblastoma is a tumor of ameloblasts that have failed to induce odonotoblastic differentiation. As enamel formation by enameloblast occurs only after odontoblastic differentiation begins, pure ameloblastomas are characterized by the absence of enamel organ formation. In humans, it is important to distinguish odontoma and ameloblastic odontoma as the latter behaves more aggressively and in a manner akin to ameloblastomas.

253. Haibach H, Burns TW, Carlson HE, Burman KD, Deftos LJ. Multiple hamartoma syndrome (Cowden's disease) associated with renal cell carcinoma and primary neuroendocrine carcinoma of the skin (Merkel cell carcinoma). *Am J Clin Pathol* 97:705–712, 1992.

254. Ferrari A, Bisogno G, Macaluso A, et al. Soft-tissue sarcomas in children and adolescents with neurofibromatosis type 1. *Cancer* 109:1406–1412, 2007.

255. Kruse-Losler B, Gaertner C, Burger H, Seper L, Joos U, Kleinheinz J. Melanotic neuroectodermal tumor of infancy: systematic review of the literature and presentation of a case. *Oral Surg Oral Med Oral Pathol Oral Radiol Endod* 102:204–216, 2006.

256. Jiang X, Iseki S, Maxson RE, Sucov HM, Morriss-Kay GM. Tissue origins and interactions in the mammalian skull vault. *Dev Biol* 241:106–116, 2002.

257. Warwar RE, Bullock JD, Shields JA, Eagle RC. Coexistence of 3 tumors of neural crest origin: neurofibroma, meningioma, and uveal malignant melanoma. *Arch Ophthalmol* 116:1241–1243, 1998.

258. Picquet J, Valo I, Jousset Y, Enon B. Primary pulmonary meningioma first suspected of being a lung metastasis. *Ann Thorac Surg* 79:1407–1409, 2005.

259. Vogel KS, Klesse LJ, Velasco-Miguel S, Meyers K, Rushing EJ, Parada LF. Mouse tumor model for neurofibromatosis type 1. *Science* 286: 2176–2179, 1999.

260. Donovan PJ, de Miguel MP. Turning germ cells into stem cells. *Curr Opin Genet Dev* 13:463–471, 2003.

261. Matsui Y, Zsebo K, Hogan BL. Derivation of pluripotential embryonic stem cells from murine primordial germ cells in culture. *Cell* 70:841–847, 1992.

262. Schneider DT, Schuster AE, Fritsch MK, et al. Multipoint imprinting analysis indicates a common precursor cell for gonadal and nongonadal pediatric germ cell tumors. *Cancer Res* 61:7268–7276, 2001.

263. Teilum G. Classification of endodermal sinus tumour (mesoblatoma vitellinum) and so-called "embryonal carcinoma" of the ovary. *Acta Pathol Microbiol Scand* 64:407–429, 1965.

264. Allegrucci C, Thurston A, Lucas E, Young L. Epigenetics and the germline. *Reproduction* 129:137–149, 2005.

265. Okamoto K, Kawakami T. Epigenetic profile of testicular germ cell tumours. *Int J Androl* 30:385–392, 2007.

266. Yang SY, Fugmann SD, Schatz DG. Control of gene conversion and somatic hypermutation by immunoglobulin promoter and enhancer sequences. *J Exp Med* 203:2919–28, 2006.

267. Codrington R, Pannell R, Forster A, et al. The Ews-ERG 4 fusion protein can initiate neoplasia from lineage-committed haematopoietic cells. *PLoS Biol* 3:e242. The stem cell theory of cancer holds that the progenitor cells of each cancer is a stem cell belonging to an ancestral lineage of the cells comprising the tumor. Most of the cells in a tumor are merely descendants of the progenitor stem cell. Killing the differentiated descendants of the tumor stem cell has no therapeutic value. Killing the progenitor stem cells lurking in the tumor will result in a cure. The stem cell theory may have merit, but this study would indicate that you need not travel very far up the ancestral tree to find a cell capable of producing a clone of malignant cancer cells.

268. Young RH, Kurman RJ, Scully RE. Proliferations and tumors of intermediate trophoblast of the placental site. *Semin Diagn Pathol* 5:223–237, 1988.

269. Therasakvichya S. Gestational trophoblastic disease in 2005. *J Med Assoc Thai* 88(Suppl 2):119–123, 2005.

270. McGrath J, Solter D. Completion of mouse embryogenesis requires both the maternal and paternal genomes. *Cell* 37:179–183, 1984.

271. Barton SC, Surani MAH, Norris ML. Role of paternal and maternal genomes in mouse development. *Nature* 311:374–376, 1984.

272. Stevens LC, Little CC. Spontaneous testicular teratomas in an inbred strain of mice. *Proc Natl Acad Sci USA* 40:1080–1087, 1954.

273. Brinster RL. The effect of cells transferred into the mouse blastocyst on subsequent development. *J Exp Med* 140:1049–1056, 1974.

274. Mintz B, Illmensee K. Normal genetically mosaic mice produced from malignant teratocarcinoma cells. *Proc Natl Acad Sci USA* 72:3585–3589, 1975.

275. Papaioannou VE, McBurney MW, Gardner RL, Evans MJ. Fate of teratocarcinoma cells injected into early mouse embryos. *Nature* 258:70–73, 1975.

276. Kahan BW, Ephrussi B. Developmental potentialities of clonal in vitro cultures of mouse testicular teratoma. *J Natl Cancer Inst* 44:1015–1029, 1970.

277. Edwards WD, Berman JJ, Rice JM, Townsend MW. Differentiation of a primary chemically induced rat nephroblastoma in organ culture. *Dev Growth Differ* 21:519–525, 1979.

278. Miura K, Mineta H, Yokota N, Tsutsui Y. Olfactory neuroblastoma with epithelial and endocrine differentiation transformed into ganglioneuroma after chemoradiotherapy. *Pathol Int* 51:942–947, 2001.

279. Bolande RP. Spontaneous regression and cytodifferentiation of cancer in early life: the oncogenic grace period. *Surv Synth Pathol Res* 4:296–311, 1985.

280. Sonnenschein C, Soto AM. Somatic mutation theory of carcinogenesis: why it should be dropped and replaced. *Mol Carcinogen* 29:205–211, 2000.

281. Berman JJ. *Biomedical Informatics.* Sudbury, MA: Jones and Bartlett Publishers, 2007. This describes the fundamental issues and questions in the field and reviews the different types of biomedical data resources and open source tools needed to fully use biomedical data.

282. Berman JJ. *Perl Programming for Medicine and Biology.* Sudbury, MA: Jones and Bartlett Publishers, 2007.

283. Berman JJ. *Ruby Programming for Medicine and Biology.* Sudbury, MA: Jones and Bartlett Publishers, 2008.

284. National Committee on Vital and Health Statistics. Report to the Secretary of the U.S. Department of Health and Human Services on Uniform Data Standards for Patient Medical Record Information. Retrieved July 6, 2000, from http://www.ncvhs.hhs.gov/hipaa000706.pdf. The Committee on Vital and Health Statistics recommends that the United States have designated national healthcare vocabularies. The committee indicated that such vocabularies would assure that that data shared across systems would be comparable at the most detailed level.

285. National Library of Medicine. Unified Medical Language System (UMLS). http://www.nlm.nih.gov/research/umls/umlsmain.html. The UMLS is the largest curated medical vocabulary in existence. The UMLS metathesaurus is a composite of over 100 medical nomenclatures and is available at no cost from the U.S. National Library of Medicine.

286. FAQs: Inclusion of SNOMED CT in the UMLS. United States, National Library of Medicine, National Institutes of Health. http://www.nlm.nih.gov/research/umls/Snomed/snomed_faq.html.

287. Berman JJ. Developmental lineage classification and taxonomy of neoplasms. http://www.julesberman.info/neoclxml.gz. This is the web location for the latest version of the Developmental Classification.

288. Fritz A, Percy C, Jack A, et al. International Classification of Diseases for Oncology (ICD-O). Lyons, World Health Organization, Third Edition, 2001.

289. Berman JJ. Modern classification of neoplasms: reconciling differences between morphologic and molecular approaches. *BMC Cancer* 5:100, 2005. http://www.biomedcentral.com/1471-2407/5/100. A new classification can sometimes resolve certain questions that arise in alternate classifications.

290. Berman JJ. Tumor taxonomy for the developmental lineage classification of neoplasms. *BMC Cancer* 4:88, 2004.

291. Nogales FF, Buritica C, Regauer S, Gonzalez T. Mucinous carcinoid as an unusual manifestation of endodermal differentiation in ovarian yolk sac tumors. *Am J Surg Pathol* 29:1247–1251, 2005.

292. Berman JJ. Doublet method for very fast autocoding. *BMC Med Inform Decis Mak* 4:16, 2004.

293. Berman JJ. Nomenclature-based data retrieval without prior annotation: facilitating biomedical data integration with fast doublet matching. *Silico Biol* 5:0029, 2005.

294. Lacy LW. *OWL: Representing Information Using the Web Ontology Language.* Cheshire: Trafford Publishing, 2005.

295. Ceusters W, Smith B, Goldberg L. A terminological and ontological analysis of the NCI thesaurus. *Methods Inf Med* 44:498–507, 2005.

296. Harris MA, Clark J, Ireland A, et al. Gene Ontology Consortium: the Gene Ontology (GO) database and informatics resource. *Nucl Acids Res* 32:D258–D261, 2004. http://nar.oxfordjournals.org/cgi/content/full/32/suppl_1/D258. GO can be thought of as three classifications merged into an ontology: molecular functions, biological processes, and cellular components. Molecular function describes molecule-level activities such as a specific catalytic or binding action. An example is "kinase activity." Biological process describes coordinated cellular activities (e.g., respiration and cell death) that involve macromolecules. Cellular component describes subcellular locations of molecular activities and structures.

297. Smith B, Kumar A. On controlled vocabularies in bioinformatics: a case study in the gene ontology. *BIOSILICO: Drug Discovery Today* 2:246–252, 2004. http://ontology.buffalo.edu/bio/Compositionality_in_GO.pdf. The Gene Ontology is an open-access nomenclature and classification of macromolecules, cellular processes, and functions.

298. Smith B, Williams J, Schulze-Kremer S. The ontology of the Gene Ontology. *Proc Ann Symp Am Med Inform Assoc* 609–613, 2003. http://ontology.buffalo.edu/medo/Gene_Ontology.pdf. Barry Smith is one of the most important thinkers in the realm of biomedical ontologies. His comments on the Gene Ontology are particularly salient at this time because they focus on some of the inherent complexities and unintended consequences of ontologies.

299. Smith B, Ceusters W, Klagges B, et al. Relations in biomedical ontologies. *Genome Biology* 6:R46, 2005. http://genomebiology.com/2005/6/5/R46. The relationships in ontologies connect classes and instances (e.g., is_a, part_of, derived_from, located_in) and need to be formally defined in a manner that permits their broad use in different biological ontologies. This article discusses the importance of relationship terms in ontologies.

300. Guo Z, Zhang T, Li X, et al. Towards precise classification of cancers based on robust gene functional expression profiles. *BMC Bioinformatics* 6:58, 2005. The authors suggest a novel method of analyzing gene expression arrays of cancers by first grouping the expressed genes (in the array) into Gene Ontology functional classes.

301. Kochanek KD, Murphy SL, Anderson RN, Scott C. Deaths: final data for 2002. *Natl Vital Stat Rep* 53: 53:1-115, 2004. Retrieved October 12, 2004, from http://www.cdc.gov/nchs/data/nvsr/nvsr53/nvsr53_05.pdf. Contains summary data for the leading causes of death in the United States.

302. Prostate, lung, colorectal, and ovarian cancer screening trial (PLCO). http://www3.cancer.gov/prevention/plco/. PLCO is an example of a large, long, and expensive clinical trial.

303. Leaf C. Why we're losing the war on cancer: and how to win it. *Fortune Magazine*, March 22, 2004.

304. Jacoby RF, Cole CE, Hawk ET, Lubet RA. Ursodeoxycholate/sulindac combination treatment effectively prevents intestinal adenomas in a mouse model of polyposis. *Gastroenterology* 127:838–844, 2004.

305. Horning SJ, Rosenberg SA. The natural history of initially untreated low-grade non-Hodgkin's lymphomas. *N Engl J Med* 311:1471–1475, 1984.

306. Food and Drug Administration's Modernization Act. http://www.fda.gov/cder/guidance/105-115.htm.

307. O'Shaughnessy JA, Kelloff GJ, Gordon GB, et al. Recommendations of the American Association for Cancer Research Task Force on the treatment and prevention of intraepithelial neoplasia. Treatment and prevention of intraepithelial neoplasia: an important target for accelerated new agent development. *Clin Cancer Res* 8:314–346, 2002.

308. National Cancer Institute. *SEER Cancer Statistics Review* 1975–2004. http://seer.cancer.gov/csr/1975_2004/results_merged/topic_lifetime_risk.pdf.

309. Tsai SP, Lee ES, Hardy RJ. The effect of a reduction in leading causes of death: potential gains in life expectancy. *Am J Public Health* 68:966–971, 1978.

310. Puska P. Successful prevention of non-communicable diseases: 25 year experiences with North Karelia Project in Finland. *Public Health Med* 4:5–7, 2002.

311. Khurana V, Bejjanki HR, Caldito G, Owens MW. Statins reduce the risk of lung cancer in humans: a large case-control study of US veterans. *Chest* 131:1282–1288, 2007.

312. Jemal A, Murray T, Ward E, et al. Cancer statistics, 2005. *CA Cancer J Clin* 55:10–30, 2005.

313. Minucci S, Pelicci PG. Histone deacetylase inhibitors and the promise of epigenetic (and more) treatments for cancer. *Natl Rev Cancer* 6:38–51, 2006.

314. Sakr WA, Haas GP, Cassin BF, Pontes JE, Crissman JD. The frequency of carcinoma and intraepithelial neoplasia of the prostate in young males. *J Urol* 150:379–385, 1993.

315. Guileyardo JM, Johnson WD, Welsh RA, Akazaki K, Correa P. Prevalence of latent prostate carcinoma in two U.S. populations. *J Natl Cancer Inst* 65:311–316, 1980.

316. United States Court of Appeals for the Eighth Circuit, No. 06–2286, Washington University, Appellee v. William J. Catalona, M.D., Appellant. http://www.ca8.uscourts.gov/opndir/07/06/062286P.pdf.

317. Department of Health and Human Services. 45 CFR (Code of Federal Regulations), 46. Protection of Human Subjects (Common Rule). *Federal Register*, Volume 56, p. 28003–28032. Retrieved June 18, 1991, from http://www.hhs.gov/ohrp/humansubjects/guidance/45cfr46.htm. Although HIPAA (the Health Insurance Portability and Accountability Act) privacy regulations seem to get all the attention, the common rule sets the basic principles for protecting patients from research risks, mandating the activities of institutional review boards, and using human tissues in support of medical research. It is essential reading for anyone involved in human subject research.

318. Department of Health and Human Services. 45 CFR (Code of Federal Regulations), Parts 160 through 164. Standards for Privacy of Individually Identifiable Health Information (Final Rule). *Federal Register*, Volume 65, Number 250, Pages 82461–82510, December 28, 2000. http://aspe.hhs.gov/admnsimp/final/PvcPre01.htm. This is the much-feared HIPAA rule for protecting patient privacy.

319. Loeb LA. **Mutator phenotype** may be required for multistage carcinogenesis. *Cancer Res* 51: 3075–3079, 1991.

320. Trosko JE. From adult stem cells to cancer stem cells: Oct-4 gene, cell–cell communication, and hormones during tumor promotion. *Ann NY Acad Sci* 1089:36–58, 2006.

321. Ames BN, Durston WE, Yamasaki E, Lee FD. Carcinogens are mutagens: a simple test system combining liver homogenates for activation and bacteria for detection. *Proc Natl Acad Sci USA* 70:2281–2285, 1973.

322. Goldberg P. An old drug's 21st century makeover begins with 84-fold price increase. *Cancer Letters* May 13, 2005.

323. Berenson A. A Cancer Drug's Big Price Rise Is Cause for Concern. *New York Times*, March 12, 2006.

324. Vanchieri C. When will the U.S. flinch at cancer drug prices? *J Natl Cancer Inst* 97:624–626, 2005.

325. Committee report. Classification and nomenclature of morphological defects (Discussion). *Lancet* 305:513, 1975.

326. Izeradjene K, Combs C, Best M, et al. Kras(G12D) and Smad4/Dpc4 haploinsufficiency cooperate to induce mucinous cystic neoplasms and invasive adenocarcinoma of the pancreas. *Cancer Cell* 11:229–243, 2007.

327. Cabelof DC, Ikeno Y, Nyska A, et al. Haploinsufficiency in DNA polymerase beta increases cancer risk with age and alters mortality rate. *Cancer Res* 66:7460–7465, 2006.

328. Thomas RM, Berman JJ, Yetter, RA, Moore GW, Hutchins GM. Liver cell dysplasia: a DNA aneuploid lesion with distinct morphologic features. *Hum Pathol* 23:496–503, 1992.

329. Yu Y, Xu F, Peng H, et al. NOEY2 (ARHI), an imprinted putative tumor suppressor gene in ovarian and breast carcinomas. *Proc Natl Acad Sci USA* 96:214–219, 1999.

330. Knudson AG Jr, Hethcote HW, Brown BW. Mutation and childhood cancer: a probabilistic model for the incidence of retinoblastoma. *Proc Natl Acad Sci USA* 72:5116–5120, 1975.

331. Seidman JD, Berman JJ, Yetter RA, Moore GW. Multiparameter DNA flow cytometry of keratoacanthoma. *Anal Quant Cytol Histol* 14:113–119, 1992.

332. Barosi G, Bergamaschi G, Marchetti M, et al. JAK2 V617F mutational status predicts progression to large splenomegaly and leukemic transformation in primary myelofibrosis. *Blood* 110:4030–4036, 2007.

333. Lugo TG, Braun S, Cote RJ, Pantel K, Rusch V. Detection and measurement of occult disease for the prognosis of solid tumors. *J Clin Oncol* 21:2609–2615, 2003.

334. Kraus JA, Koopmann J, Kaskel P, et al. Shared allelic losses on chromosomes 1p and 19q suggest a common origin of oligodendroglioma and oligoastrocytoma. *J Neuropathol Exp Neurol* 54:91–59, 1995.

335. Dahlen A, Fletcher CD, Mertens F, et al. Activation of the GLI oncogene through fusion with the beta-actin gene (ACTB) in a group of distinctive pericytic neoplasms: pericytoma with t(7;12). *Am J Pathol* 164:1645–1653, 2004.

336. Dai C, Celestino JC, Okada Y, Louis DN, Fuller GN, Holland EC. PDGF autocrine stimulation dedifferentiates cultured **astrocytes** and induces oligodendrogliomas and oligoastrocytomas from neural progenitors and astrocytes in vivo. *Genes Dev* 15:1913–1925, 2001.

337. Yamaguchi H. Mutations of telomerase complex genes linked to bone marrow failures. *J Nippon Medical School* 74:202–209, 2007.

338. Lerat E, Semon M. Influence of the transposable element neighborhood on human gene expression in normal and tumor tissues. *Gene* 396:303–311, 2007.

339. Duan DR, Pause A, Burgess WH, et al. Inhibition of transcription elongation by the VHL tumor suppressor protein. *Science* 269:1402–1406, 1995.

340. Kibel A, Iliopoulos O, DeCaprio JA, Kaelin WG. Binding of the von Hippel-Lindau tumor suppressor protein to elongin B and C. *Science* 269:1444–1446, 1995.

341. Lewis PD, Baxter P, Paul Griffiths A, Parry JM, Skibinski DO. Detection of damage to the mitochondrial genome in the oncocytic cells of Warthin's tumour. *J Pathol* 191:274–281, 2000.

Index

C

D

F

G

H

I

M

N

O

P

Q

R

S

U

Figure Credits

Figure I.1 Age-adjusted U.S. cancer mortality, 1969–2004. Public domain image. Surveillance, Epidemiology, and End Results program, National Cancer Institute.

Figure 1.1 Fine needle aspirate of pancreas. Courtesy of Dr. Ed Uthman.

Figure 1.2 Small cell carcinoma of lung. From the University of Alabama at Birmingham Department of Pathology, © PEIR Digital Library.

Figure 1.3 A flat clump of metaplastic cells with reactive atypia. Courtesy of Dr. Jules J. Berman.

Figure 2.1 Schematic of progression from precancer to cancer. Public domain image from the National Cancer Institute, available at: http://visualsonline.cancer.gov/

Figure 3.1 Colon adenoma. From the University of Alabama at Birmingham Department of Pathology, © PEIR Digital Library.

Figure 5.1 Crown gall tumors. Courtesy of Clemson University - USDA Cooperative Extension Slide Series, Bugwood.org.

Figure 5.2 Basal cell carcinoma. Courtesy of David Lortscher, M.D.

Figure 5.3 Diagram of hair shaft. Adapted from Gray, Henry. *Anatomy of the Human Body*. Lea & Febiger Publishers, 1918.

Figure 5.4 Pulmonary carcinoid. Public domain. Courtesy of G. William Moore, MD, PhD, Pathology and Laboratory Medicine Services, Baltimore VA Medical Center, Baltimore, MD.

Figure 8.1 A single spherical object. Courtesy of Dr. Jules J. Berman.

Figure 8.2 Four spheres of the same size. Courtesy of Dr. Jules J. Berman.

Figure 8.3 A simple epithelium. Adapted from Gray, Henry. *Anatomy of the Human Body*. Lea & Febiger Publishers, 1918.

Figure 8.4 Cytologic preparation of small cell carcinoma. Courtesy of Dr. Ed Uthman.

Figure 8.5 Granular cell tumor. From the University of Alabama at Birmingham Department of Pathology, © PEIR Digital Library.

Figure 9.1 Warthin tumor. From the University of Alabama at Birmingham Department of Pathology, © PEIR Digital Library.

Figure 11.1 Age specific cancer incidence rates. Public domain chart from NIH Pub. No. 99-4649. Bethesda, MD, 1999.

Figure 11.2 Average annual incidence rates. Public domain chart from NIH Pub. No. 99-4649. Bethesda, MD, 1999.

Figure 12.1 Unrelated organisms can fly. Reproduced from Thomas, J.A. *The Outline of Science: A Plain Story Simply Told.* G. P. Putnam's Sons, 1922. Available at http://www.gutenberg.org/etext/20417

Figure 15.1 Graphic representation of the first three class tiers of the Developmental Lineage Classification of. Courtesy of Dr. Jules J. Berman.

Figure 15.2 Transverse sections through a dog's embryonic disk. Adapted from Gray, Henry. *Anatomy of the Human Body.* Lea & Febiger Publishers, 1918.

Figure 16.1 Hierarchy of Class Endoderm/Ectoderm tumors. Courtesy of Dr. Jules J. Berman.

Figure 16.2 Oral squamous cell carcinoma. From the University of Alabama at Birmingham Department of Pathology, © PEIR Digital Library.

Figure 16.3 Bowen's disease. From the University of Alabama at Birmingham Department of Pathology, © PEIR Digital Library.

Figure 16.4 A section of pancreas. Adapted from Gray, Henry. *Anatomy of the Human Body.* Lea & Febiger Publishers, 1918.

Figure 16.5 Adenocarcinoma of stomach. From the University of Alabama at Birmingham Department of Pathology, © PEIR Digital Library.

Figure 16.6 Normal prostatic tissue. Courtesy of Laura P. Hale, M.D. Ph.D., Duke University Medical Center.

Figure 16.7 Adenocarcinoma of prostate. Courtesy of Dr. Jeffrey S. Ross.

Figure 16.8 Prostatic intraepithelial neoplasia. Courtesy of Dr. Alessandro Brollo.

Figure 17.1 Hierarchy of Class Mesoderm tumors. Courtesy of Dr. Jules J. Berman.

Figure 17.2 Carcinosarcoma of uterus. From the University of Alabama at Birmingham Department of Pathology, © PEIR Digital Library.

Figure 17.3 Burkitt lymphoma. Public domain image. Author, Dr. Louis M. Staudt, National Cancer Institute.

Figure 18.1 Hierarchy of Class Neuroectoderm tumors. Courtesy of Dr. Jules J. Berman.

Figure 19.1 Hierarchy of Class Neural Crest tumors. Courtesy of Dr. Jules J. Berman.

Figure 20.1 Hierarchy of Class Germ Cell tumors. Courtesy of Dr. Jules J. Berman.

Figure 20.2 Hierarchy of Class Trophectoderm tumors. Courtesy of Dr. Jules J. Berman.

Figure G.1 Breast adenocarcinoma From *An Introduction to Human Disease, Seventh edition.* Photo courtesy of Leonard V. Crowley, M.D., Century College.

Figure G.2 Bronchogenic adenocarcinoma. *From An Introduction to Human Disease, Seventh edition.* Photo courtesy of Leonard V. Crowley, M.D., Century College.

Figure G.3 The logwood tree. Reproduced from Adolphus Ypey. Vervolg ob de Avbeeldingen der artseny-gewassen met derzelver Nederduitsche en Latynsche beschryvingen. J. C. Sepp en Zoon, 1813.

Figure G.4 Metastatic nodules in spleen. *From An Introduction to Human Disease, Seventh edition.* Photo courtesy of Leonard V. Crowley, M.D., Century College.

Figure G.5 Hodgkin lymphoma. *From An Introduction to Human Disease, Seventh edition.* Photo courtesy of Leonard V. Crowley, M.D., Century College.

Figure G.6 Hemangioma. *From An Introduction to Human Disease, Seventh edition.* Photo courtesy of Leonard V. Crowley, M.D., Century College.

Figure G.7 Malignant melanoma. *From An Introduction to Human Disease, Seventh edition.* Photo courtesy of Leonard V. Crowley, M.D., Century College.

Figure G.8 Mitosis. Reproduced from Flemming, W., 1879. *Archiv fur Mikroskopische Anatomie.*

Figure G.9 Multiple myeloma. *From An Introduction to Human Disease, Seventh edition.* Photo courtesy of Leonard V. Crowley, M.D., Century College.

Figure G.10 Normal cervix and carcinoma in situ of cervix. *From An Introduction to Human Disease, Seventh edition.* Photo courtesy of Leonard V. Crowley, M.D., Century College.

Figure G.11 Pap smear. *From An Introduction to Human Disease, Seventh edition.* Photo courtesy of Leonard V. Crowley, M.D., Century College.

Figure G.12 Philadelphia Chromosome. *From An Introduction to Human Disease, Seventh edition.* Photo courtesy of Leonard V. Crowley, M.D., Century College.

Figure G.13 Submucosa. *Anatomy of the Human Body.* Lea & Febiger Publishers, 1918.

Author Biography

Jules Berman, Ph.D., M.D., has studied cancer for the past 36 years. After receiving two Bachelor of Science degrees (mathematics and earth sciences from MIT), he entered the graduate program in pathology at Temple University, where he started his thesis work within the Fels Cancer Research Institute. He spent the final year of his graduate studies at the Naylor Dana Institute of the American Health Foundation in Valhalla, New York, before beginning his post-doctoral studies in the *Perinatal Carcinogenesis* section of the Laboratory of Experimental Pathology at the U.S. National Cancer Institute. He earned his medical degree from the University of Miami, and he completed his residence in anatomic pathology at the George Washington University Medical Center in Washington, D.C. He became Board Certified in Anatomic Pathology and in Cytopathology, and served as the chief of Anatomic Pathology, Surgical Pathology, and of Cytopathology at the Veterans Administration Medical Center in Baltimore, Maryland. While at the Baltimore VA Medical Center, he held appointments at the University of Maryland Medical Center and at the Johns Hopkins Medical Institutions. In 1998, he became the Program Director for Pathology Informatics in the Cancer Diagnosis Program at the U.S. National Cancer Institute. In 2006, he became President of the Association for Pathology Informatics. Today, Dr. Berman is a full-time writer. His previously published books, also published by Jones & Bartlett publishers, are *Biomedical Informatics* (2007), *Perl Programming for Medicine and Biology* (2007), *Ruby Programming for Medicine and Biology* (2008), and *Ruby: The Programming Language* (2008). Over the course of his career, he has co-authored hundreds of scientific contributions, and he has first-authored more than 100 publications.